Soul Full of Coal Dust

Soul Full of Coal Dust

A Fight for Breath and Justice in Appalachia

Chris Hamby

Little, Brown and Company
New York Boston London

Little, Brown and Company
Hachette Book Group
1290 Avenue of the Americas, New York, NY 10104
littlebrown.com

First Edition August 2020

Little, Brown and Company is a division of Hachette Book Group, Inc. The Little, Brown name and logo are trademarks of Hachette Book Group, Inc.

The publisher is not responsible for websites (or their content) that are not owned by the publisher.

The Hachette Speakers Bureau provides a wide range of authors for speaking events. To find out more, go to hachettespeakersbureau.com or call (866) 376-6591.

ISBN 978-0-316-29947-3
LCCN 2019952818

Book design by Sean Ford

10 9 8 7 6 5 4 3 2 1

LSC-C

Printed in the United States of America

To my parents, Elizabeth and Roger, who instilled a love of the written word and a belief in the human spirit

Contents

And here's a story you can hardly believe, but it's true, and it's funny and it's beautiful. There was a family of twelve and they were forced off the land. They had no car. They built a trailer out of junk and loaded it with their possessions. They pulled it to the side of 66 and waited. And pretty soon a sedan picked them up. Five of them rode in the sedan and seven on the trailer, and a dog on the trailer. They got to California in two jumps. The man who pulled them fed them. And that's true. But how can such courage be, and such faith in their own species? Very few things would teach such faith.

The people in flight from the terror behind—strange things happen to them, some bitterly cruel and some so beautiful that the faith is refired forever.

—John Steinbeck, *The Grapes of Wrath*

He's had more hard luck than most men could stand
The mines was his first love but never his friend
He's lived a hard life, and hard he'll die
Black lung's done got him, his time is nigh

Black lung, black lung, you're just biding your time
Soon all of this suffering I'll leave behind
But I can't help but wonder what God had in mind
To send such a devil to claim this soul of mine

He went to the boss man, but he closed the door
Well, it seems you're not wanted when you're sick and you're poor
You're not even covered in their medical plans
And your life depends on the favors of man

Down in the poorhouse on starvation's plan
Where pride is a stranger and doomed is a man
His soul full of coal dust till his body's decayed
And everyone but black lung's done turned him away

Black lung, black lung, oh, your hand's icy cold
As you reach for my life and you torture my soul
Cold as that waterhole down in that dark cave
Where I spent my life's blood diggin' my own grave

Down at the graveyard the boss man came
With his little bunch of flowers, dear God, what a shame
Take back those flowers, don't you sing no sad songs
The die has been cast now, a good man is gone

—Hazel Dickens, "Black Lung" (written for her brother, who died of the disease)

Soul Full of Coal Dust

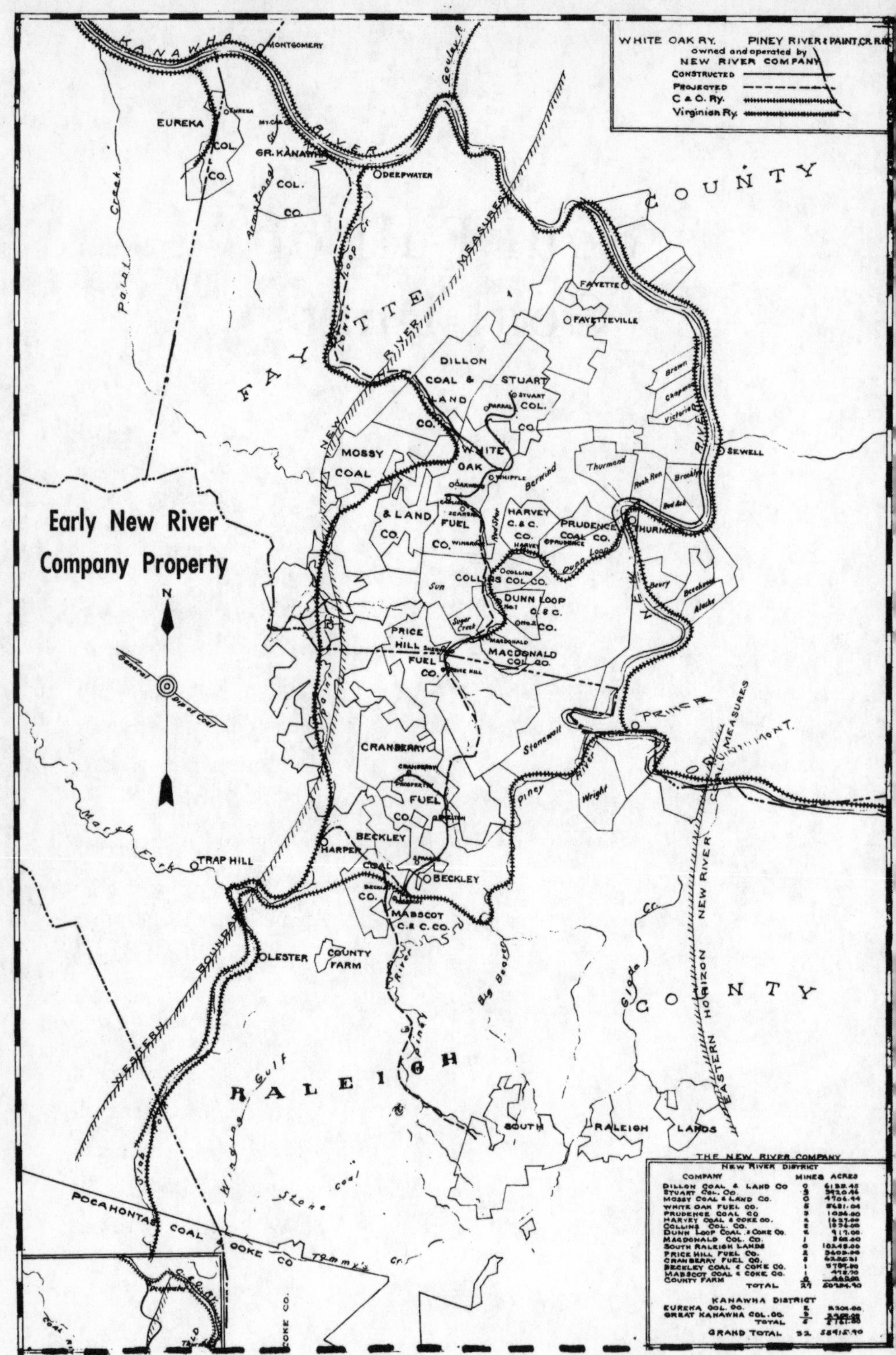

Early twentieth-century claims of the first coal barons to arrive in the area near what is today Beckley, West Virginia, a hub of the state's southern coalfields. *(New River Company, 1976)*

Prologue

January 10, 2007

John Cline looked across his kitchen table at a gaunt man with a countenance etched by a life of hard labor. The two had spoken by phone, but now that they were face to face for the first time, the man's rapidly declining health became apparent. Each breath, it seemed, required more effort than the last. John had heard such strained exertions many times before, and he knew the suffering these sounds signified.

As each man appraised the other, the faint gray light of a frigid day in the southern West Virginia coalfields fell on the smooth maple top, fashioned by John's middle son, that adorned the table John's grandfather had made out of fir. Wind gusted through the patch of land where John and his wife kept a vegetable garden in warmer months. The trees bounding his property to the south had shed their foliage, revealing the precipice preceding the sharp plunge to the Piney Creek Gorge a thousand feet below. In the distance, benches of bare earth lined a stretch of the hillside—the scars of long-finished strip-mining. A railway ran through the gorge, and sometimes when John went for walks in the trails he kept clear behind his house, he still heard the coal trains rumbling through.

This land of scarred beauty had forged both John and the man now seated in his kitchen, a longtime coal miner named Gary Fox. Though they had started life in different worlds, both had come of age amid the political tumult of the late 1960s and the historic coal miners' rebellion that had swept southern West Virginia at the time. Both had made lives for themselves and their families here in the heart of Appalachia. And both had spent decades working, straining against setbacks, building something bigger than themselves.

John was a rarity here in the coalfields: a lawyer who was willing to help coal miners navigate an abstruse legal system in pursuit of modest monthly payments and medical coverage as recompense for the disease that robbed their breath, the old scourge with the disturbingly accurate name "black lung." Most lawyers wouldn't touch these cases. They were complex, time-consuming, and fiercely contested; coal companies and their lawyers made sure of that. Success rates were low, and even after a win, a miner's lawyer had to prevail in yet another round of legal combat against the company to collect fees that barely kept the lights on.

Yet this was the only type of case John took, the only type he'd ever wanted to take. It was why, a few years earlier, he had gone to law school at age fifty-three and emerged with a load of student debt he would still be paying off long after others his age had retired.

Law was his fourth vocation. The first three—community organizer, carpenter, and rural medical-clinic staffer—might suggest the incongruous roving of a restless soul, but John saw each of them, as well as his current one, as variations of the same job, which he described as "trying to be of use."

Here in the house he'd built with his own hands in a small community on the outskirts of Beckley, the nearest thing to a big city southern West Virginia coal country had, he ran a solo practice. He had no assistants, no secretaries, no paralegals; each case was John versus the coal company. His kitchen was the de facto meeting room, and directly overhead on the second floor was his office, a small space covered in manila folders bearing the names of sick miners or their widows, each file stuffed with legal and medical arcana designed to befuddle and discourage those who couldn't find a lawyer like John. Spare shelf space held family photos, and on the wall hung a portrait of Mary Harris "Mother" Jones, the labor activist and "miners' angel" who had adopted West Virginia as her second home.

A famous quote of hers was inscribed beside her bespectacled face: *Pray for the dead and fight like hell for the living.*

John's comportment more closely resembled that of his clients than that of his fellow members of the bar. He favored corduroys or jeans and plaid button-downs. His mop of brown-gray hair and soft voice conveyed a certain boyishness. He had the sturdy frame and callused palms of a laborer.

Gary had found him only through a stroke of luck: a recommendation from a legendary doctor who had been conducting pioneering research into black lung and treating miners for more than forty years at a small clinic in Beckley. John had liked Gary from the first time they spoke on the phone about seven months earlier, but before he agreed to take the case, he needed to gauge whether they would have any shot at winning. Together in John's kitchen, the two men were now trying to figure that out.

This was one of the last stages in a system John had set up; the goal was to take as many cases as he could while making sure his one-man operation didn't get hopelessly overstretched. He typically juggled about fifty cases, their deadlines staggered to avoid everything hitting at once, and each week he fielded about ten more calls from people who hoped he'd represent them.

When a potential client called, the first step was to explain the peculiar legal gauntlet that miners and widows had to traverse. For the few lawyers like John who even considered helping a miner file for benefits, these preliminary conversations could be taxing, as they often engendered frustration for the caller—not at the lawyer but at the demoralizing realization of how the system for awarding claims functioned.

The process wasn't like going to civil court and presenting a case to a jury, and it wasn't like filing for workers' compensation. It was somewhere in between, a byzantine administrative system that combined the contentiousness of a multimillion-dollar lawsuit with the lesser payouts of a disability-insurance claim—in other words, the worst of both systems, from a miner's perspective. The fact that a miner was struggling to breathe because of black lung caused by years of inhaling coal dust didn't necessarily translate into a win; he had to clear the high bar of proving "total disability." There were no settlements or partial awards; each case was an all-or-nothing fight to the end.

After explaining these daunting realities to the caller, John would pull out a piece of paper, run through a list of basic questions, and jot down the answers, with key dates and facts underlined: age, years of mining experience, specific job duties, smoking history, past medical diagnoses or legal claims. Sometimes it was clear from the start that the caller had no case, and John would explain why. But, he'd tell the miner, the disease might worsen, even without any more dust invading the lungs, so keep monitoring it with periodic X-rays and lung-function tests. Don't hesitate to call back.

If the basic information pointed to a potentially viable claim, John explained what would happen next. The miner had to fill out forms and submit them to the U.S. Department of Labor, which processed these claims. That would unleash a flood of paper from bureaucrats, companies, insurers, and lawyers, and the miner would have to undergo two different medical exams—one by a doctor chosen by the miner from a government-approved list and one by a doctor chosen by the coal company.

If the case cleared the first stage—a determination made by a Labor Department claims examiner, who, at the time, turned down about 85 percent of claims—the coal company's lawyers almost certainly would appeal the decision to an administrative law judge under the Labor Department umbrella. These lawyers would then send copies of the miner's X-rays, CT scans, lung-function tests, and any other medical records to an established network of impeccably credentialed physicians—including some at renowned institutions—who, for a fee, would review them and submit reports that often poked holes in or outright blew up the miner's claim. A miner might be able to afford an additional report or two, but the medicolegal arms race was inevitably one-sided. If a judge nonetheless upheld the award of benefits, the company lawyers could appeal to a higher administrative court, then to a federal appeals court, then possibly to the U.S. Supreme Court.

As the process dragged on, miners sometimes withered and died, leaving their widows to fight on if they could. All of this battling was over monthly payments set by law at just over a third of the salary of an entry-level government employee—in 2007, it was $876.50 for a miner and his wife—plus medical expenses for treatment related to black lung.

Word had spread through the coalfields that companies would rather spend stacks of cash fighting each case to the bitter end than pay the modest benefits to their former employees. No wonder, then, that many miners didn't bother with the system or that most lawyers didn't want to wait for an uncertain payout that might amount to fifteen thousand dollars for five years of work, if they were fortunate.

John had gone through these daunting realities with Gary when he'd first called in June 2006. Gary wasn't discouraged; it was just another trial for a man who'd endured more than his share. He had worked in coal mining for more than thirty years, but his breathing had finally gotten so bad that he thought he had no choice but to retire early and file for benefits, he told John. He'd call back when he did.

John wrote all of this down and stuffed the paper inside a manila folder labeled DOL PHONE CONTACTS. This was where logs of the many calls with potential clients resided until John had enough information to decide whether to take the case. If he eventually did agree to represent the miner, he'd grab a new manila folder, write the client's name on the cover in black Sharpie, put the phone logs inside, and place this fresh case file among the others, which filled shelves lining the office walls, covered the small sofa by his desk, and spilled onto the steps leading up to a small attic where old case files sat packed in boxes and shoved back in a crawl space.

Gary called again in November after he'd filed the forms that set the wheels in motion at the Labor Department's office in the state capital of Charleston, then again in December after he'd been examined by the doctor he'd chosen. Now the paperwork was starting to arrive at the house in Beckley where he and his wife, Mary, lived: a notice that his former employer was contesting his claim, an intimidatingly worded list of twenty-seven questions seeking personal information and medical records dating back decades.

John took it all down; the phone log grew longer. Gary's case was not an obvious loser, he thought. It was time they met.

Now, on this chilly day in January, the two men whose lives and legacies would become intertwined sat together at John's home in the small community of Piney View, about twenty-five minutes from Gary's house. They went over the stack of files Gary had brought. These papers supported John's impression of Gary: He was an extremely sick man. The results of his breathing tests were dismal, and doctors were treating him for what they believed was advanced black lung. Gary's physique, always lanky, was now all sharp angles. Protuberant tendons accentuated his long neck. A dimple on his left cheek popped below his soft green eyes. He'd shed more than thirty pounds over the past couple of years, despite his efforts to keep the weight on.

But John also felt a growing sense that something more was going on. It turned out that this was not the first black lung–benefits claim Gary had filed. He had tried in 1999. After a months-long, ultimately fruitless search for a lawyer, he had gone it alone against the company and lost. Poring over the reports from that initial claim, John saw that many of the same facts that underpinned Gary's current claim had been documented in his previous one. Even then, eight years earlier, doctors had seen evidence of the advanced stage of black lung that, under the rules of the federal program, was supposed to result in an award of benefits. The purpose of these rules was to ensure that miners with this advanced stage got out of the dust immediately, giving them at least a chance to avoid the worst ravages of the disease. In Gary's case, however, that hadn't happened.

Something in Gary's 1999 claim must have gone very wrong. As John read through the files, he got an idea of what that might be.

For about fifteen years, John had been trying to unravel what he believed was a systematic scheme to defraud sick miners orchestrated by the law firm of choice for many titans of the coal industry. Sure enough, this was the firm that had found a way to tank Gary's earlier claim and was now fighting his current one. The case file was filled with reports from the prominent doctors at major national institutions whom the firm frequently enlisted as experts. And all of it was on behalf of one of

Appalachia's biggest and most notorious coal companies, run by a baron often portrayed in critical press accounts as a sort of dark lord of the coalfields.

John scribbled a note to himself and, when the two-hour meeting was over, tucked it inside a new manila folder with GARY FOX written in black Sharpie on the front. In time, this small slip of paper would spawn multiple motions and reply briefs, decisions and appeals, documents that filled one folder, then another, then another. It would lead to fraught decisions and sleepless nights, elation and disappointment. And all of this would have great implications not just for Gary and his family but also for countless other miners and their families.

Gary's fight would be, in many respects, the culmination of a battle that had started forty years earlier, when a historic grassroots uprising had extracted long-overdue promises from the government to virtually eradicate black lung and provide compensation to those already afflicted with it. Despite the significant strides advocates had made over the years, these dual pledges had become a mirage for far too many miners and their widows.

This had not happened by chance. The coal industry and the elite professionals it enlisted in its cause had systematically undermined the law in courtrooms, legislative chambers, and dusty mine tunnels. Companies had found ways to dodge the rules intended to prevent the disease, and they had persuaded sympathetic lawmakers to gut some of the law's existing protections and stall or kill proposed improvements. Nationally prominent doctors had found a lucrative market in supplying the reports and testimony that helped doom miners' claims. And top-notch lawyers had become maestros of the benefits system, adept at orchestrating a cacophonous symphony that left many miners disoriented and that denied them what they were due.

At the same time, a small but scrappy coalition of advocates from various walks of life—clinic workers, lay representatives, lawyers, doctors, miners—had been fighting back, pushing reforms in the hopes of making the promises of prevention and compensation realities. John had long been a crucial animating force at the heart of this coalition. With his temperament and experience, he found equal comfort and satisfaction in drafting proposed legislative changes, organizing potluck fund-raisers, sparring with coal companies' lawyers, and traveling to DC with miners to lobby Congress.

The coalition had achieved some reforms, but the coal industry, with its allies in political office and in the medical and legal communities, had beaten back the most sweeping changes that the miners' advocates sought.

Now, Gary seemed to be the embodiment of all the ways that the pledges of disease eradication and fair compensation remained unfulfilled. But John saw something else in Gary. He suspected there was something appalling somewhere in the records of Gary's case and in his lungs, something that might reverberate well beyond Gary's individual claim.

About a week after meeting with Gary, John notified the Labor Department that he was taking the case. Years later, he mused, "I had no idea what I was getting into."

Part One

The breath of a coal miner stricken with advanced black lung disease often has a distinct and haunting timbre. It is a hoarse gasp, a desperate attempt to draw in the oxygen that the body needs to sustain itself, an audible representation of the losing battle being waged every minute of every day inside scarred and shriveled lungs.

The first time I heard this sound was in the spring of 2011. I was a cub reporter at the nonprofit investigative news outlet the Center for Public Integrity, based in Washington, DC. I had the good fortune to be assigned to a team focusing on the environment, labor, and public health led by editor and reporter Jim Morris. With his beard and shaved head, Jim could come off as gruff, an impression he didn't discourage, but he had a contagious compassion for working people. He both guided me and left me largely to my own devices—a reporter's dream.

I began to learn about a part of America I hadn't known. I grew up in Tennessee, but I'd been ensconced in a middle-class suburb. Now, reporting on labor, I found myself drawn to gritty industrial towns and rural communities. I spent hours driving around chemical plants and oil refineries, marveling at the mazes of twisting metal and flashing lights. (More than once, this ended with my getting pulled over and questioned by plant security.) I donned comically large coveralls and toured a sugar refinery. But what I found most compelling were the people—factory workers, men and women who'd lost loved ones in accidents, or survivors whose lives had been forever altered. So many of them shared traits that I'd not seen often when writing about other subjects (especially politics): candor, quiet strength, gratitude that someone cared about their stories.

It was May 2011 when Jim received an embargoed copy of a report on the worst coal-mine disaster in forty years: the 2010 explosion at Massey Energy Company's Upper Big Branch mine in southern West Virginia that killed twenty-nine men. In coal country, Massey was notorious. The company's cantankerous, mustachioed CEO, Don Blankenship, was a throwback to a past era of coal barons. Under his leadership, Massey was fabulously productive, but this economic success was accompanied by a prodigious tally of environmental and safety violations. The Upper Big Branch blast was the worst in a series of accidents at the company's mines.

The report had been commissioned by the West Virginia governor and written by a team of investigators led by former Mine Safety and Health Administration chief Davitt McAteer. Blankenship had claimed that the explosion was an unforeseeable catastrophe that had resulted from a sudden, unexpected release of methane from the coal seam. McAteer's report determined that, to the contrary, the disaster was a result of company management's disregard for basic, time-tested safety practices. The mine wasn't properly ventilated, leading to a dangerous buildup of explosive gases, the report found, and because the company had allowed coal dust to accumulate, the blast spread quickly throughout the mine.

"The story of Upper Big Branch is a cautionary tale of hubris," the report concluded. "A company that was a towering presence in the Appalachian coalfields operated its mines in a profoundly reckless manner, and 29 coal miners paid with their lives for the corporate risk-taking."

Jim and I knew that other news outlets had also received an embargoed copy of the report. We assumed that their stories would focus on the findings related to the explosion's cause, so we discussed whether there was anything noteworthy in the report that other publications might overlook. It turned out that there was: on page 32 out of 120, a seven-paragraph sidebar described a grim discovery that doctors had made when examining the bodies of the miners killed in the blast.

Of the twenty-nine victims, twenty-four had enough lung tissue for pathologists to examine during autopsies. Of those, seventeen had black lung. The victims ranged in age from twenty-five to sixty-one. This astronomical rate of disease accompanied by the relative youth of some of the miners whose lungs bore the hallmarks of it was "an alarming finding," the report said.

Until I read that page, I'd thought, as many Americans probably do, that black lung was a historical relic, the sort of medieval scourge that advances in engineering and medical science surely had eradicated long ago. And in fact, by the twenty-first century, that should have been the case. In 1969, Congress enacted a law imposing tough new limits on the amount of disease-causing coal dust allowed in mine air, a provision that was supposed to have virtually eliminated black lung.

More than forty years later, however, the centuries-old miners' curse remained. The law had led to significant improvements, but since the late 1990s, the once-encouraging trend line had flipped: rates of disease had been rising. Government researchers were tracking the resurgence, trying to understand the cause of the dismaying reversal. Worse, as these scientists were documenting, there was an even larger increase in the prevalence of the most severe stage of black lung, and it seemed to be taking a nasty new form, striking younger miners and progressing more quickly.

This, Jim and I decided, would be the focus of our story. Over a few days, I read up on the disease, the regulatory system, and the recent research. I spoke with doctors who said there was no justification for the continued existence of an entirely preventable ailment, with coal-industry representatives who expressed sympathy for miners but cautioned that enhanced regulation could cripple a vital American industry, and with miners who were the largely unseen victims.

Singular mine catastrophes—fires, explosions, cave-ins—garnered media attention and prompted reform proposals by lawmakers, but black lung continued to exact a little noticed but much larger toll. Between 1995 and 2004, more than three hundred coal workers died in accidents; meanwhile, black lung claimed about ten thousand.

I noted all of this in the article that we published when the embargo on the report was lifted, but I had a lingering sense that I had only begun to see the contours of a more far-reaching story. I was slowly being drawn into a world that had captivated many before: the coalfields of central Appalachia, with southern West Virginia at its core. The region possesses a magnetism that defies explanation. It is, alternately, the butt of uninformed jokes and a romanticized embodiment of the American spirit. One of the first explorers to arrive in this rugged territory found it "pleasing tho' dreadful." Generations of writers have celebrated its beauty and lamented its poverty.

Fascinated as I was, though, I didn't anticipate that I would spend the better part of the next eight years reporting and writing about the men and women who lived and worked in these mountains—their hopes and disappointments, their struggles and triumphs. The revelation contained in that one page from the Upper Big Branch report would lead me to rural clinics where nurses and doctors grappled with an unseen epidemic, to homes tucked in hollows where miners battled breathlessness to make it from their front porches to their mailboxes and dragged oxygen tanks wherever they went. It would lead me to law offices and courtrooms where overmatched but inspired advocates challenged the coal industry in order to secure compensation and medical coverage for ailing miners in a bizarre legal system that had become dizzyingly complex and lopsided, to government offices housing reams of largely unexamined and nonpublic records that, once organized and pieced together, revealed startling injustices. Most important, it would lead me to Gary Fox and John Cline.

1

Gary

Frank Ratcliff was struggling. Just outside the mouth of the Birchfield mine in the heart of the southern West Virginia coalfields, he was straining to load wood posts into a scoop machine that would haul them underground.

He'd been back on the job for only a few days after a six-week recuperation from hernia surgery. His doctor had urged him to wait at least a couple weeks more before returning to work as a coal miner, but he needed money. Plus, his bosses didn't look kindly on absenteeism, whatever the reason. On most days he worked as a roof bolter, a skilled position commanding higher pay and respect. He was pretty sure his current cumbersome task was a company-imposed penance for his sin of missing work.

As the afternoon waned and quitting time for the day shift neared, Ratcliff fumbled and grimaced, flipping the posts over end to end along the ground, advancing them toward the scoop a few feet at a time. Across the yard, some of the men on the evening shift chatted as they waited for their turn to chisel away at the rich black seam below.

Looking up, Ratcliff saw one of them walking toward him, approaching fast thanks to the long strides that came with his rangy, six-foot-two-inch frame. Ratcliff knew the man, but not well. They'd both started recently at the Birchfield mine. His name was Gary Fox and, like Ratcliff, he was a roof bolter. Gary knew about Ratcliff's recent surgery; he'd filled in occasionally during Ratcliff's abbreviated convalescence.

"You got no business doing this; get out of the way," Gary said, brushing Ratcliff aside. "I'll load this for you." When Gary finished, he carted the posts underground, then worked his own shift.

"After that," Ratcliff recalled later, "me and Gary was awful good buddies. I respected him for that. I never forgot it."

Both men performed one of the most difficult and dangerous jobs a miner could do. After a machine with a spinning drumhead studded with carbide-tipped teeth chewed through the coal seam, the roof bolter was the first person to venture into this freshly cut area, which, at Birchfield, was usually an opening about twenty feet wide, twenty feet deep, and six feet high. The immense weight of the mountain above had, over millennia, pressed decaying plant matter into an energy-rich black mineral, and those same forces now pressed on the small cavity where the roof bolter set to his task: using a machine to drive metal bolts through rock, pinning together unstable layers to render a cave-in less likely. Only then did the rest of the crew enter the new section; they all depended on the roof bolter to do his job quickly and well. The price of a mistake was measured in crippling injuries or death.

Gary was, by coworkers' accounts, a damn good bolt man—"Probably the best I've ever seen" in the eyes of one fellow Birchfield miner. In moments of candor, Ratcliff would admit that Gary likely had him beat. Gary was fast and steady, drilling through sandstone or shale with ruthless efficiency, and he never seemed to let up. From time to time, he could be demanding and quick to anger. "He could be hateful," one coworker recalled. "Not hateful," he said on further reflection, "just hard. Bolt men work real hard, and they're particular how you do stuff."

Gary's occasional flashes of frustration passed without lingering effects. In the mines, you pretty much couldn't hold a grudge; you depended on your colleagues to alert you to the multitudinous dangers. If you spotted a loose hunk of rock on the mine roof and didn't grab a slate bar and chip it down right then, it'd be on your conscience when it later fell and maimed your buddy. There tended to be a camaraderie of the sort found in foxholes.

The men shared a certain pride in the job, a notion of the dignity that came with hard work that brought value to the larger society. The fruits of their labors kept the lights on for millions of Americans. Some of the mines in the area yielded the high-grade metallurgical coal used to forge the steel backbones of modern American cities.

Gary and his coworkers spent more time here in the damp grit of a dusty tunnel than they did with their families. The standard workweek was fifty-four hours—nine-hour shifts Monday through Saturday—but overtime was routine. Shifts often stretched to ten, twelve, or fourteen hours. Many of the men were happy for the extra pay that came with this, but, practically speaking, overtime was not optional.

Gary and his coworkers had been "Mabenized." Birchfield was one of a handful of mines under the corporate parent Maben Energy, a company run by a quintessentially West Virginian would-be coal baron named H. Paul Kizer. Raised in a coal camp, Kizer was the son of a coal miner who had died of black lung, and he spent time working in the mines himself. After graduating from Marshall University, where he'd been a lineman on the football team, he held various jobs, including a stint as a federal mine-safety inspector. In the early 1980s, he began acquiring coal mines of his own, and by the time Gary went to work for him in 1987, he had dozens, with around a thousand miners on his payroll.

Kizer was also at the center of two tabloid-worthy scandals. First, there was the murder charge. A Boone County man had discovered his girlfriend in bed with Kizer, and he'd walloped the coal man. The next day, prosecutors said, Kizer asked an assistant to arrange for a bit of payback, which he'd done. The two-man muscle-for-hire team was supposed to rough up the angry boyfriend, but one of them shot him dead. Kizer's assistant was convicted of voluntary manslaughter, but a jury acquitted Kizer.

Then there was his testimony that helped put a former West Virginia governor in prison. Kizer characterized himself as the victim of a shakedown by then governor Arch Moore, who Kizer said had not so subtly implied that if the coal boss wanted his application for a state insurance refund to go through, he'd need to give the governor a 25 percent cut.

Acquiescing, Kizer had paid Moore more than five hundred thousand dollars out of his roughly two-million-dollar refund. Moore ended up pleading guilty to five federal corruption-related charges and serving just short of three years behind bars.

Amid all this, Kizer traveled the coalfields in a company helicopter and sometimes visited his mines. He'd go underground, walk the tunnels, talk to the men. Whatever his public persona, he knew coal mining.*

"Mr. Kizer was an interesting individual," said Dale Birchfield, the president of the company that bore his name. (Kizer named numerous Maben Energy subsidiaries for the officials who ran them, Birchfield said.) "They could probably make a movie about him—good things and bad things. But he was certainly good for this area at the time. A lot of people were able to work and stay in their homes because of him."

At a time when mining jobs were scarce in the area, Kizer was hiring. Miners who went to work for him quickly learned that keeping that precious job meant doing things the Maben Energy way. The company gave employees bumper stickers that said I'VE BEEN MABENIZED.

Among the miners at Birchfield, the slogan meant different things to different people. Some proudly put the sticker on their car or truck bumper; it meant they had a job and worked damn hard. Scoop operator Gene Stewart's take: "Mr. Kizer, I'm thankful that he opened the mines. If it hadn't been for him, the economy around here would have been a lot worse."

To other miners, it was more like a brand on a cow's hide. Ratcliff's take: "Paul Kizer owned you, and he pretty well owned your family too. That's 'Mabenized' for you."

Whatever else the slogan meant, there was a consensus among the men, a sort of common understanding: If you liked having a paycheck, you worked when and how the company told you and didn't complain. When Jerry Bailey started at Birchfield, he recalled, a company official told him, "You can make good money, and your wife and daughter will be spending it because you'll be here all the time."

---

* H. Paul Kizer died in 2017, so I was unable to speak with him.

At Birchfield, you had a starting time; quitting time was whenever the boss said. Occasionally, you'd work from sunset to sunrise, go home and get three or four hours of sleep, and head back to work before lunch. If a boss called you at home and asked you to come in, the only correct response was yes. Maben mines were unionized, but given the job scarcity at the time, the company's miners generally felt the union offered little real protection. Gary Hairston, a union representative who worked at another Maben mine, said that when he filed grievances on behalf of a miner, the company usually responded with threats of retaliation, and the miner almost always ended up dropping his complaint.*

To some miners, this was just the way of the world; a person ought to be satisfied with hard work and good pay. Others felt like they were toiling under an iron hand, and they resented it. But economic realities tended to check their otherwise fiercely independent tongues. "If you was lucky enough to get a job and you had a family to feed, you done what you had to do to feed your family, sacrificed your body," Ratcliff said.

Gary Fox didn't complain, but he didn't put the company's sticker on the bumper of his truck either. Day after day, he'd climb in the mantrip and take a thirty-minute ride down to the dark caverns where black dust floated everywhere and coated everything. The haze was especially bad for roof bolters like Gary. They had front-row seats for the spectacle of the hulking mining machine ripping through coal, and their own machinery gouged out openings in the mine roof. Sometimes, it was difficult to see through clouds of dust. They'd guzzle water from gallon jugs they lugged underground in an attempt to clear the grit from their throats. Some men wore respirators at times, but the devices were bulky and uncomfortable. They got in the way, might not work right if a man had a beard, and were, at best, an imperfect last line of defense against the thick fog that the machines unleashed.

---

* The president of Birchfield Mining at the time, Dale Birchfield, said that the company did not punish miners if they declined extra work. Usually, he said, miners volunteered to work overtime, happy for the additional pay. He also said that as far as he knew, the company did not retaliate against employees for filing grievances.

After a day underground and another thirty-minute ride on the mantrip, Gary would emerge from the mouth of the mine covered in black dust. Unlike some mines, Birchfield didn't have a bathhouse, just a changing room with a toilet and a place to hang up soiled work clothes. For much of his six-year tenure with the company, Gary carpooled with scoop operator Stewart and fellow roof bolter Bailey, commuting an hour and a half each way. Lots of workers made similarly long drives to mines tucked in rural hollows; you went where the jobs were, largely at the mercy of both the vagaries of global energy markets and the predilections and ambitious gambles of coal bosses such as Kizer.

After finishing a shift, Gary and his two riding buddies would meet up at the vehicle of whoever's turn it was to drive that week. When it was Gary's, they piled into his black pickup. He had different models over the years but eventually settled on the Toyota Tacoma. He'd buy a cheap one, run the hell out of it, get under the hood himself to keep it going well past its natural life, then start the cycle anew.

Gary would steer the truck onto the two-lane blacktop of Route 85, entering the rural county's circulatory system: road, river, and rail line twisting in tandem—arteries bringing metal and men, veins sending fuel and waste. He drove south, tracing the western flank of the Coal River Valley, a sort of inverted V formed by the Big and Little Coal Rivers, which flowed north and converged near the state capital of Charleston.

The road snaked through densely forested mountains, still tracing the path of the river, which gathered strength from creeks fed by mountain headwaters. Narrow roads shot off alongside these creeks into hollows where people lived in small homes and trailers deep in the mountain's shadow. For generations, people here had worked the mines, grown potatoes and beans, hunted turkey and deer, and foraged for mulberries, ramps, and ginseng. In small riverside communities, families set aside some of the crops from their vegetable gardens to can for winter, and miners congregated at the general store, identifiable by their dust-covered clothing studded with reflective bands. Belt lines poked out from openings in the sides of mountains, conveying coal to nearby tipples, where it was sorted and loaded into trucks or railcars that rattled through towns.

On and on, the landscape of the coalfields unfolded before Gary in rhythmic cycles: sugar maples, white oaks, creek; house, church, general store; belt line, coal pile, railyard. It is a region steeped in beauty and blood, an expanse of rugged terrain that is the heart of Appalachia and the embodiment of its contradictions—breathtaking vistas alongside the scars of faded industrialism, a wealth of natural resources but continued isolation and poverty, indefatigable optimism despite a legacy of neglect and repression.

It's here where eighteenth-century land speculators staked their claims and clashed with American Indians, where one nineteenth-century prophet of the United States' manifest destiny saw the work of divine hands in what lay beneath the mountains: immense deposits of coal "scattered by the hand of the Creator with very judicious care, as precious seed, which, though buried long, was destined to spring up at last, and bring forth a glorious harvest." This bounty helped fuel the Allied victories in World Wars I and II, and it powered the industrialization that brought America prosperity. The riches of the earth and the sweat of the people here helped make the United States a global superpower.

Yet it's here where prospectors conned mountaineers out of their share of the riches. Where, in the pursuit of these riches, coal bosses sent thousands to their deaths in one needless disaster after another and dominated every corner of community life like feudal landlords. Where one species of hired gun suppressed resistance on the ground with deadly force and another suppressed it from afar in courts and legislative chambers.

And still, it's here where average people have banded together, defied disheartening defeats, and won striking victories with dogged grassroots activism. Because of the allure of southern West Virginia's buried treasure and its defiant mountaineer ethos, the area has been the scene of historic clashes between rank-and-file agitators and the coal industry that so controlled their lives. Some of the bloodiest battles in the history of American labor have played out here.

Gary's commute took him through the mountain towns that, in 1921, had supplied many of the thousands of miners who, in the throes of a bitter unionization drive, had taken up arms and fought in what became known as "the Battle of Blair Mountain." From Boone County, the miners had marched west for a deadly shootout with about three thousand hired guns and company-aligned volunteers led by the sheriff of neighboring Logan County. By various tallies, anywhere from twenty to fifty men died before President Warren G. Harding sent in federal troops, putting an end to what is often called the largest armed insurrection since the Civil War.

More than six decades later, these small communities nestled in hollows and valleys were beginning to feel the sting of the Appalachian coal industry's decline—a trend that would only accelerate in coming years as production continued to shift to surface mines in the West, as fracking yielded a glut of cheap natural gas that increasingly displaced coal as the fuel of choice for power generation, and as market forces and, to a lesser degree, environmental regulations led to the closure of some coal plants, crimping demand. Precipitous declines in coal jobs would hollow out hamlets throughout central Appalachia. Poverty rates would rise. These same communities would be ravaged by a flood of prescription opioids pushed by drug companies and distributors along with a related wave of potent illicit drugs. West Virginia would come to experience the highest rate of overdose deaths in the country, driven in large part by the hard-hit counties of the southern coalfields.

Navigating the mountainous terrain made for white-knuckle driving even for Gary, who was accustomed to the narrow roads with sharp curves that wound through much of the region. The fastest route home was across Bolt Mountain, a notoriously treacherous trip. The ever-present fog was sometimes so thick Gary could barely see farther than his headlights. Near the summit, which was more than three thousand feet high, hazardous weather conditions were amplified. Two inches of snow at the base could become eight inches at the peak. The two-lane road that had been cut into the mountain offered little room to maneuver around common impediments—falling rocks or the massive coal trucks that made heavy use of the road. "Just getting to work was more dangerous than anything we ever did in the mines," Stewart said, only half jokingly.

On these long rides, Gary and his commuting buddies sometimes traded stories, especially about their families; often, the two who weren't driving would get some much-needed sleep. Back at their meeting place outside Beckley, they'd part ways and head home for a few precious hours of rest or time with loved ones.

Finally, after at least nine hours underground and three hours on the road, Gary would pass from rural terrain into the commercial landscape of small-town America: all-night diners and fast-food joints, chain hotels and grocery stores. Then into planned communities and cul-de-sacs. And finally, home.

Inside the two-story house, he would peel off his clothes and place them in a pile separate from the rest of the laundry. Once a week, his wife, Mary, took them to a laundromat; she didn't want the coal dust clogging up their home washing machine.

In the shower, he tried to scrub off the finely ground coal dust that had found its way into nearly every cranny and crevice of his body. He had a brush for the dust embedded under his fingernails. In front of the mirror, he pulled at the skin below one eye and, with a towel, wiped away the coal dust that had collected at the lash line. He did the same thing for the top lid, then repeated the process on the other eye. Some men didn't bother with these thin black rings that folks in the coalfields called "miner's mascara," maybe out of laziness, maybe just pride in the job.

But Gary wouldn't have it. Mining was his job, not his life. Many men were brimming with stories of underground exploits, and in their spare time, they grabbed beers with work buddies. Not Gary. When he was underground, he was all business; he refused to bring the job home. It wasn't that he didn't like the work—he just had more important things to do than rehash the day's labors. Whatever spare time he did have was spent with Mary and their adolescent daughter, Terri; he wasn't going to waste that time carrying on about the mines. And he certainly wasn't going to go out to eat with them looking like he was wearing makeup.

From the medicine cabinet Gary grabbed pills to loosen the joints in his arthritic hands. His body ached from a lifetime of wounds. While serving in Vietnam, he'd been in a truck accident, and sheet metal had shattered his foot. Falling rock in a mine had crushed his shoulder. His hands locked up from years of drilling metal bolts through layers of sandstone and shale.

Six days a week, sometimes seven, Gary repeated these rituals in a meticulous attempt to scrub his world of coal dust. But over the years, it had seeped inside him, lodged itself so deeply that there was no scrubbing it away. It was amassing in his lungs, and each day underground added to that load little by little. Gradually, these fine black specks would stain and scar what they touched; specks would reach toward other specks until they coalesced and turned chunks of tissue into useless black masses.

But Gary was focused on what came with the dust: the best paycheck and health insurance around. In southern West Virginia, the clearest path to earning a good living had long been coal mining. It was coal mining that would allow him to provide the kind of life he had in mind for Mary and Terri.

His own childhood spurred him on. Mary, unlike his mother, would not have to hustle for work or worry about scraping together the next meal. Terri would go to college, something no one in his family had gotten the chance to do. He'd sworn to himself that, when he had a family of his own, he'd seal them off from a world of worry, want, and black dust. So, every day, he returned to that world himself.

Coal suffused Gary's life from the day he was born in 1950. While many Americans were buying new homes in suburbs and TVs to watch Lucille Ball and Joseph McCarthy, the feudal coal-camp system endured in the southern West Virginia town where Gary grew up. Even as the United States emerged from a coal-fired war as a global superpower, Gary's family continued to strain under a weight that had squeezed the area for more than half a century.

Gary was the third of four boys born to Edward and Lorraine Fox and raised on the edge of Skelton, a company town that is part of Beckley today. At the time of Gary's childhood, it was a fading remnant that reflected King Coal's imprint throughout the region.

Skelton was the creation of coal baron Samuel Dixon, who had left Yorkshire, England, at age twenty-one in 1877 to pursue the riches of the

West Virginia coalfields. He'd arrived at a time when life was changing dramatically in the region. Up to that point, the frontier lifestyle had persisted in many areas. People lived off the crops they could grow and the game they could kill (and maybe a little homemade whiskey they could sell). But a place so rich in resources could evade the gaze of the titans of burgeoning American industrialism for only so long.

At the start of the twentieth century, the region was undergoing a transformation as outsiders began arriving. Railroads extended their reach here so they could feed the appetite for coal of industrial plants in the East. Land speculators fanned out across the coalfields. These newcomers eventually gave locals a rude introduction to a couple of themes that reappear throughout West Virginia's history: control of natural resources by capitalists from afar and a tilt in the scales of justice toward these moguls and their top-flight lawyers.

The land speculators convinced many locals to sell their mineral rights—the fossil fuels that might be beneath their land—for a song. Often, these rights ended up in the hands of out-of-state energy companies. Only later did many locals realize that the "broad form deeds" they had signed also gave these companies the right to get at those minerals however they chose, including by setting up machinery on the land and boring into the earth. Landowners who fought back didn't have much success, as the courts typically sided with the companies.

The region was soon teeming with coal camps. Mining families—many of them recent European immigrants, African-Americans making the small step up from Southern sharecropping, and other vulnerable members of society—spent their entire lives under the thumb of the coal company. Children might be born in the company hospital, grow up in a run-down company shack, go to work before their tenth birthday in a company processing plant, toil for decades underground in the company's mines, and, when their bodies were used up, be buried in a company cemetery. Workers got paid not in cash but in scrip—the company currency, redeemable only at the company store for goods at jacked-up prices that kept the coal-camp denizens in poverty and squalor, perpetually on the verge of starvation.

Dixon had his eye on the bounty buried near the breathtakingly beautiful New River Gorge. The Chesapeake and Ohio railroad had put these riches within reach when it extended its lines through the gorge. The laying of this track was supposedly the scene of the competition that spawned the legend of John Henry, the steel-driving man who, as the story goes, bested a machine and died with a hammer in his hand. Dixon eventually bought many of the coal deposits in the area and consolidated everything under the umbrella of the New River Company.

In 1906, Dixon built a three-story mansion on a hill overlooking Skelton, which he'd named after a village in his native Yorkshire. From this perch, he could survey his domain: a town of single-story homes with white siding and metal roofs that the company rented or sold to workers and their families; the tipple; the company store.

The heavy demand for coal during World War II was a boon for the company and the region, but prosperity didn't last. After the war, mining jobs grew scarce, and thousands of West Virginians went north in search of work in manufacturing centers such as Detroit and Chicago, some commuting every weekend and others leaving for good. From 1950 to 1970, the state lost 13 percent of its population. The southern coal regions were hit even harder, with some counties losing half of their citizens during this period.

The hard years came to Skelton around the time Gary was born. The area's mines sometimes slowed or stopped production temporarily. The New River Company started selling what had been company houses, and scrip became a thing of the past. But the town and the mines limped on.

The Fox brothers—Ernie, Freddie, Gary, and Jerry, each separated by about two years—barely knew their father, Edward. A longtime coal miner, he died when they were just kids, and even when he was alive, he mostly was laid up in a hospital ward for tuberculosis patients. At the time, many miners with lung disease were diagnosed with TB, and there was little acknowledgment among doctors—especially coal-company doctors—that the real cause, or at least a contributing factor, was the men's jobs. The brothers always suspected their father's death had a lot to do with the dust he inhaled for years underground.

Lorraine found a way to scrape by, eking out a living and raising the boys more or less on her own. She worked as a maid for better-off people in the area and always managed to put food on the table and keep the boys in decent clothes. It might be bean soup for three nights in a row and hand-me-down shirts, but the brothers learned not to complain. "I don't know how she done it," Ernie recalled. "I saw her not eat so everybody else could."

The family had an up-close view of the industry in action every day. When they looked out their window, they saw the company's tipple, a tall wooden structure where coal was loaded onto railcars. Interspersed around the area were piles of coal and rock and the portals that led to underground tunnels. There was hardly a place in town where coal hadn't left its stain.

"Everything was black," said Jerry, the youngest Fox brother. The coal dust blew over from the tipple or off the piles and coated homes, clothes, whatever it touched. Lorraine didn't have the luxury of a washer or dryer, so she hung wet clothes in the backyard and hoped the wind wouldn't change direction. If it did, "you'd have to do your washing over," said Earl Waddell, who'd lived near the Fox family.

The brothers spent much of their childhood outside, playing football or baseball or wandering through the woods behind the house. The neighborhood kids played together, sometimes atop a dusty pile of waste rock from the mines. "We were running around all snotty-nosed and black as coal and didn't realize we didn't have nothin'," Waddell said. "We were happy just playing and carrying on."

When the mines had temporary halts in production, the Fox boys occasionally ventured into the portals to get a glimpse of the tunnels. Even as kids, though, they had a healthy respect for the hazards. "You don't mess with anything down there," Jerry said. "You know that there's danger there somewhere."

When the brothers were teenagers, their mother remarried, and they got two older stepbrothers along with a stepfather who worked on power lines for the local electrical company. It was in high school, Ernie remembered, that he and his brothers started to realize that other families were better off. They noticed little things—their classmates had nicer clothes and money to go out to eat.

The Fox brothers, especially Gary, dived into work; college wasn't really an option. "You grow up and you don't have nothing to speak of, it kind of drives you," Ernie said. Ernie worked his way up to a spot as a manager at a farm-supply store. Freddie participated in the government-backed program Job Corps, learning a variety of skills and working as a mechanic and a carpenter. After a few years in mining, he worked odd jobs. Jerry started out driving a truck and managing a warehouse for a lumber company, then he went into business by himself, building decks and doorways for houses and selling furniture and crafts.

Gary was particularly eager to get to work. He dropped out of high school, though he eventually earned a GED. In 1968, he got a job at Burger Boy Food-O-Rama, a popular joint selling fifteen-cent hamburgers. The following year, he started at the Acme grocery store. He was making money, building for the future. Once he'd begun working, he wouldn't stop until he had to, in the final years of his life.

On Memorial Day weekend of 1969, students from Beckley College unwound at their usual drinking haunt, the Rathskeller. Mary Lynn Smallwood was two years too young to be there—she was sixteen—but her older sister, Gloria, had sneaked her in. Gloria and her friends, all of them classmates at the college, had a dare for the tagalong: Go steal the pitcher of beer sitting in front of that guy across the bar.

Mary was game, and she walked over and gave it a shot. The guy stopped her. "If you want to drink this beer," he said, "you're going to have to sit here and drink it with me."

She did. He was "tall, slim, and good-looking," Mary thought. He introduced himself: Gary Fox.

They got to talking, and they didn't stop for hours. "From that time on, they were Mutt and Jeff," recalled Richard Roles, who was dating

Gloria at the time and later married her. Gary and Mary often went on walks hand in hand, Gary standing about a foot taller than Mary.

At the Smallwoods' home in Beckley, Gary became a fixture. The patriarch of the family, Wallace, was a mechanic and foreman at a dairy company; he died not long after Mary and Gary started dating. Mary's mother, Frankie, raised six children, three from a previous marriage and three with Wallace: Gloria, then Mary, then John.

Frankie immediately took a liking to Gary. "Gary was her pride and joy," Gloria said. "My mother would have done anything for him." She wasn't sure why; there was just something quietly endearing about the clean-cut young man and his curiously chivalrous courtship of Mary.

"There wasn't a time that we went out that he didn't bring me something," Mary remembered with a smile. "Flowers, stuffed animals—even if it was just a small pack of chewing gum, he never showed up at my house without bringing me something."

Activities included the usual small-town fare. There was an ice-skating rink, a Dairy Queen, and a drive-in movie theater. Teenagers would climb up onto the roof of the Freewill Baptist church nearby and get a glimpse of the film. ("Actually, we watched everybody making out," Gloria confessed, laughing.) Mary and Gary enjoyed spontaneous adventures, but they were also content simply being in each other's company. Taciturn with just about everyone, Gary found a confidante and companion in Mary. "We could just sit and talk," Mary said. "Or we didn't have to talk. We would just sit. But we was always together."

Their budding relationship, however, soon faced a complication. In March 1970, at age nineteen, Gary was drafted and shipped to Fort Jackson, South Carolina, where he learned to fight and shoot an M16. A few months later, he was in Vietnam driving a truck for the army. He wrote home to his mother and exchanged letters regularly with Mary. "She'd go to the post office all the time," Mary's brother John remembered.

After being overseas for almost a year, Gary was in a truck accident, and sheet metal crushed his foot. The injury was so bad that the army brought him home and sent him to Walter Reed Army Medical Center in Washington, DC. There, doctors surgically realigned the fractured bones in his foot and inserted metal rods. After a two-month stay to recuperate, he returned to Beckley, still in a cast and on crutches. He would receive a few medals for his service and an honorable discharge.

He and Mary picked up right where they'd left off, but she noticed something had changed in him, something that would endure for decades. "When he came back, he was a little disturbed. Seeing all that killing and stuff," Mary recalled. "His attitude toward things changed. He was just hostile, like he didn't know how to laugh. He was so calm and everything before he went over. It changed him." Gary's brothers saw it too, but he wouldn't discuss what had happened in the Southeast Asian jungles with anyone—not Mary, not his mother, not his brothers—for years.

Still, Gary and Mary resumed their inseparability, and less than a year after Gary returned home, they got married in a small ceremony at the church whose roof offered teenagers a view of movies and necking love-birds. They chose the date with deliberation: May 25, 1972—just before Memorial Day weekend, almost three years to the day after Mary took the beer-pitcher dare and met the man she'd love for the rest of her life.

Gary had gone back to work at the Acme grocery when he'd returned from Vietnam. Before long, he began driving a truck for various companies, from Coca-Cola to local oil, concrete, and lumber businesses. Mary started at Beckley College but didn't finish. She worked a few jobs off and on. The young couple lived in a trailer on the outskirts of Beckley. Money was tight. Gary was thinking about building a family. He wanted to have a child, buy a home, and supply all the comfort and opportunity that had been out of his reach when he was growing up. All of that would be hard to do on trucking wages. So in 1974, he decided to go underground.

Mary didn't know much about mining at the time. No one in her family had done it. She just knew what everyone else in the area did: "You couldn't beat the insurance or the pay around here." Only later did she learn what went on deep in the earth, all the things that came with that paycheck and insurance coverage. "If we knew, would he have worked in the mines in the first place, if we knew? No," she reflected. "I don't care what we'd had. No, it's not worth your life. The suffering that you do, and there's nothing they can do for you."

2

The Mines

The American coal industry has a virtually unmatched record of death, disease, and destruction, and for more than a century, many of King Coal's worst excesses have been concentrated in central Appalachia, with southern West Virginia at its core.

The essence of a coalfield adage dating to the industry's early days, when mules hauled out of the mine what men had extracted, has remained true for generations of miners: "Coal operators always cared more for the mules than the men. A man they just paid for the work he did; a mule they had to *buy*."

The hard-won, incremental reforms achieved by miners and their allies over the decades since have been battles in the ongoing war to put the man on par with the mule in the eyes of company bean counters. Make the companies pay for a portion of the damage their jobs wrought on men's bodies, and they'd take a bit more interest in preventing that damage in the first place. Or so the thinking has gone. But coal companies have proved endlessly adaptive, adept at wriggling out of one legal vise after another.*

---

* Representatives of the industry's primary trade group, the National Mining Association, declined my requests for an interview about the many issues, both historical and contemporary, raised throughout this book. "We disagree with the premise of your book, which appears to be already fixed," a spokesperson said.

Anyone who buys the stereotype of the simple-minded mountaineer who must make his living with brute labor because he is incapable of more high-minded pursuits surely hasn't spent much time talking with coal miners. Yes, the work is backbreaking, but it also demands a wealth of knowledge, keen intuition, and skill. The miner is, in essence, part geologist, part engineer, and part all-purpose handyman with at least a passing familiarity with principles of aerodynamics and hydrology. The self-effacing miner describes all of this as simply "running coal." But the job is anything but simple, and the conditions are almost unfathomable to the uninitiated. The miner may earn his pay with the sweat of his brow, but he survives by the acuity of his wits.

For the underground miner, the possibility of death or crippling injury is ever present, and in few workplaces are there so many different ways death might come. At any given moment, the miner could be asphyxiated, crushed, drowned, burned alive, or blown up. Though surface mines account for the greater share of production in the United States today, underground mines employ more workers.

The underground miner goes to work in the middle of a layer cake, each layer telling the story of a past epoch in the Earth's history. Descending through sheets of sandstone and shale, the miner is traveling back in time to those precious eras when plants died in conditions that forestalled decomposition, allowing them to retain much of the solar energy they'd captured. Over millennia, layer after layer piled atop, pressing down and squeezing the layer of plants into the black rock that humans later discovered and burned to release the trapped energy that has powered the modern world.

The techniques people developed to extract this precious mineral were often crude. For hundreds of years, mining amounted to little more than hacking at the layer of coal with sharp tools and blasting it with explosives. In the United States, methods remained surprisingly rudimentary well into the twentieth century.

As mining boomed in Appalachia, starting in the late 1800s, the dominant method of removal was room-and-pillar, and the technique remains one of the two primary approaches to underground mining today.

After the company bored a hole in the earth and reached the coal seam, miners pushed forward in parallel, hollowing out an organized network of tunnels. Each miner got a "room": an opening that might be twenty feet wide and was only as high as it needed to be. The thickness of a coal seam varies, and companies didn't want to waste time cutting out the rock above and below it. As a result, a miner might spend the day standing in a six-foot-high opening or crawling in a three-foot-high opening, a rite of passage known as working in "low coal."

In that room, the miner attacked the untouched seam, an area called the face. During the late 1800s and early 1900s, this meant cutting a few feet of coal from the base using a pick, drilling holes in the face, and inserting explosives. The miner then retreated to relative safety to detonate the blast, and if the undercutting worked, the coal would shatter and fall. Finally, he'd shovel the coal into cars to be hauled out of the mine.

In the early 1900s, a miner might earn about two dollars on a good day. He[*] had to buy most of his own tools, and the company paid him only for the weight of the coal he produced. If his car had "dirty" coal—a load containing some waste rock—the company docked his pay. The company also wouldn't pay him for his time performing what many miners called dead work—the tasks that kept the mine in working order and kept the miner alive. As the miner moved forward, chipping and blasting away at the face a few feet at a time, he had to do something to prevent the increasingly unstable roof from caving in. So he set up wood beams, hoping this would hold off the weight of the mountain above, the same pressure that had squeezed plant matter into the black rock he was removing. It was a little like siphoning out the fudge in the

* The use of masculine pronouns throughout this book is not meant as a slight to the women who work in the mining industry. Rather, it is an attempt to avoid awkward and confusing syntax and reflect the fact that the workforce in underground mines has long been predominantly male. A 2012 government survey estimated that about 96 percent of coal miners were male. Well into the twentieth century, some states had laws barring women from working underground. Nonetheless, many women have proven themselves in the mines, and in numerous bitter labor strikes, miners' wives have been the driving force on picket lines, persisting amid hardship and winning concessions. Since the 1970s, the number of female miners has increased, as women have in many cases overcome discrimination and harassment.

middle of a layer cake and hoping a few toothpicks would keep the whole thing intact.

After a bit of progress, the miner cut an opening in the side of the room, connecting the area where he was working with that of the man advancing in parallel in the next room over. What was left behind were pillars of coal to support the roof. Thus, the grid of rooms and pillars emerged. Often, after a company finished with an area, it would have miners remove the pillars on the way out, lest any of the coal go to waste. As miners retreated, they allowed the roof to collapse behind them in a sort of controlled cave-in—a precarious practice still used in some mines today that has often proved deadly.

By the 1940s, many mines had upgraded their equipment, replacing picks and shovels with machines that undercut the coalface and loaded and hauled away the spoils. Some miners worried technological advances would put them out of a job, but the opposition to mechanization received a death blow in the 1950s when powerful new machines called continuous miners appeared on the scene. A worker could drive this behemoth up to the coalface and use its spinning drumhead lined with drill bits to rip through the seam at a pace once unthinkable. The United Mine Workers embraced the new technology, and production skyrocketed.

But while technology brought stunning advances in production, it didn't bring comparable advances in safety and health. Miners continued to die at an alarming clip.

There is, of course, an inherent danger to the work. Aside from a full-fledged cave-in, huge chunks of rock can break loose and cause crippling or fatal injuries. Underground water sources can burst into and flood a tunnel. Miners know well the three deadly gases: choke damp, also known as black damp (carbon dioxide); white damp (carbon monoxide); and fire damp (methane). The first slowly robs the mine air of oxygen, causing suffocation. The second poisons the miner by preventing oxygen from reaching the tissues. The third is perhaps the most feared; as methane seeps from coal seams, it can fuel massive explosions. Another fuel for fires and explosions is the coal itself; blasting and gouging at the face produces tiny dust particles that are themselves combustible. In a nightmare scenario, the two fuels team up: a spark triggers a methane explosion, the blast wave lofts accumulated dust into the air, and a wall of flame races through the network of tunnels, feeding off the tiny coal particles as it goes.

Over the years, Gary Fox suffered a few injuries, but he managed to avoid the deadly accidents that claimed the lives of some of his coworkers. In the two years before Gary took his first mining job at Itmann Coal Company, which was owned by the industry-leading Consolidation Coal Company, there were ten deaths at the complex of a few mines where Gary would work. Three men were crushed by falling rock, two were struck by equipment, and five died in a methane explosion. The blast victims ranged in age from twenty-four to forty-four and left behind five wives and seven children. While Gary worked at Itmann, two more miners died after being hit by machinery. A year after Gary started at Maben Energy's Birchfield mine, a roof fall killed a thirty-three-year-old miner.

As a roof bolter, Gary had to train his eyes and ears on the roof and intuit what the mountain was telling him. He listened for faint creaking or popping sounds, though the hum and grind of the machinery usually made it hard to hear much. He watched for "workin' top"—small pieces of rock dribbling off, cracks forming. Bolt men were especially wary of "kettle-bottoms." These fossilized remnants of tree stumps fell out in one big, deadly hunk, leaving behind a space in the roof that looked like an inverted kettle. For a few months during Gary's time at the Birchfield mine, the crew bored through the earth below a river, and water dripped ominously from cracks in the roof. Sometimes, when Gary drilled an opening in the top, a geyser shot down into his face. He'd spend the entire shift in water up to his waist, and when he emerged outside on cold days, his work pants froze. By the time he made the short walk to the changing room, his body was half Popsicle.

Among miners, there has long been an acceptance of risk, almost a pride in it. Just as the region's early inhabitants managed to survive in a landscape too unforgiving for many settlers, coal miners earned their living in a workplace so harsh, only the hardy could last. Some have always been convinced that there's no way to operate a coal mine without occasional accidents and a constant haze of disease-causing dust—or at least, no way to operate it *profitably*. Many mine owners have been more than happy to nurture that belief, and they've found pencil-pushing bureaucrats who don't know the first thing about coal mining to be an easy target, right up there with the environmentalists hell-bent on shutting down every mine in the country.

But the dangers of mining are hardly as unpredictable or impossible to address as many coal operators have led the public to believe. Some of the remedies that lessen the risk are well known and fairly simple—setting up roof supports, keeping plenty of air flowing to ventilate the tunnels, spreading dust from rocks that aren't combustible to nullify the coal dust. But doing these things takes time—time that miners aren't producing coal.

Many of the worst disasters in the long record amassed by the industry have resulted not from acts of God but from failures of men. The toll of preventable catastrophes is staggering. More than one hundred thousand miners have died in coal-mine accidents in the United States since 1900. The worst of these came at Monongah, West Virginia, in 1907, when an explosion killed three hundred and sixty-two. The number of deaths each year didn't dip below one thousand until 1946.

But even this record of violent death pales in comparison with the toll from the disease that slowly steals miners' breath. For the poor souls taken by this scourge, there are no news stories commemorating the anniversary of their sacrifice, no public apologies to grieving widows, no rallying cries of "Never again!" They simply suffocate in a slow-motion disaster that plays out over years in homes tucked deep in mountain hollows. They are trapped in that moment just before death when a miner hears the crackle of coal dust igniting and realizes he is powerless to escape the flames. For the diseased miner, this moment is slowed to a crawl, advancing by torturous milliseconds month by month, year by year. The quick deaths of lost coworkers from a fireball, blast wave, or falling slate can start to seem merciful.

The response to this plague by many coal companies and some in the medical and legal professions has been nothing short of a national disgrace.

* * *

In 1831 a doctor in Scotland named James C. Gregory sliced into the chest of a dead coal miner and was struck by what he saw: the man's lungs were almost entirely black, and they contained large cavities filled with black fluid. The miner had been a patient of Gregory's who had come seeking treatment for terrible chest pain and shortness of breath. Gregory wrote up a case study arguing that the reason for his patient's respiratory problems was the coal dust he'd inhaled on the job.

Across the Atlantic, in Pennsylvania coal country, Dr. John Carpenter reported similar findings to colleagues in the county medical society in 1869, and others in the United States followed with further reports. At the 1881 annual meeting of the Colorado State Medical Society, physician and president of the society H. A. Lemen presented a paper on a patient, a longtime coal miner, who coughed up a pint of inky black liquid each day. To drive the point home, he announced to the audience: "The sentence I am reading was written with this fluid. The pen used has never been in ink."

This coal miners' curse was so well known that it even seeped into popular culture. In French author Émile Zola's 1885 novel *Germinal*, the protagonist, a young man, arrives in a coal-mining town and meets an old miner who struggles to carry on a conversation amid coughing fits: "When finally, with one deep rasping scrape of the throat, he had finished coughing, he spat by the foot of the brazier, and the earth turned black." When the young man asks the miner if he's hacking up blood, the old man replies: "It's coal... I've got enough coal inside this carcass of mine to keep me warm for the rest of my days. And it's five whole years since I was last down the mine. Seems I was storing it up without even knowing."

By the early 1900s, there was broad recognition within much of the medical community that a distinct and deadly disease was afflicting coal miners. The symptoms were well documented: chronic coughing that sometimes produced black sputum, ever-worsening breathlessness. So was the pathology: blackened, scarred lung tissue. Articles and textbooks situated the disease within a family of illnesses called pneumoconioses, which were caused by breathing various kinds of dust—in this case, coal.

And doctors recognized that it was common among miners and that the damage it wrought appeared to be irreversible.

But government agencies in coal-mining states and at the federal level largely balked at supporting the kind of research that could have yielded a definitive understanding of how the disease worked and how prevalent it was. Meanwhile, there was little progress made in response to some medical professionals' calls for controlling the amount of dust miners had to breathe. State and federal agencies often viewed it as their job to promote the industry's economic prosperity, and inspectors with limited authority tended to focus on preventing explosions and other well-known calamities, not on the slow death of dust disease.

Miners might not have been versed in the burgeoning medical literature, but it seemed quite clear to them that their breathing problems were caused by the dust they sucked in day after day. Yet for the most part, they were left to fend for themselves. People living in coal towns throughout Appalachia often had little choice of how to earn a living. Many started as breaker boys, some of them as young as eight, separating the waste rock from the coal in a dust-filled processing plant. Over the years, a boy could work his way up the ladder to better-paying jobs—tending the doors that regulated ventilation to the mine, leading the mules that hauled cars loaded with coal, and, by his twenties, mining the coalface, the highest rung reachable for most. If he was lucky and escaped fatal or career-ending accidents, he could work in this position for years. But the descent back down the wage ladder was almost inevitable. Age and disease sapped productivity, and the miner found himself demoted to the lesser-paying jobs he'd done before. Eventually, he might end up back as a breaker boy, pulling down the meager wages he had as a child. An 1884 folk song captured the indignity: "It's twice a boy and once a man, is a poor miner's life."

Health benefits and pensions were almost nonexistent. Miners tried to stick it out for as long as they could, many turning to folk remedies. Some men wore handkerchiefs to cover their mouths or chewed tobacco, thinking this would catch the dust particles. Tradition held that the best way for a man to clear the dust out of his system was to drink a shot of whiskey with a beer chaser. Merchants peddled nostrums that often contained some mixture of liquor and herbs.

Then, despite what appeared to be a growing consensus in the early decades of the 1900s that inhaling coal dust could be harmful, something remarkable happened: The disease seemed to vanish. Some coal companies and industry-aligned doctors mounted a sustained disinformation campaign during much of the first half of the twentieth century that proved quite effective.

Their main strategy was to attack the cause of this supposed disease, which went by a variety of names, including miners' asthma and anthracosis. Unlike some types of dust that were recognized as agents of illness in workers, coal dust was "relatively harmless," a pair of prominent radiologists wrote in a 1926 monograph.

Sure, coal dust might blacken miners' lungs and sputum, but that didn't mean it was doing any harm. In fact, companies and doctors contended, a bit of coal dust might actually be a blessing in disguise. Repeating a common claim, Illinois physician M. C. Carr wrote in a 1905 article published in the *Journal of the American Medical Association* that coal miners were "practically immune from tubercular infection." In 1915, the industry publication *Coal Age* proclaimed that "the atmosphere of the mine is now vindicated even though its healthfulness has not yet been extolled." The same year, the magazine published an article titled "The Long Life of Coal Miners" and a third piece that contended, "There are no healthier men anywhere than in the mining industry."

If any miners really were sick, there was some other cause. Maybe it was booze or tobacco. Maybe it was laziness or a weak-minded "fear of the mines." Responding to a 1901 paper that found widespread illness among miners, one industry official maintained, "There is nothing in mining that makes it insanitary, and any insanitary conditions which may exist are doubtless closely related to the rum shop."

The second part of this deflection strategy was to blame everything on a different type of dust: silica. The mineral is commonly found in the quartz that pervades layers of rock in the Earth's crust, and its dust can cause the disease known as silicosis. Silica dust has long been recognized as a serious danger to workers in a range of industries, from mining and quarrying to construction and manufacturing. By scapegoating silica, the industry largely avoided addressing the amount of coal dust in the mine air and effectively made it impossible for most coal miners to qualify for workers' compensation.

For its part, the federal government did little to correct the record or protect the hundreds of thousands toiling underground. In the 1920s, the U.S. Public Health Service conducted a series of studies that found high rates of respiratory disease in coal miners, and some of them pointed to coal dust as the cause. But many of the key findings remained under wraps or were softened at the insistence of coal companies or officials at the Bureau of Mines, the federal agency that was charged with policing mines but that lacked enforcement authority and had a reputation as more of a cheerleader for the industry. The research that did make its way into the medical literature did little to dislodge the view that coal dust was benign.

In the 1930s and 1940s, a reinvigorated United Mine Workers union took the fight to the political arena, pushing for state compensation laws. But here, too, the industry's disinformation campaign proved effective. The laws that states including West Virginia passed allowed compensation only for silicosis.

While delay and denial reigned in the United States, major advancements were happening in Great Britain. Doctors there began amassing evidence of a disease separate from silicosis, and the government commissioned a large-scale study of Welsh coal miners. The results, published in 1942, identified a distinct and disabling disease—coal workers' pneumoconiosis—caused by coal dust alone. The next year, the government enacted a law that made miners with this ailment eligible for workers' compensation.

It would take another twenty-five years for the United States to reach this same point, and change would come only after rank-and-file miners themselves rose up and demanded it.

3

John

The man in the photograph looked directly into the camera lens, fixing readers of the *Fayette Tribune* with a gaze that was intense but somehow unthreatening, almost playful. His sandy-brown hair, parted in the middle, fell just over his ears. He wore blue jeans and a sweater-vest over a plaid shirt with the sleeves rolled up, and he sat in a rolling chair with his hand resting on the edge of his desk. In the background, papers, binders, and file folders filled the small office he shared with a coworker.

Above the picture, the headline read "John Cline—the Miner's Friend." The photo was taken by a retired coal miner who had become a skilled photographer and avid chronicler of the area's coal heritage. His wife, herself the daughter of a coal miner, had written the accompanying article, which described the work John was doing in this basement office of a rural community health clinic.

The New River Health Association was an oasis in a medical desert. Tucked against a hillside creek in the sleepy coalfield town of Scarbro, the clinic was named, like so many things in the area, for the nearby waterway that had carved one of the longest and deepest gorges in all of the Appalachian Mountains. The facility sat within what had once been the domain of the New River Company; it had used the land where the clinic now stood as a place to dump waste rock. Coal baron Samuel Dixon had named Scarbro for a town in his native Yorkshire, England, just as he had with Skelton, the company town a few miles south where Gary Fox had grown up.

The main branch of the clinic was a much-needed source of primary care for people in the area. It had a sliding scale of fees based on income and family size, and it didn't turn away those who couldn't pay at all. The clinic basement was home to the New River Breathing Center, one of a few sites spread across coal-mining states that received a federal grant to diagnose and treat working and retired miners with black lung.

The reputation of the breathing center had filtered out along coalfield grapevines: Folks there, it was said, were on the coal miners' side. They wouldn't look down on you on account of the way you dressed and talked, probably because lots of people who worked there were themselves the daughters, sons, sisters, or brothers of coal miners. Coal miners practically ran the place. They'd helped start the clinic, and they made up a large portion of its board.

Miners traveled to the breathing center from all over West Virginia; some even came from Virginia and Ohio. They didn't just want to know how their lungs were; if the tests showed impairment, they wanted help trying to win benefits payments. That was what many doctors balked at doing, not wanting any part of the legal mess. The staff at New River didn't have those qualms. In fact, that was exactly what John had been hired to do, and he'd taken to the task with alacrity.

John's title was benefits counselor. The job called for what the nurse who supervised him dubbed "cross-training"—a little medicine, a little law. John had possessed no expertise in either, but he'd been happy to learn. He'd gotten certified to perform the breathing tests used to detect impaired lung function, and he'd learned on the job how to help a miner file a benefits claim. He had developed a certain medicolegal fluency and an easy rapport with the miners who came into the clinic. Spirometry—which evaluated pulmonary function—could be a downright unpleasant test, especially for someone with dust-laden lungs, so John liked to start by explaining exactly what was being measured and why it was necessary. It tended to defuse any tension.

"One of the main things our lungs do," he'd begin, "is take air in and get it back out, like a bellows. But coal dust can choke off that flow. We need to know how much air your lungs can get in and out and how fast." He'd nod to the machine in front of them in the small exam room and say, "That's what this measures."

John then had the miner blow as hard as possible into a tube. As the air flowed from the tube into the main apparatus, a cylinder rose up, lifting a pen that was marking the changes on graph paper attached to a rotating drum. After a few repetitions of this, John would show the miner the arc of his breath and explain what it meant. A sharply rising arc was a good sign; a nearly flat arc meant the miner could barely blow out a candle. John would run some calculations by hand to get a few scores.

Sometimes, the news was good—the scores were in the normal range; the miner should come back in a couple of years for another check-in. If the news was bad, John might offer to set up an appointment with the clinic's doctor and help the miner file a benefits claim. When John had started at the clinic, that usually meant a claim under an occupational-disease program that was part of West Virginia's workers' compensation system; it offered lump-sum payments based on how disabled someone was and allowed for further payments if later tests showed worsened breathing problems. These payments, however, weren't a long-term solution for miners who were permanently disabled.

Under the state system, it was difficult for miners to qualify for the ongoing payments that came with a finding of total disability, and even if a miner cleared that high bar, the payments eventually would end. The better option for miners who became so ill they could no longer work was the federal black lung–benefits program. If a miner could prove he was totally disabled by black lung, he was entitled to monthly payments and coverage for medical care related to the disease for the rest of his life, and if he died, his widow could qualify to continue receiving payments.

The federal program, however, was contentious and complicated. It was a coalfield axiom that to win, "you gotta have one foot in the grave." John would help a miner file a federal claim if he wanted to, but not many did. Even in these cases, John's job as a benefits counselor was to help get the claim started and maybe provide some guidance along the way. Getting more involved in what could be a lengthy, intricate legal process seemed prohibitively time-consuming, not to mention a little crazy. A bit of cross-training wasn't preparation for a marathon.

John's plate was already so full, it had become a sort of running joke among the other clinic staffers. "He would leave work, and I would tell whoever I was with, 'Watch, he'll be back in a minute,'" said coworker Susie Criss. "You could go, 'Five, four, three'—he would run back in because there was something else he wanted to take care of before he left. He would do that about three times before he would actually pull out of the parking lot."

Yet he somehow never seemed to be in a hurry. He sometimes picked up patients, drove them to the clinic, then took them back home again. He encouraged them to call him anytime about anything, whether it was a health problem or a family issue. One retired miner occasionally caught a ride with John just for the companionship. He didn't have much money and sometimes would ask John to file a benefits claim for him. John had to explain each time that his breathing-test scores simply didn't qualify; there was no point. Still, John liked him, and when he'd show up at the clinic selling honey from the bees he kept at his home, John always bought a jar.

John craved social interaction. As a kid walking home from school, he'd sometimes stop at the local pub, sidle up to the bar, and order a water just so he could chat with the bartender. Because of his soft and deliberate manner of speaking, John could seem almost shy at first. Really, though, he was choosing his words with care; it was almost as if he were analyzing his thoughts as he had them.

George Bragg, a retired miner who joined the board of the New River clinic, noticed John's interactions with miners and assumed John was a local, probably the son of a coal miner. John's boss, Brenda Halsey Marion, knew John's background. The Cline family was middle-class, from the Northeast, and white-collar. Whatever drove John was something different.

"He could relate for some reason to the blue-collar worker," said

Marion, herself the granddaughter and niece of miners. "He could relate to people from the coal camps that didn't have much, trying to bring up kids, not much money, not much of a house or supplies. For some reason, he had that, and if you ever met his dad or his mom, you would understand."

John's path to Appalachia began in Queens, New York. His parents, Crawford and Catharine, had met through their work at American Airlines, and they had John in 1945. Crawford soon took a job in Washington, DC, and the family lived in a nearby Virginia suburb. Before long, John had two younger sisters, Ann and Cathi.

These big cities weren't really home, though. John's father had moved for work, but much of his family remained in East Aurora, a village in western New York, near Buffalo. John was a boy of five when his father decided they were moving back there.

East Aurora was, in many respects, a quintessential slice of small-town America, and it largely still is. Even today, visitors can stroll down Main Street, pop into the crowded five-and-dime for some sponge candy, catch a show at the single-screen movie theater with a marquee of neon lights, or head over to Wallenwein's restaurant and bar to try the area's signature delicacy: beef on weck (sliced roast beef on a kummelweck roll, a bun seasoned with salt and caraway seeds). The toy company Fisher-Price was born here, and its headquarters remain today. Annual traditions such as the Kiwanis Club's chicken barbecue have endured for decades, and the community theater group is one of the country's oldest.

John's paternal great-grandfather came to East Aurora in the early 1900s. A Baptist minister from Canada, he laid down the family's American roots. One of his children, John's grandfather Herbert, married a woman from a local farming family, and he managed a few area businesses and served as a member of the village board and president of the Kiwanis Club. He also had his own woodworking shop, and John's periodic visits exposed him to the craft that generations of Cline men had practiced.

The other side of the family tree was of similar stock. John's maternal great-grandfather, James Byron Brooks, fought for the Union in the

Civil War, then earned degrees from a Methodist seminary, Dartmouth College, and Albany Law School. Settling down in Syracuse, he made a name for himself teaching and practicing law. Brooks and his wife, a seminary teacher named Caroline Jewell, adopted an orphan, Elizabeth Brooks Lyon, who would have a profound effect on the lives of John and his siblings. She graduated Phi Beta Kappa from Syracuse University and taught Greek and Latin. She and her husband had a girl named Catharine, John's mother.

In 1951, Crawford moved the family back to East Aurora so his mother-in-law—"Grandma Lyon" to the Cline kids—could live with them. Crawf and Kay, as friends called them, soon had a fourth child, Andy. They were a singing family with a wide-ranging repertoire, everything from folk songs to show tunes, religious hymns to Bob Dylan. Behind their house they had a barn where they raised chickens and a garden where they grew rhubarb. Like many people in the area, they and their neighbors would make their own ice-skating rink in the winter by flooding a flat area in the yard and letting the frigid air from Lake Erie do the rest. Crawf ran an insurance business, selling the locals policies for their houses and cars, and Kay stayed at home to take care of the kids. They lived comfortably but modestly. Both Crawf and Kay had grown up during the Great Depression, and it had left an indelible imprint.

In high school, John was a solid athlete, finding his place on the defensive side in football. "He was small, but he liked to hit," Andy recalled. "He was not afraid of contact at all."

High-school friend and football teammate Tim Burke still laughs about the time John got in a fight—who started it and why is lost to the years—outside the movie theater on Main Street: "John and this guy go behind the movie theater and duke it out. Can you imagine what happens after that? They both come walking out, and John and this guy are best buddies after punching each other. When I look back on it, that really is John. He's a tough guy, but he's still really friendly."

As the big brother, John looked out for his siblings. When Cathi was smarting after a breakup, John had the right salve. "He took me into Buffalo to see the Beatles movie," she said. "That was a big adventure.

I think he had a way of knowing when people were hurting and then finding a way to try and alleviate that."

Perhaps most important for John and his siblings' future, the Cline household was a virtual lyceum. Grandma Lyon kept a collection of books, and she voraciously consumed all manner of newspapers and magazines, from the *New York Times* and the *Christian Science Monitor* to the left-wing magazine *Ramparts* and the newsletter from the Highlander Center, a grassroots organization pushing for social justice in Appalachia. They would read and discuss everything from Plato to George Orwell to the reviews of the latest Beatles album.

The evening news provided grist for many discussions. The Vietnam War was raging, and the family was captivated by the civil rights movement. East Aurora was almost entirely white, so Kay joined a program mentoring black youth from nearby Buffalo. Crawf worked with a nonprofit to combat racial discrimination in home sales and lending practices.

Crawf and Kay were active in the Methodist Church, but they didn't force belief on their children. They sought out a church in line with the Cline family tradition—less fire and brimstone, more New Testament compassion. John was confirmed in the church as a teenager. He didn't particularly care for the sermons, he later said, but "I was very much interested in learning about the life of Jesus and trying to understand what that meant or what He was trying to teach."

Cathi recalled, "We were taught as kids that, as much as we were meant to be happy, it was also a part of our obligation to be involved and to work for justice for other people also."

Ann went on to become an elementary-school teacher, working much of her career in Chicago public schools. She and her husband took in two foster children, in addition to raising their biological daughter, and helped set up a progressive church focused on organizing social welfare efforts. They eventually retired to southern West Virginia, where Ann would work part-time as a supervisor for student-teachers.

Cathi became a supervisor in the New York State government office charged with collecting unpaid child support. She and her husband got involved in a refugee-resettlement program. At the same time, she served as a lay preacher with the United Church of Christ, often at a church in inner-city Albany. After retiring, she attended seminary, was ordained as a minister, and spent a year working in an AIDS-ravaged village in Kenya teaching women practical skills. She then moved to Houston, where she coordinated community organizing and social programs at a church.

Andy spent almost four decades teaching high-school math and coaching baseball at schools in Ohio and Massachusetts, eventually retiring to a quiet seaside town in Maine. Like John, he carried forward the generations-old Cline tradition of woodworking, spending a lot of time in the small shop built onto his garage. "It's really kind of in our heritage and in our blood to want to build things out of wood," he said.

That heritage was in John's blood too, and as a teenager, he found himself fascinated by the construction projects he'd walk past on the way to high school. Crawf suggested he ask a local contractor about a summer job. He did, and at sixteen, he was shoveling muck and applying tar to walls as part of a crew building houses. In 1963, John enrolled at Allegheny College, a small liberal arts school in Meadville, Pennsylvania. He was on the football team, but he suffered an injury before the season even started. He took to drinking heavily, and his grades tanked. Partway through his second year, he washed out and went home.

Back in East Aurora, John resumed working on houses. Pretty soon, however, a letter from the Selective Service System arrived at the Cline home. Without the deferment that came with being a college student, John now faced the possibility of being drafted and sent to Vietnam. He decided to give school another shot and headed off to Bryant and Stratton Business Institute in Buffalo. In 1967, he earned an associate's degree in business administration, but he still felt adrift, uncertain of what he wanted to do with his life. He was looking to do something useful, but the what, where, and how were inchoate notions floating around in his head.

He read something about a program under the auspices of the federal government's War on Poverty called Volunteers in Service to America, VISTA, a sort of domestic version of the Peace Corps, and decided to sign up. After being trained, he settled on an assignment in southern

West Virginia, and in February 1968 he took up residence in a two-room cabin in a narrow mountain hollow in one of the poorest counties in one of the poorest states in the country. He rented the cabin from a neighbor, a retired coal miner, with whom he shared an outhouse.

Though John hadn't known it when he selected the assignment, the region was on the cusp of one of the most unlikely and shocking labor rebellions in American history. The movement would lay the foundation on which John, his colleagues at the New River clinic, and countless miners would later stand, and it was this movement that they, in their own way, would seek to revive.

For years, anger had been spreading among rank-and-file miners over the refusal by government, industry, and their own union even to acknowledge the toll of black lung, let alone prevent it and provide compensation to its victims. As John got an introduction to the landscape and the people of rural Appalachian coal towns, that anger boiled over into a grassroots uprising that would capture the nation's attention and assume a preeminent spot in the canon of coalfield lore.

4

Wildcat

In the early-morning hours of a frigid Wednesday in November 1968, explosions ripped through the underground tunnels of the Consolidation Coal Company's no. 9 mine in Farmington, West Virginia. The blasts shook homes located miles away and launched plumes of smoke from mine shafts. When the dust settled, seventy-eight miners were dead in what amounted to the worst mining disaster in almost two decades. Efforts to recover all of the bodies would drag on for years until the company finally gave up and sealed the mine, entombing nineteen of the dead.

In the immediate aftermath of the explosions, television crews descended on the small mining town in northern West Virginia, and they stayed for days as further explosions shook the earth and emergency responders battled out-of-control fires underground. The major networks broadcast nightly updates, and Americans got to see the destruction, the harrowed first responders, and the women and children pleading with the company to continue searching for their husbands and fathers. The human toll of a broken regulatory system became impossible to ignore.

When Governor Hulett C. Smith and United Mine Workers president Tony Boyle arrived in Farmington, they seemed to defend the company, which was known as Consol and was one of the nation's leading coal producers. Smith said he was confident it was taking "the most improved safety measures" and added that "mining is a hazardous profession." Boyle, sporting a sharp suit with a rose in his buttonhole, called Consol "one of the better companies" and said, "As long as we mine coal, there is always this inherent danger."

In fact, evidence would show that the disaster had been far from unavoidable; rather, Consol managers had failed to follow well-established safety standards. If miners needed further confirmation that their government and their union weren't going to stand up for them, this was it.

Among those who came to Farmington and *did* express outrage was a Charleston-based doctor named I. E. Buff. The outspoken physician-activist seized the opportunity to direct the attention of news crews to the groundswell of miners demanding that companies and the government admit that coal miners' lung disease was real and deserving of compensation.

By that point, a movement had been spreading gradually throughout the year, and it was starting to gain traction. The Farmington disaster brought the issue a sense of immediacy and a national spotlight. With the union unwilling to support the movement, rank-and-file miners found other allies: a trio of doctors, led by the bombastic Buff, and a group of youthful volunteers versed in grassroots activism who had arrived as part of the federal government's War on Poverty, which included the VISTA program that John Cline joined.

The cause proved to be a draw for many youngsters. Among them was Craig Robinson, who eventually became a colleague and friend of John Cline's. A native of the Buffalo area, Robinson arrived in southern West Virginia in 1968 just a few months before the Farmington blast. As a student at Oberlin College, he had become fascinated with Appalachian culture after attending a presentation by miners who described their fight for health benefits. Following graduation, he signed up as a VISTA and requested to be sent to Kentucky or West Virginia.

Not long after moving to the coalfields, he connected with a local association of disabled miners and widows who were pushing for changes to state law that would allow miners with lung disease to qualify more easily for workers' compensation. He started attending meetings and talking with miners about their breathing problems and their fruitless attempts to win disability claims.

Meanwhile, Buff began advocating for compensation for miners sickened by coal dust, and beyond the ears of the Boyle-controlled union hierarchy, frustration swelled among rank-and-file miners.

In a fortuitous convergence, Robinson and a few other reformers attended a gathering at a Kanawha County union hall where Buff was holding court. By then, the sixty-year-old cardiologist had connected with two like-minded doctors and formed a group that eventually became known as Physicians for Miners' Health and Safety. They were fine-tuning what many referred to as their "dog-and-pony show." Seeing them in action at the union hall was "revelatory," Robinson said. "That inspired us."

Robinson forged an alliance with the trio, and in counties throughout the coalfields, VISTAs joined with local unions to arrange gatherings in schools, churches, courthouses, and community centers at which the doctors would speak. Their mission was a radical one: Tell miners the truth about what was happening to their lungs. They would provide the facts—mixed, of course, with more than a little rabble-rousing—that would empower miners to help themselves.

The doctors demystified the disease, dragging it out of the abstruse medical morass that was the province of company doctors and into the light of plain language. They eschewed medical jargon and instead used a term that was simple, evocative, and horrifying: *black lung.*

"You all have black lung, and you're all gonna die!" Buff would roar to audiences of miners. Then he'd pull out two hats for a sort of Jekyll-and-Hyde routine. Playing the role of a state legislator, he would don a white hat and speak reassuringly to miners. Then he'd switch to the black hat and describe the more sinister conversation he said that lawmakers engaged in with coal executives.

Next up would be Hawey Wells, a pathologist from Morgantown, West Virginia, who had performed autopsies on many coal miners. He'd hold up a piece of blackened lung tissue removed from a dead miner and crumble it between his fingers. That, he'd tell the miners, was what would become of their lungs too. "You're crazy if you let them do this to you," he'd say.

After the two showmen finished, it would be Donald Rasmussen's turn. A soft-spoken and methodical pulmonologist with bright red hair and a thick red beard, he'd describe what the medical evidence showed about black lung, drawing on laboratory test results from the many coal miners he had treated at a clinic in Beckley.

Buff could trade rhetorical barbs with anyone, but Rasmussen's sober science backed up Buff's bluster. Rasmussen turned out to be the linchpin of the whole movement. Buff would die five years later, and Wells would turn to other pursuits, but Rasmussen would remain the miners' tireless champion for the rest of his life. In time, he would become a key counselor to John Cline and a resource for Gary Fox.

Rasmussen grew up in Colorado and earned his medical degree from the University of Utah in 1952. From 1955 to 1962, he served in the army, working at military hospitals in California, Colorado, and Texas as a specialist in chest diseases. After completing his army service, he was looking for someplace to practice when he saw an ad in the *Journal of the American Medical Association* that read, "Doctors needed in Beckley, West Virginia, at the Miners Memorial Hospital." He recalled, "I came in October 1962 just to look around, and I never left."

He was impressed by the facilities and staff, and, he later said, "I also liked the idea of caring for the coal miners." He had no experience with coal miners' lung disease, but he soon became interested and began studying his patients' peculiar affliction.

Many of them presented a puzzle: When they exerted themselves, they quickly became breathless, but when they took the standard tests used to identify lung disease, their results were basically normal. Sometimes their X-rays didn't show significant scarring of the lungs. Rasmussen had seen something similar in Colorado, so he started running an additional test on his patients. He drew blood while they were exercising, and analyses confirmed what he'd suspected: Their lungs weren't transferring enough oxygen to their blood. These miners weren't malingerers, as some doctors contended; they were really sick.

Coal companies and their allies in the medical establishment vilified the trio of Buff, Wells, and Rasmussen. Two county medical societies passed resolutions declaring that the three physicians were spreading alarmist messages untethered to fact. One doctor who led the charge in Cabell County insisted: "There is no epidemic of devastating, killing and disabling man-made plague among coal workers.... It is my opinion that the false prophets and deluded men who present their hypotheses concerning technical medical problems to the general public without discussion or presentation should be condemned."

The harsh reaction laid bare the real points of debate: Was there a particular disease afflicting coal miners, and, if so, who got to define it? Much of the medical, legal, and political fighting that has endured from the time doctors first noticed breathless miners with coal-stained lungs until today boils down to answering these questions.

Many coal-industry officials and their allies in the legal and medical professions have used the same playbook over and over: Deny, then contain. First, they insisted coal dust was harmless and actually protected miners from tuberculosis. When that position became untenable, they moved to containment: Some coal miners might have gotten sick because of their work, but only the ones who were exposed to a lot of silica from ground-up rock.

In 1968, the industry again found itself in a crumbling fortress, and, with Buff, Wells, and Rasmussen leading the charge, thousands of coal miners were banging at the gate. The scientific evidence that lung disease caused largely or entirely by coal dust was rampant had reached critical mass. A Public Health Service study estimated that one in ten working miners and one in five nonworking (mostly retired) miners had a coal dust–related lung ailment, and the surgeon general soon would estimate during a congressional hearing that one hundred thousand current or former miners had the disease. The Department of Health, Education, and Welfare had proposed a limit on the amount of dust allowed in mine air, saying this would help prevent the disease.

In the face of all this, the industry abandoned the fort, retreated, and dug in anew. The containment strategy this time: Coal dust could cause a specific disease, coal workers' pneumoconiosis, but this condition resulted in significant disability only when it reached an advanced stage, known as the complicated phase or progressive massive fibrosis.

Once again, the industry had found a way to limit its liability, this time by excluding two groups of sick miners: those who had the simple form of the disease and those who had a manifestation of dust disease that defied easy classification.

Doctors in the industry-aligned medical establishment mocked the label *black lung* as some sort of bogeyman, the brainchild of populist instigators. One coal boss worried that he and his colleagues "might end up paying... for lung infections caused by other health factors." The industry invoked what remains its favorite scapegoat today: smoking.

Buff fired back: "It doesn't matter what the damn thing is called. The man can't work. He's disabled." Buff was certainly guilty of exaggerating for dramatic effect on many occasions, but his larger point was correct. It is now widely accepted that inhaling coal dust affects people in different ways. The appearance of the disease on X-rays can vary, as can the type of breathing impairment revealed by testing. Even today, doctors are discovering previously unrecognized variations of lung disease that appear to be caused by breathing coal dust.

In reality, then, there is not *one* disease that afflicts coal miners; it is a *family* of diseases, all with the same cause. The industry tried to disown most members of this family; Buff and the miners embraced them all.

As the fight for recognition of the disease intensified, rank-and-file miners created the Black Lung Association, which would become a powerful and lasting force. They wanted to hire a lawyer to help draft legislation, so they needed to raise ten thousand dollars. Some of the money came from their elected president, a local union leader named Charles Brooks, who mortgaged his house. The rest came from donations solicited from other local unions.

The coalition was now assembled: the rank-and-file miners, led by the Black Lung Association; the trio of doctors; and the young activists, led by people like VISTA Craig Robinson. They took the show on the road, and the movement expanded throughout the coalfields.

When the state legislature held a hearing on various proposals for workers' compensation reform, miners packed the galleries and spilled into the corridors, some of them carrying signs that read NO LAW, NO WORK. Still, the legislation remained stuck in committee.

The breaking point came on February 18, 1969. At a mine in Raleigh County, more than two hundred workers walked out. The news quickly spread, and within just a few days, more than ten thousand miners in neighboring counties had joined the protest. It was a wildcat strike; the United Mine Workers leadership didn't approve. The UMW accused the Black Lung Association of dual unionism—a charge considered akin to treason—and forbade local unions to donate to the organization.

Undeterred, the miners made plans they thought would force the legislature to act.

On February 26, two thousand miners and their allies from across the state converged on Charleston. After a rally at the Municipal Auditorium, they marched down Kanawha Boulevard, the main road running through the city, toward the capitol building. Some wore hard hats; others carried signs that said NO LAW, NO COAL.

Marching alongside them was a shaggy-haired twenty-three-year-old who had come to West Virginia from western New York as a foot soldier in the War on Poverty: John Cline. He wasn't part of the group organizing the black lung rallies. The march was just something he'd heard about, and it appealed to his sensibilities and notions of justice.

John aspired to do something useful for others, but he was still figuring out what that meant. He seemed to have found at least a piece of what he was looking for in a two-room cabin up a mountain hollow in the heart of Boone County, where he had arrived as a VISTA in February 1968.

In a letter to his sister Cathi, John wrote of the experience: "The involvement I have had this year has opened a door to learning and growing without limits. . . . I feel like I am slowly on a rampage—applying myself to every situation and learning from it, because of it."

John's brain was churning, grappling with issues both timely and eternal: racial discrimination, war, poverty, religion. In regular dispatches to his family, he mixed observation with introspection. He ruminated on marriage, the consolidation of the coal industry, and the seeds of rural poverty. He quoted French philosopher Albert Camus and American advocate for radical pacifism David Dellinger. For Christmas, he asked, without a whiff of irony, for "maybe a couple of T-shirts and a better world for everyone."

The war in Vietnam often occupied his thoughts. Before moving to West Virginia, after much reading and deliberation, John had reached a consequential conclusion: The draft was discriminatory, and war was never justified. As a VISTA, he was likely to qualify for a deferment. But to John, that was beside the point. Even accepting a deferment amounted to participating in a system he thought was unjust.

Shortly before arriving in the coalfields, John had been drafted. He'd gone to his scheduled induction proceeding and told the government he refused to serve. Soon, John became one of the more than twenty-five thousand men indicted on draft-related charges during the Vietnam War era. His case would remain unresolved during his VISTA stint.

John and about a dozen other VISTAs spread out in the hollows, renting spare living space from locals. They became familiar with a type of poverty that had tended to remain hidden. People lived up winding dirt roads in isolated crevices of the area's hills. Families packed into small shacks and survived largely on staples they could grow, such as beans and potatoes. The bulk of the coal-mining jobs that once offered decent wages had disappeared thanks to increased competition from other fuel sources, such as oil and natural gas, and the introduction of new machinery in the mines that allowed operators to produce more coal with fewer workers. Many coalfield residents had gone north in search of factory jobs.

The VISTAs spent their days talking to people on front porches, helping them with errands and chores, and asking them what community problems they'd like to see fixed. With the VISTAs' help, community members formed a small advocacy group that successfully lobbied for additional federal funding for free or reduced-price school lunches.

John enjoyed the company of neighbors who had spent their lives in the mines, some of whom were disabled by old injuries or lung disease. One old-timer told stories of fighting in the infamous Mine Wars decades earlier. John's landlord and neighbor, a retired miner, took John on hikes in the woods and showed him where to find ginseng and the spongy, edible mushrooms that mountaineers called Molly Moochers.

Meanwhile, John took a leap that he would later view as impetuous. In May 1968, he married fellow VISTA Pat Richards, who had grown up in California. She, too, wanted to do something to alleviate the area's endemic poverty, and she was drawn to John's quiet intensity. "I knew he could help me become a better person," she later said. The two wed within a few months of meeting.

When the VISTAs' allotted tenure in Boone County ended, in the fall of 1968, John and Pat decided to stay and try to continue their organizing work.

Support for the War on Poverty, however, was on the decline after a backlash primarily from conservatives and machine politicians. Some in the coalfields considered the whole thing to be another patronizing, failed "rediscovery" of Appalachian poverty, but in retrospect, chroniclers of the tumultuous time have credited the effort as a foundation for future successes. Labor historian Alan Derickson described the poverty warriors as a "vital component" in the "explosive mixture" that led to reforms, writing, "The VISTAs brought the attitudes, tactics, and ideals of the civil rights and anti-war movements to the coalfields."

There is a long tradition of fatalism among many in Appalachia. The weight from generations of oppression can feel impossible to shed. Some believe hardship is a test set before them by God. But with the infusion of the 1960s protest spirit and the victories won by groups like the school-lunch advocates, some West Virginians began to feel their lot was not preordained. Boone County resident Shelva Thompson, who went on to lead efforts for welfare reform, put it simply: "We started fighting back."

The region was like a mine saturated with methane; with one small spark, there'd be one hell of a conflagration. The Farmington disaster proved to be that spark. As John marched with the miners down Kanawha Boulevard toward the capitol building in February 1969, the fiery insurgency had engulfed much of the state. By the following week, more than forty thousand miners had walked off the job in the largest political strike in U.S. history. The wildcatters had shut down virtually every coal mine in West Virginia, crippling the state's biggest industry.

Just hours before the scheduled end of the legislative session, lawmakers hammered out the final details of a bill, and all but two legislators voted in favor. Governor Arch Moore signed the bill into law, and the miners went back to work.

The final legislation included much, though certainly not all, of what the Black Lung Association had sought. The mere recognition of black lung as a disease that qualified for workers' compensation payments was a breakthrough, but justice remained out of reach for the vast majority of disabled miners who had already retired and would be unable to file claims.

Nonetheless, the reach and importance of the black lung uprising in West Virginia extended far beyond the state compensation law. The West Virginia revolt "sent out shock waves that proved decisive" in the more far-reaching changes still to come, Derickson wrote. The rebellion spread, and Ohio and Kentucky soon passed their own compensation laws. But state-by-state reform was time-consuming and inconsistent, and it seemed inevitable that thousands of disabled miners would remain left out.

Beyond these practical concerns, many workers and advocates were becoming increasingly skeptical that after-the-fact compensation for the disease would prod the industry to solve the more fundamental problem of preventing it in the first place. One line of thinking held that forcing companies to pay for the damage caused by mines filled with clouds of dust would lead them to clean up the workplace out of economic self-interest. But this would work only if the compensation system was efficient and free from the influence of coal companies and their network of sympathetic physicians and government administrators, something many activists and miners doubted would happen in coal-dominated states such as West Virginia.

In light of all this, the black lung movement began to adapt, refocusing on national policies that would not only provide a more comprehensive compensation system but also cut down on the amount of dust that miners had to breathe, attacking the disease at the source.

These once-unthinkable goals now seemed like realistic possibilities. The audacious coalition of rank-and-file miners, sympathetic doctors, and volunteers had captured the nation's attention. "The wave of wildcat strikes by West Virginia coal miners makes it plain that action on mine health and safety cannot wait," the *New York Times*'s editorial board wrote in February 1969. Another article penned by the board that same month said: "The crusade to clean up the mines and make them safer must get priority in the statehouses and in Congress. The Black Lungers have a claim on the conscience of a nation in which coal remains a vital fuel."

For the movement the insurgents had started in the West Virginia coalfields to succeed in Washington, DC, the miners would need a champion in Congress. In the aftermath of the Farmington disaster, they had gotten one.

Congressman Ken Hechler, who represented a stretch of western West Virginia, had arrived at the Farmington mine shortly after the initial blasts. At first glance, his résumé didn't appear to be that of a likely ally for the black lung insurgents. Born in Long Island, New York, he had earned a PhD in history and political science from Columbia University, taught at Princeton University, then served on the staff of President Harry Truman. He was in his forties by the time he moved to West Virginia to teach at Marshall College, a public institution near the Ohio border.

After unseating the Republican incumbent in 1958, Hechler established himself as one of the most liberal members of Congress, evincing a penchant for populist causes and fiery rhetoric. By the fall of 1968, he already had called for strong mine safety and health legislation. When he saw the devastation in Farmington and met the widows of the fallen miners, the cause took on a sense of urgency.

He began a public crusade for stronger laws not just to prevent catastrophes like Farmington but also to end the scourge of black lung. As the insurgents in West Virginia fought for a state compensation law, he helped put their cause before a national audience in television interviews and press statements.

On February 6, 1969, as the newly formed Black Lung Association hammered away at the state legislature, Hechler introduced a bill in Congress, and the national fight over coal mine health and safety was on. Others in Congress introduced their own versions, and hearings soon began.

Though some within the coal industry adopted a hard-line stance against any new laws, the largest companies and the national trade association took a different tack. After Farmington and the black lung uprising in West Virginia, outright opposition to regulation seemed untenable. Instead, the leading voices in the industry publicly supported some form of legislation but objected to specifics, all the while working behind the scenes to weaken the proposals.

Taking aim at the most potentially onerous measures, the industry objected in particular to the provisions intended to prevent black lung. At the time, there was no standard for how much coal dust was allowed in the air miners breathed. Companies had to control dust only to the extent necessary to prevent fires and explosions. In December 1968, the Department of Health, Education, and Welfare recommended a dust concentration of no greater than 3.0 milligrams per cubic meter of air—a standard that had been imposed years earlier in Great Britain and had substantially lowered the incidence of disease.

Hechler's bill would make that level the law. The industry warned that such a strict limit would be either impossible to achieve or financially ruinous. It didn't oppose a standard for the amount of dust allowed in mine air altogether; it just sought to weaken the standard or put off regulation until further study was done. The National Coal Association sent a letter to every member of Congress warning that an overly stringent law could "jeopardize the public welfare by bringing on a nationwide power and steel shortage" and that "the nation will face power blackouts." Companies were more comfortable with the looser standard of 4.5 milligrams per cubic meter of air proposed by the administration of Richard Nixon.

The United Mine Workers' stance was more complicated. Worried that the health provisions meant to address black lung would be so expensive and controversial that they could tank the entire legislative effort, the union favored splitting the single bill into two: one to address health hazards and one to address safety hazards such as explosions and cave-ins. UMW leaders were willing to sacrifice the former to ensure they at least got the latter.

Hechler wouldn't bend, and he called on the reform movement's greatest strength: the voices of the miners themselves. He brought the faces and stories of rank-and-file miners and the Farmington widows to Washington, injecting the spirit of the black lung insurgency into committee hearings and the halls of Congress. Buff brought a toned-down version of his dog-and-pony show to the nation's capital, performing his white hat/black hat routine before a Senate subcommittee and passing around a plastic bag containing a dead miner's blackened lung.

Soon, the debate broadened to include what would become the most controversial piece of the entire legislation: a federally administered program to provide black lung compensation and medical benefits. Some congressional Republicans and the Nixon administration raised a pair of grave concerns: first, that establishing a federal workers' compensation program for coal miners would set a dangerous precedent that could spread to other industries and undermine the consensus that workers' compensation was a matter for the states to handle, and second, that the program would prove ruinously expensive for the federal government, as it would involve the U.S. Treasury picking up the tab for at least a few years.

The bill that made it through both houses of Congress in October provided compensation for any proven case of totally disabling pneumoconiosis, which was defined as "a chronic dust disease of the lung arising out of employment in an underground coal mine." Miners who had the complicated form of the disease would automatically get benefits, and those with the simple form could qualify if they proved it had caused total disability. The bill's language, however, left it unclear whether miners with disease variations that manifested as emphysema and chronic bronchitis would qualify for benefits.

The concerns over costs and states' rights led to a complex compromise intended to keep federal involvement temporary. The Social Security Administration would process and pay approved claims through 1972; states were supposed to set up their own compensation programs and take over administration of claims in 1973, with coal companies on the hook for paying benefits. If a company went out of business, the federal government would pay.

In the fight over the provisions meant to prevent black lung, Hechler prevailed. The standard of 3.0 milligrams per cubic meter of air that Hechler had fought so hard to keep would become effective immediately, and within three years, companies would have to comply with an even tougher standard of 2.0 milligrams.

Though the sections related to black lung aroused the most contentious debate, they were only part of a far-reaching bill. The legislation created strong standards requiring companies to make a host of improvements—such as better roof control, ventilation, and methane detection—to prevent accidents such as explosions and cave-ins. It gave federal regulators real enforcement authority, empowering inspectors to issue citations that carried stiff fines and temporarily shut down mines where conditions posed an imminent danger.

The bill, however, needed Nixon's signature. He was threatening to veto the entire thing because of his concerns about the black lung–benefits program. On December 22, 1969, the *New York Times* editorial board weighed in again, chastising Nixon for his veto threat: "Given the epochal character of the law in reversing decades of neglect on the part of industry, the union and officialdom, it would be astonishing if the President did veto it."

As the new year approached, the situation became urgent. Even if Nixon didn't veto the bill, it would die by pocket veto on January 1. Hechler and the rank-and-file miners weren't about to let the hard-won legislation collapse so close to the finish line. Hechler chartered two small planes to fly seven of the Farmington widows to Washington so they could try to meet with Nixon. Meanwhile, word had reached the president that twelve hundred miners in West Virginia had walked off the job in a wildcat strike, demanding he sign the bill.

Nixon declined to see the Farmington widows, but on December 30, he signed the bill. There was no ceremony. He issued a statement calling the law "a crucially needed step forward in the protection of America's coal miners." But he devoted the bulk of the statement to criticizing the black lung–benefits provisions. He was signing the legislation despite having "reservations," and he emphasized that the benefits program was "temporary, limited, and unique and in no way should it be considered a precedent for future Federal administration of workmen's compensation programs."

The moment, nonetheless, was a landmark. The rank-and-file miners, barnstorming doctors, youthful activists, and crusading lawmakers had taken on the coal industry, the medical establishment, and opponents in the West Virginia statehouse, the U.S. Congress, and the White House, and they'd won what remains one of the strongest occupational health and safety laws ever passed.

The law resounded with noble intent: "The first priority and concern of all in the coal mining industry must be the health and safety of its most precious resource—the miner." The provisions forcing companies to limit the amount of dust miners breathed were designed "to permit each miner the opportunity to work underground during the period of his entire adult working life without incurring any disability from pneumoconiosis or any other occupation-related disease."

But holding the government to those promises was a complicated and never-ending task. Some coalfield activists soon became disillusioned with the law's implementation. The rules established by the Social Security Administration, which was responsible for processing and paying claims during the program's first three years, were too strict, miners and their advocates contended.

Black Lung Association leaders revived their protest tactics and pushed for further reforms. As it happened, John Cline's path intersected again with the movement roiling the coalfields.

The specter of the Vietnam draft still hung over John, but by 1970 his thinking on how best to express his pacifist beliefs had changed. Rather than refusing to comply with the process, as he had previously, he applied for conscientious objector status and volunteered to do civilian service work to fulfill his obligation. "I feel the CO position still shows I oppose war but also want to be of service to others," he told the draft board in August 1970.

Family and friends submitted letters attesting to the sincerity of his anti-war convictions. Two high-school classmates of John's—one of them a self-described hawk and the other a member of the army who was about to deploy to Vietnam—wrote that they had debated John numerous times but that he had remained firm.

John's parents related their own internal conflict after John had told them of his planned refusal to serve. After questioning John and discussing alternatives with him, they wrote, "We came at least to understand John's position and to respect him for it and his willingness to accept the consequences, hard as they could be for him and for those who loved him.... John's father has said that he has known men of greater and of less abilities than John, but he has never known any man more determined to use what abilities he has for others."

The Selective Service approved John's application and reclassified him as a conscientious objector. Later, the agency allowed some of his previous work to count toward his service obligation and permitted him to finish out his time working for an organization named Designs for Rural Action, which was composed of young activists who, like John, had come to the coalfields as part of the War on Poverty.

The group joined with the Black Lung Association in its drive for improvements to the benefits program. From an office in Charleston, the allies helped organize the work of chapters throughout the coalfields, staged media events to keep the issue in the public eye, and spread information and inspiration to the chapters through a newsletter called the *Black Lung Bulletin,* which mixed practical advice on filing for benefits with a heavy dose of militant calls to organize and protest.

John's job was, as he put it, "office flunky." His duties included taking each edition of the *Black Lung Bulletin* to the printer, then hauling the copies to the Black Lung Association chapter on Cabin Creek, where he and the members bundled them for mass mailing.

In 1971, debate in Congress over proposed reform legislation echoed that of 1969, with mostly the same supporters and opponents making mostly the same arguments. Members debated, cajoled, and bargained against a backdrop of rank-and-file demonstrations and threatened strikes. Black Lung Association members regularly carpooled to the nation's capital, where they held press events and walked the hallways of congressional offices to lobby legislators.

On May 10, 1972, Congress passed a reform bill, but rumors swirled that President Nixon might refuse to sign it. His concerns were much the same as they had been in 1969: the high costs and possible undermining of states' rights. On May 19, members of the Mingo County chapter of the Black Lung Association made good on the strike threats. The walkouts would spread, chapter leaders warned, if Nixon didn't sign the bill. In another echo of the 1969 fight, Nixon signed the legislation but said he was doing so with "mixed emotions."

The law was another resounding victory for the black lung movement. It eased the standard for proving total disability and allowed widows to prevail if they could show that their husbands had been totally disabled by black lung at the time of their deaths; previously, widows had been compelled to prove that black lung *caused* their husbands' deaths. The law also established a critical provision known as the fifteen-year presumption: if a miner had worked at least fifteen years and had a totally disabling respiratory disease, it was up to the company to show that the miner's impairment was caused by something other than black lung.

By 1972, states were supposed to have set up their own programs to administer claims in which companies would pay benefits. Because no state had yet done so, the 1972 amendments gave them another eighteen months. States would continue to ignore the directive, however, and later amendments would make the benefits system a permanent federal program. The Labor Department would process new claims, and the miner's most recent employer would be on the hook for payments if a claim was successful.

Labor advocates heralded the 1969 and 1972 laws as historic achievements that had broken down the wall of denial and neglect that had surrounded workplace illness for decades, but their hopes that the reforms would spread to other industries were largely dashed.

In 1970, Congress passed the Occupational Safety and Health Act, but its provisions were significantly weaker than the coal-mine law enacted a year earlier. The new law tasked one agency, the Occupational Safety and Health Administration, with policing virtually all U.S. workplaces other than mines, but it gave these beat cops less-potent weapons than the mine regulators had. Today, the process of issuing a new regulation to deal with a specific hazard is so onerous and time-consuming that the agency has been able to address only a small fraction of the toxic substances that sicken and kill workers. Enforcing the relatively few rules that exist is notoriously challenging for the perennially understaffed, underfunded agency.

Nor did the coal-mine law spawn a host of other federal benefits programs. With a few exceptions, workers' compensation remains the province of the states, whose systems range from solid to hopelessly porous.

Even the 1969 and 1972 laws would not be the cure-alls some had hoped. Conditions for miners improved substantially, but political victories are never permanent. As miners' advocates would learn, statutes can be amended, regulations can be rewritten, enforcement budgets can be slashed, and protections can be gutted by adverse court rulings.

Decades after the original black lung insurgency, Rasmussen would pause between appointments with coal miners at his clinic in Beckley and look back at the movement that had drawn him into the role of reluctant hero, the dramatic victories it had won, and the successes and unfulfilled pledges in the years since. "In 1969, I publicly proclaimed that the disease would go away before we learned more about it," he mused. "I was dead wrong."

5

Footer

Almost twenty years after the wildcat strikes and watershed legislation of 1969, John Cline was trying to make sense of what he was seeing every day in the basement clinic in Scarbro. Why, he wondered, were so many miners coming in with serious lung problems, and why were they having such trouble winning benefits? By now, the disease was supposed to be well on its way toward obsolescence.

John was far from the only one to notice these problems. Concerned clinic workers, miners, and union leaders across the region were trying to revive the Black Lung Association. After having seemingly won the battle over prevention and compensation years earlier, the association had largely receded from the political scene. Now, miners and their advocates were attempting to resurrect local chapters or, in some instances, start new ones.

John, his colleagues at the New River Breathing Center, and area miners, many of them clinic board members or patients, eagerly joined the effort. In early 1988, they formed the Fayette County Black Lung Association. The New River clinic was a natural home base for the chapter; it was one of the more than fifty sites across fourteen states that received federal funding to diagnose and treat miners with black lung. The clinic's cofounder and director was a veteran of the original battle for disease recognition and compensation.

Craig Robinson—the VISTA who had forged an alliance with the fledgling Black Lung Association, helped arrange the dog-and-pony shows of the barnstorming doctors, and collaborated on draft legislation—had taken a job with the United Mine Workers' occupational health department in 1973. Tasked with establishing community health clinics in underserved areas, he had visited local union houses throughout the coalfields and offered funding and help getting a clinic started.

After a few years, however, the union, buffeted by complex politics and funding troubles, had discontinued the effort. At the time, Robinson had been in the early stages of talks with local union leaders in the communities near the New River Gorge in Fayette County, and, thanks to an outreach effort at Harvard Medical School, he had identified a doctor who seemed interested.

The prospect of working in a rural medical center had appealed to Dan Doyle, a recent graduate in his second year of residency in family medicine at the University of Massachusetts Medical School in Worcester. A native of South Bend, Indiana, he had graduated from Notre Dame in 1968, then spent a summer working in a rural community in Bolivia. He'd later taken two years off in the middle of medical school to work on a War on Poverty project in Roxbury, a predominantly African-American community in Boston.

He had little interest in becoming a highly paid specialist or an academic. "I wanted to be on the front lines of primary care," he said. "I wanted to be in an underserved place."

Doyle had visited the coalfields twice and met with Robinson, who still wanted to start a clinic in Fayette County despite the loss of union backing. The two had decided to go for it. Doyle finished his residency and moved with his wife to West Virginia.

"We had no money," Robinson recalled. "We didn't have a place. We had nothing, actually, just kind of an idea."

The Robert Wood Johnson Foundation had supported the effort with a grant, and a few area miners found a site for the clinic, a small warehouse and shop that backed up to a hillside creek. Robinson, Doyle, and the miners had renovated the building themselves, and the clinic opened in 1978 with a handful of staffers and some bare-bones equipment.

Soon the clinic outgrew the warehouse, and the miners found a new site, this one just down the road. People from the community streamed into this larger, better-equipped clinic. The sliding fee scale proved to be a lifeline for many patients, and the clinic treated some financially strapped people for free. A significant portion of the clinic's patients, naturally, were current or former coal miners, and Doyle began to learn, through reading and hands-on experience, about black lung.

The federal government program offering grants for black lung clinics seemed to be a natural expansion of New River's offerings, and Robinson and the board had applied, received the grant, and built an addition to the clinic to house what became known as the New River Breathing Center. They hired Brenda Halsey Marion, then a young nurse, to head up the new effort.

Marion had grown up in the area, and her grandfather and uncle had worked in the mines. She had previously worked at Appalachian Regional Hospital in Beckley, where she gained experience treating miners with black lung and learned from Dr. Don Rasmussen.

She took the job at the New River clinic in 1983. Her duties included helping to perform pulmonary function tests and advising miners who were pursuing benefits claims.

Four years later, Marion had a chance to hire a benefits counselor. One of the applicants was someone she'd seen around but didn't know well. He was a carpenter, and the company he ran with his father had built an addition to the clinic. She recalled his attentiveness; he had seemed to be taking in every detail about the clinic's work. For some reason, he now appeared to be enthralled by the idea of advocating for coal miners. She decided to give John Cline a shot.

After finishing his stint with Designs for Rural Action, John looked for work that might provide greater financial stability. He had a family to consider. In 1972, Pat gave birth to the young couple's first son, Jesse. Their second, Matthew, arrived two years later.

John decided to try his hand at the skill that seemed to have been encoded in his family's genes for generations and that had been awakened in John during his early childhood days in his grandfather's woodworking shop straightening bent nails with a hammer. He began doing carpentry in the area, and he eventually got a job with a housing-construction program started by a local college that had received funding to build homes for low-income people.

John saw a need for more affordable housing in the southern coalfields, and he sometimes discussed the shortage with his father, who shared an interest in the subject. The two decided to form a business, and in 1975, Crawf and Kay moved from New York to West Virginia.

Cline Building Corporation focused almost exclusively on a niche market where demand far outstripped supply. Virtually every home John and Crawf built was for people whose modest income qualified them for flexible, low-interest loans through the federal Farmers Home Administration. The program allowed people in rural areas who otherwise wouldn't be able to get a mortgage to purchase a home.

Crawf handled most of the business side of the operation, and John spent his days on the construction site, overseeing the crew and working as a carpenter himself. John put a premium on quality, and he paid particular attention to the early stages of construction. One of the first tasks was excavating a footer and having concrete poured to form the foundation on which the rest of the house would stand. Lots of contractors hired someone to come in and dig it with a backhoe. John thought this was too imprecise. It was critical that the footer be square and level; if it wasn't, the whole crew would pay for it for the rest of the job. So John and the crew dug footers by hand. "Nobody does that anymore; there weren't too many people that did it when John was doing it," noted carpenter Daniel Richmond, who, along with his brothers, worked on about thirty Cline homes. "Anybody that'll take the time to dig a footer by hand, it's right."

John believed that crews were happier and more productive if there was camaraderie and a bit of fun on the site. John might have been the boss, but he didn't mind some humor at his own expense, recalled Arland Griffith, who installed the plumbing for about fifty Cline homes. Exhibit A: the Clivus Multrum, a contraption that earned John some good-natured ribbing. One of the houses the Clines built was a new place for John and his family, and John decided to install this unusual appliance in it. "It looked like an outside toilet inside," Griffith said years later, his tone still incredulous that such a thing could exist.

John explained: "It's an indoor outhouse, really.... The one we had was about nine feet long on a thirty-degree angle. The very top was the chute where the human waste went. It would gradually, glacier-like, move its way down and dehydrate, and you're supposed to get usable manure out of it at the end. There wasn't any odor.

"We did put in an exhaust fan for the fruit flies," he added with a wry grin.

Richmond, reflecting on the many bosses he'd had and the many homes he'd helped build over a long career, concluded, "If I was starting over and John was in the business, I'd get a job with him, and I'd stay there till I died."

Many of the homes that the Clines built were in Piney View, a community of about a thousand people just outside Beckley. One of the first houses they erected there went to Donna Meadows, who was caring for her infant son while her husband did various jobs at a local hospital. Crawf and John went over floor plans with the couple, stretching the twenty-five-thousand-dollar Farmers Home Administration loan as far as they could. They built a one-level ranch house with three bedrooms in six weeks.

"I was amazed by that," Meadows said. "I was living in a brand-new house for eighty-one dollars a month, built the way I wanted. Oh, I was just dancing around. I thought I was something, and I thought I could do anything."

John's new home was also on one of these Piney View lots. The nurses, truck drivers, and coal miners for whom he'd built houses were now his neighbors.

John also found a new focus for his organizing efforts in his new community. He helped establish the Piney View Improvement Association, which successfully lobbied state and local government to bring paved roads and improvements to the school building. He was part of an organization that convinced the state government to block the building of a landfill near the community. He was active in the local parent-teacher organization and the 4-H club. He helped start an organization that put together concerts and art shows featuring the work of locals. John's neighbors jokingly dubbed him the "Mayor of Piney View."

His marriage, however, had been rocky for some time, and the discord now reached a breaking point. He and Pat divorced. They shared custody of Jesse and Matthew. Pat eventually remarried, and in 1987, she moved to South Carolina. Jesse, then fifteen years old, went with Pat, and Matthew, thirteen, stayed with John.

Crawf was getting ready to retire, and John was lonely and a bit lost. "I was feeling the need to do something more—social organizing," John said. Though the Clines had built mostly low-income housing, they did take other projects from time to time, and one of these was building an addition to the New River clinic in Scarbro. John heard that the clinic's breathing center had an opening for a benefits counselor. Intrigued, he applied.

The new job drew him back into the sphere of miners' advocates that he had briefly inhabited before. This time, he would stay for good.

Over the years, with experience and self-analysis, John had figured there were two ways he could, as he sometimes put it, "be of use": first, by applying whatever skills he had to help people tackle their individual problems, and second, by helping communities address larger problems using his penchant for networking and activism. At the New River clinic, the first part of John's effort to be useful came naturally with the job. He helped miners get the care they needed—doctors' appointments, breathing machines, hearing aids, and so on—and guided them if they wanted to apply for benefits payments for their lung disease. The fledgling Fayette County Black Lung Association afforded him ample opportunity to apply his networking abilities.

Once a month, the association met in the breathing center. It started small. "Ten members were in attendance," the minutes from one 1988 meeting recorded, and "$9 in dues was collected." Soon, though, the number of names on the sign-in sheets reached twenty or thirty for an average meeting; John's name was always among them.

These evening gatherings often included prayers and updates on members who were in the hospital. Miners would sip coffee from Styrofoam cups and share stories of their own struggles to win benefits claims. It was an outlet to vent frustration and learn from the experiences of others.

Increasingly, the federal black lung–benefits program became a topic of discussion at meetings. The initial claims approval rate was hovering around 4 percent. Many members regarded filing a claim as futile, and those who tried were mostly unsuccessful. Other chapters had similar concerns, and leaders agreed that fixing the problems with the federal benefits system called for a lobbying campaign. They developed priorities and started raising money.

While the long-term push for reform continued, however, there were many miners and widows who needed immediate help with their benefits claims. As benefits counselor, John helped with claims through the state system that provided lump-sum payments, but some miners coming into the clinic were totally disabled and needed the long-term monthly payments and medical coverage that came with federal benefits. John helped them file the paperwork and offered some guidance, but actually representing them was not part of his job description—for good reason. Just about everyone thought it was crazy to wade into that muck. Just about everyone, it turned out, except John.

Part Two

I worked in a coal mine for twenty-three year.
Seems to me I breathed enough bad air
To kill me and three and four more.
But I'm gonna get my black lung pension
And that's for sure.
'Cause I coughed all night last night
And the night before.

I said, Woman, I believe I'll walk off
Down about the store.
I got down there and I bought a jar of coffee.
I said, Merchant, I would buy more
If you'll wait on me
'Till I get my black lung pension.
I said, I'm gonna get it
And that's for sure.
'Cause I coughed all night last night
And the night before.

I come back home
And I got mad and I walked the floor.
I said, Woman, when I get my black lung pension
Remind me to trade at another store.
I said, I'm gonna get it
And that's for sure.

'Cause I coughed all night last night
And the night before.

I get up early every morning.
I don't know how many trips I made to the door,
Runnin', peepin', lookin'
To see if that mailboy
Is gonna bring my black lung check up to the door.
When he don't bring it
I say, Woman, we just have to wait a little more,
But I'm gonna get it
And that's for sure.
'Cause I coughed all night last night
And the night before.

One mornin' the mailboy pulled up to my door
Read me a notice to go back to the doctor once more
To see if I had the black lung for sure.
I went back there for a breathin' test,
And that lady stood up there.
She just kept hollerin'
More and more and more.
I got done, I said, Listen here, lady,
I'm gonna get that black lung pension
And that's for sure.
'Cause I coughed all night last night
And the night before.

I done waited eleven months
And just a little more.
I done wore out one knob on my door,
Runnin', peepin', lookin'
To see if that mailboy is gonna bring my black lung check
Up to my door.

I says, I'm gonna get it
And that's for sure.
'Cause I coughed all night last night
And the night before.

—Walter Brock (retired coal miner), "Black Lung Paycheck"

In the waiting room of a small medical clinic in Beckley, I sat alongside a handful of miners, each of them here to undergo the medical exam paid for by the Labor Department that would become part of their black lung–benefits claims.

"They say this doctor—what he tells you is true," one of the miners told me. "If he says you've got black lung, you've got it. He's not no company man, and he's not no worker man."

The miner introduced himself as Thomas Marcum. He was fifty-nine years old, with sandy-brown hair just starting to gray. He'd driven two and a half hours from eastern Kentucky to be here, having decided to eschew the services of government-approved physicians closer to home. His two brothers, both coal miners, had done the same, he said. It was well worth the trip.

Before long, a man dressed in blue scrubs shuffled down the hallway with the aid of a walker to greet us. This was Dr. Don Rasmussen.

By this point, April 2012, more than four decades had passed since the historic coal miners' rebellion that had so depended on Rasmussen's expertise. His once-maligned views on black lung now were widely accepted. At age eighty-four, he still had a full head of white hair that retained streaks of red, and he was still evaluating and treating coal miners.

I had come here to learn from the man whom many regarded as a sort of godfather of black lung medicine. It had been almost a year since I'd read the report on the Upper Big Branch mine disaster and been surprised to learn that black lung was surging back. Since then, a nagging and obvious question had stuck in my mind: Why?

It turned out I was not the only person with that question. Howard Berkes, a correspondent for National Public Radio who had reported extensively on Upper Big Branch, also had been intrigued by the page in the report about black lung. We had worked together before. As a small nonprofit investigative news organization, my employer, the Center for Public Integrity, sometimes partnered with radio or TV outlets to extend the reach of our stories. We'd do at least some of the reporting together, then release stories at the same time online and over the airwaves.

After talking in late 2011, Howard and I realized that we both had been mulling over a deeper examination of the causes behind the disease's resurgence. We decided to team up, and, along with NPR producer Sandra Bartlett, we headed for the central Appalachian coalfields, hoping to find answers.

One of our first stops was the Morgantown office of the National Institute for Occupational Safety and Health, where we met with the government scientists who were documenting the uptick in black lung and trying to make sense of it. The agency was in charge of the X-ray surveillance program that tracked the disease, and since detecting the rising rates in the early 2000s, researchers had intensified their efforts.

We talked for hours with the scientists in a lab where an X-ray light box displayed films of diseased lungs and a plastic cube housed a shriveled black slab that, apparently, had once been a human lung. Epidemiologist Scott Laney described, with obvious alarm, what he and his colleagues were finding. The surveillance data had indicated a steady decline in disease prevalence from the 1970s until the late 1990s, but since then, rates had been rising. This reversal was even more striking for cases of complicated pneumoconiosis. "I think any reasonable epidemiologist would have to consider this an epidemic," Laney told us.

Howard, Sandra, and I headed south from Morgantown and traversed the coalfields, talking with miners, clinic workers, and lawyers. Finally we arrived here at the tucked-away clinic that doctors, lawyers, and miners alike seemed to regard as a sort of mecca.

Gracious and soft-spoken, Rasmussen introduced himself and invited us to watch him at work, provided it was okay with the miners. Each said it was.

I followed Marcum through the various stages of the examination process, including what some in the black lung community called "the Rasmussen test." This was the test that he'd started performing on miners when he'd arrived in West Virginia in the 1960s even though most other physicians didn't do it, the one that revealed impairment that eluded other tests. It was now a standard part of Labor Department evaluations. Called an arterial blood-gas study, it measured the lungs' ability to transfer oxygen to the bloodstream.

When it was Marcum's turn, he stepped onto a treadmill; he wore a cumbersome array of equipment: ten cables running from pads stuck on his chest, an armband to monitor his blood pressure, and a mouthpiece connected to headgear. Earlier, a respiratory therapist had inserted a catheter in his wrist so that while he was walking, she could draw blood with relative ease—two vials at the beginning and two more every couple of minutes. Rasmussen sat in a chair next to the treadmill, taking notes as a computer spit out numbers and squiggly lines. He periodically raised the incline. When he brought the treadmill to a stop after ten minutes, Marcum was out of breath and twelve vials of blood lighter.

Marcum peeled the sticky pads off his chest, and as he waited for the results, we sat and talked. He'd gone underground at age seventeen, just after graduating from high school, and he'd worked twenty-seven years in mines near the West Virginia border. He'd been diagnosed with simple pneumoconiosis, but his impairment was relatively mild. (Rasmussen's tests confirmed that was still the case.) Compared to the other men in his family, he said, he was fortunate. His father and two younger brothers also had black lung, but they were worse off. As the exam wrapped up, Marcum invited me to his upcoming family reunion, and I gratefully accepted.

A few months later, I met the rest of the Marcums at their annual gathering in the shadow of Dewey Dam at Jenny Wiley State Resort Park in Prestonsburg, Kentucky. Over fried chicken, coleslaw, potato salad, and an array of creamy desserts, we chatted about their time in the mines and the price that had come with it. Middle brother Donald—age fifty-one, with complicated pneumoconiosis—said he sometimes had to pause and bank extra air to make it through the prayers he delivered at church events. The youngest—James, age fifty, with category 3 simple pneumoconiosis—had the most severe breathing impairment. His doctor had already brought up the possibility of going on oxygen.

The patriarch—Ray, age eighty-three—had worked during a different era of mining and had been receiving benefits for decades. But all three Marcum brothers went underground after the dust limits mandated by the 1969 law had taken effect—rules that were supposed to ensure that miners like them could work a full career and retire free from disabling lung problems.

After meeting the Marcums, I followed up with Rasmussen, who had evaluated all three brothers. "They're young; they're not smokers," he said. "These guys should be hale and hearty." Yet all of them, he said, had impairment and X-ray evidence of disease. "This should not be."

Donald and James in particular exemplified a trend that Rasmussen said he'd been seeing among the miners coming to his clinic over the past fifteen years: younger men with more severe disease. "It's disturbing," he said. "We had such great hopes in 1969. We thought we had nipped it in the bud. Now we come to find out that it hadn't worked."

The pieces of evidence seemed to be coming together in a troubling mosaic. The observations of Rasmussen and other doctors with whom I'd spoken, the accounts of lawyers and clinic workers, the research from government epidemiologists, the stories of the individual miners I'd met—all of it indicated that black lung was again ascendant. I also began hearing grave concerns about the benefits system, a safety net that was supposed to ensure ailing miners were not left without desperately needed treatment and financial support. I had come across seemingly inexplicable outcomes—miners losing their claims despite having evidence of disability, significant abnormalities on their X-rays, and a diagnosis of black lung by their own physicians.

Miners and advocates shared theories that might explain this state of affairs, all of them tracing back to coal companies and the handsomely remunerated legal and medical professionals they employed. Each theory, stripped of the gloss imparted by jargon, amounted to base human behavior—coercion, gamesmanship, outright fraud.

I was being drawn deeper into a world in which I had considered myself merely a passing visitor. Each discovery raised new questions. As it happened, a fortuitous connection—one I almost missed—set me on a path that, after considerable time, would lead to some answers.

On the same day that Howard, Sandra, and I had spent at Rasmussen's office, we had been scheduled to meet with a lawyer and a few miners he represented. We became so invested in seeing miners through the entire evaluation process at the clinic, however, that we'd been loath to leave and had canceled the appointment at the last minute.

We went on to finish our stories on the resurgence of black lung, after which Howard and Sandra were pulled into work on other subjects. As we finalized those articles, though, I once again had the gnawing sense that there was another, larger story to pursue. I decided to reach out to the lawyer we'd stood up. I assumed he would be annoyed, but when I called, he didn't seem to be.

On my next reporting trip to the central Appalachian coalfields, I added a stop in Beckley to meet John Cline.

6

"Go Jump"

John Cline's unplanned initiation into the federal black lung–benefits system began on a bus ride to Kentucky in 1990. A few members of Congress had introduced legislation to reform the program, and they were holding a hearing in one of the mining towns in the eastern part of the state. John and others from the Fayette County Black Lung Association planned to be there.

As the group's rented bus wound through the central Appalachian coalfields en route to the hearing, retired miner Ralph Manning approached John and told him the basics of the situation he was facing: He had previously filed a benefits claim and lost. He was now sixty-eight, and his breathing was so labored that he could no longer work. He wanted to keep fighting for what he thought he was owed. But the United Mine Workers District 29, the union's stalwart outpost in southern West Virginia for half a century, had recently stopped providing in-house legal representation for federal black lung–benefits claims, and Manning hadn't been able to find another lawyer. He asked John if he would take a look at the files from his case.

Though this level of involvement in a claim wasn't technically part of John's job, he was intrigued. He remembered from his work with the Black Lung Association twenty years earlier that retired miners and folks from coal towns had, with some training from a public-interest law firm, become effective lay advocates. Already, he was helping miners with state claims, and both the state and federal systems were supposed to be similar workers' compensation programs, more or less, John thought. How different could they really be?

Sure, he told Manning; bring the documents by the clinic.

The next Monday, Manning and his wife arrived at the breathing center with two cardboard boxes full of papers, and John quickly began to worry that he had underestimated the magnitude of what he'd promised Manning. That night, he took the files home to peruse, and his concerns were confirmed.

He spent every night for the rest of the week trying to make sense of the stacks of documents. The volume and complexity of the medical reports were staggering, even to someone who regularly administered pulmonary function tests. There were more than eighty different X-ray interpretations, the majority of them negative readings (meaning that no disease was seen) submitted by the company that had employed Manning. The legal filings evidenced an administrative system that seemed so specialized and technical as to be almost indecipherable. Worse, there was a notice from the Labor Department that said Manning owed the agency more than thirty thousand dollars.

As John leafed through the file on Manning's claim, he became baffled as to why the longtime coal worker had lost. Multiple breathing tests had shown his disability, yet a judge had determined that Manning didn't have the very disease for which his own doctors were treating him. There had to be some reason for this, but to John, it wasn't clear where to look within the peculiar language and rules of this administrative legal process.

In 1970, when John had hauled copies of the Black Lung Association's *Black Lung Bulletin* to the printer, the enemies had been Social Security Administration bureaucrats. Winning a case hadn't been easy, but it had been doable. The system Ralph Manning and his wife—and now John—were trying to navigate seemed to be something entirely different.

Two fundamental changes had transformed the workers' compensation–style system of the program's early days into the complex litigation reflected in the mess of papers Manning brought to the clinic: The eligibility rules had become tougher, and coal companies, rather than the government, were now responsible for paying claims, pitting miners against a far more aggressive opponent.

The 1972 amendments had not worked out as planned. Because state governments had not set up benefits systems that met federal requirements, in 1974, it fell to the Labor Department to administer claims. Under the new, more stringent eligibility criteria applied by the department, miners whose X-rays didn't reveal the most advanced form of the disease had a much harder time qualifying, and the breathing-test scores that counted as evidence of total disability were lower, meaning that miners had to be even sicker to win their claims.

When Congress debated the 1972 law, supportive legislators had expressed concern that the Social Security Administration's benefit-approval rate, which was less than 50 percent, was too low. After its first three years of administering the benefits program, the Labor Department had approved just 8 percent of claims.

Once again, many miners felt the legislative victory they'd won had been snatched away. The Black Lung Association again descended on DC to hold rallies and lobby members of Congress. The legislation that President Jimmy Carter signed into law in 1978 so liberalized the program, however, that it triggered a conservative backlash.

The law substantially loosened eligibility standards and allowed many miners and widows with previously denied claims to refile under the less stringent rules. The process it set up proved to be messy and expensive, and the number of claims, approval rate, and costs of the program spiked. Thousands of previously denied claimants refiled and won. The approval rate shot back up to 50 percent.

This short-term spike, however, cost miners dearly for decades to come. At the request of Representative John Erlenborn, an Illinois Republican and critic of the original law creating the benefits system, the General Accounting Office reviewed claims processed under the new, more lenient standards and found that the vast majority of awards were based on medical evidence that "was not adequate to establish disability or death from black lung." Miners and their advocates acknowledged that there were problems with the system, including isolated incidents of fraud, but they contended that the flaws cut both ways, contributing to both unwarranted awards and unjust denials.

By the time the GAO issued its findings in 1981, a course correction already appeared to be under way. New Labor Department rules that had taken effect the previous year had tightened eligibility standards, and the claims approval rate was bound to drop substantially.

But as a new decade introduced an era of slashed government regulations and budgets, concerns about the benefits system's costs grew. A companion law passed at the same time as the 1978 amendments had sought to end the program's drain on the U.S. Treasury by creating a trust fund, financed by a tax on companies for each ton of coal produced, that would pay the miner if his former employer folded or went bankrupt. This fledgling trust fund, however, proved unable to keep up with the wave of approvals. During its first three years in existence, the fund spent $1.9 billion on benefits payments and administrative expenses, far exceeding the $700 million it collected in taxes on coal companies.

The liberalized benefits system accompanied by the trust fund's ballooning deficit looked like exactly the sort of runaway entitlement program that President Ronald Reagan's administration was targeting. For some miners and widows, the administration said, the program had become almost an "automatic pension." It was an echo of denunciations by some industry officials and conservative commentators, critics who had used terms such as *racket* and *gravy train*.

The administration's statements infuriated miners, and United Mine Workers president Sam Church warned, "If we have to, we will close down every coal mine in this country." The rhetoric was reminiscent of strikes of years past, but in 1981, the threat no longer had the same bite. The union—and the labor movement in general—had lost prestige and clout. Meanwhile, the business community had organized trade groups, formed political action committees, and captured much of the sway over lawmakers that had once been enjoyed by labor.

The union staged a rally in Washington in which more than six thousand miners and advocates from seventeen coal-producing states marched to the gates of the White House; there they chanted "Black lung kills!" and waved signs reading STOP BLACK LUNG MURDER... NOT BLACK LUNG BENEFITS. The lobbying, however, was not enough to convince Congress or the administration to change course.

In December 1981, both the Republican-controlled Senate and the Democratic-controlled House approved a bill by large margins, and Reagan signed it into law a few days later. Starting January 1, 1982, miners and widows would have to try their luck in a drastically different system. The new law didn't just roll back the most liberal provisions in the 1978 law; it erased some of the hard-won gains in the 1972 law and even narrowed some of the protections in the original 1969 law.

For miners, one of the most important changes was the elimination of the fifteen-year presumption—the critical provision stipulating that if a miner had worked at least fifteen years in dusty conditions and had evidence of a totally disabling breathing impairment, it was, in essence, up to the company to prove his disability wasn't caused by black lung.

Widows also faced new hurdles. In the original 1969 law, Congress stated that its purpose was to compensate widows whose husbands had died from black lung or had been totally disabled by the disease when they died. The Reagan amendments excised the latter qualification and required widows to prove that their husbands had died because of black lung. Now, a miner might be totally disabled by black lung, but if he was killed in a car accident, his widow likely wouldn't receive benefits.

The industry also took a hit. In an attempt to restore the trust fund's solvency, the new law doubled the tax on each ton of coal produced, setting it at one dollar for underground mines and fifty cents for surface mines.

In the end, even the miners' usual allies grudgingly supported the legislation. Both the union and the industry endorsed the law, and some coal-state representatives in Congress, including all of West

Virginia's, voted in favor of it. The union and politicians normally sympathetic to miners said their reluctant backing of the law was an acknowledgment of political realities and an attempt to stave off more devastating cuts.

Republicans, led by Erlenborn, "could have made things worse," Congressman Nick Rahall, a Democrat whose district included Beckley, said soon after the law's passage. A legislative staffer for the United Mine Workers said the decision to support the bill came after union officials were warned that continued opposition could lead to elimination of the benefits program entirely. "Of course getting benefits will be more difficult," he said, "but our options were very limited."

The more stringent Labor Department rules issued in 1980 already had cut the approval rate from the anomalous 50 percent during the previous two years to 11 percent, roughly in line with what it had been before the 1978 law. Under the criteria in the Reagan amendments, approvals plunged to 4 percent by 1988.

Meanwhile, miners and widows increasingly faced not only tighter standards but also a more combative opponent in many cases. As the obligation to pay successful claims transferred from the trust fund to coal companies, miners had to contend with lawyers from high-powered private firms who had a seemingly boundless budget to gather reams of expert medical evidence and litigate ad nauseam. The transition began in 1974, but because of the most recent amendments, the bulk of claims were still the government's responsibility until well into the 1980s.

Lawyers representing coal companies, unlike those at the Labor Department who had opposed miners' claims under the old system, rarely accepted an initial award of benefits as the end of the process. During the program's first three years, companies appealed 97 percent of initial approvals, a General Accounting Office report found. Company appeals had become more or less "automatic," and the system was plagued by "excessive litigation," a 1977 Labor Department report concluded. Companies weren't just quibbling with individual cases they thought claims examiners got wrong; they were fighting pretty much everything. The chief medical officer for one company conceded to the *Wall Street Journal* in 1975, "We've fought some claims we shouldn't have from a medical standpoint."

This became a defining feature of the system going forward and a deterrent to miners and lawyers as they considered whether to bother filing a claim.

The system grew even more costly, complicated, and slow. A 1990 GAO report found that backlogs were increasing at the appeals levels. If a miner did win at the outset, he could expect an appeal to an administrative law judge, which would tack on an average of three years, then an appeal to the Benefits Review Board, which would delay the proceedings by another two and a half years.

The plunging approval rate and the increasing complexity and delays proved to be more than enough to scare off most lawyers who would even consider representing miners. "I quit taking them a couple of years ago," one lawyer told the *Charleston Gazette* in 1987. "It got to be so time-consuming, and I found that when I did win the cases, the fee I got was not commensurate with the risk I took. It does leave a lot of poor people without representation." A lawyer in eastern Kentucky put it more bluntly: "I hate the federal black-lung act. I will not handle any federal black lung cases.... I just think the whole scheme is a hoax because they've regulated away the act that they passed in 1969."

In 1988, Harold Hayden, an official for the United Mine Workers district in southern West Virginia, testified at a congressional subcommittee hearing in Beckley that the district's one lawyer was currently juggling about seven hundred and fifty federal black lung claims, and "on June 7 of this year, our attorney is scheduled to be in five different places on the same day, a feat even Merlin the Magician would have a hard time pulling off."

One of the hundreds of cases that the attorney, Raymond Smith, had on his plate was Ralph Manning's. Two years later, Manning's claim ended in defeat. He wanted to try again, but by then, the union had discontinued its in-house handling of federal black lung cases. That's when

Manning hauled his two boxes of documents to the basement offices at the New River clinic and placed his hopes with John Cline.

In 1990, as John tried to figure out what he'd gotten into, a congressional subcommittee heard an assessment of the dismal state of the benefits program from longtime public-interest lawyer Howard B. Eisenberg, who had worked as the chief public defender for Wisconsin, represented inmates challenging prison conditions, and served on a panel investigating sexual abuse by Catholic priests in Milwaukee. Now he ran a legal clinic at Southern Illinois University that, among other things, represented miners in federal black lung claims.

"I can say without hesitation that the most unfair process I have ever run into is that which is found in the federal black lung system," he told the subcommittee members. "It defies due process of law, it defies reason and it is just simply unreasonable."

John was learning for himself what Eisenberg meant as he slowly made his way through Manning's files at home evening after evening. The voluminous record was a primer on how the system worked. John traced all that he'd missed over the years since his early involvement with the Black Lung Association—the intricate regulations and the various attempts by administrative bodies and federal courts to interpret them. Briefs and decisions contained excruciatingly detailed analyses of virtually every piece of medical evidence in the record. John began to put together an image of the decade-long legal struggle Manning had endured.

Manning had worked more than thirty years, most recently doing construction at coal preparation plants, which did the initial processing, crushing, and sorting of what came out of the mines before it was sent to market. It was not an underground job, but the plants tended to be dusty. Prep-plant workers might breathe as much coal dust as their colleagues below the surface, which was why the law made them eligible for benefits as well.

Manning had filed for benefits in 1980. His scores on breathing tests were bad enough to qualify as evidence of disability under the Labor Department standards, and doctors had pegged the cause as black lung.

Manning's former employer sent him for an exam by a doctor of its choosing—something that, under the law, miners must agree to undergo. That physician opined that Manning's breathing problems were not totally disabling and that he didn't have black lung. A Labor Department claims examiner evaluated the evidence and issued a decision awarding benefits, but the company appealed, sending the case to an administrative law judge.

After years of delays, the judge reversed the initial decision and denied Manning's claim in 1988. Manning appealed, and in 1990, the Benefits Review Board upheld the denial.

That's when Manning encountered what the Labor Department called overpayment collection. If a miner won at the initial level, he was immediately entitled to monthly payments and coverage for medical care related to black lung. But under the law, the company didn't have to pay if it appealed. Instead, the government trust fund would pay the miner's benefits until the case was ultimately resolved. If the miner ended up winning in the end, the Labor Department tried to recoup the money from the company, but if the company prevailed on appeal, the agency turned to the miner for repayment.

A miner could apply for a waiver, and the department was supposed to consider whether he could afford to pay and whether collecting would be "against equity and good conscience." But a report released by the General Accounting Office just as the Labor Department was trying to collect from Manning noted that the department didn't adequately inform miners of this right.

In Manning's case, the problem was compounded by another shortcoming the GAO documented in the report: the backlog in the administrative court system left miners and widows waiting for years to get a decision. By the time the review board had decided to uphold the judge's reversal of the initial award, the government trust fund had been paying benefits to Manning for more than eight years. Thus, the roughly thirty thousand dollars that the Labor Department now said he owed.

Finally, John felt he understood the basics of Manning's claim. Given the evidence that showed Manning's health had worsened, John felt he could reapply for benefits, something that was possible because of the specialized doctrine that evolved as the law attempted to address the distinct complexities of black lung.

The law loves finality. If you sue someone, lose, appeal, and lose again, that's usually the end of the process; you don't get to sue over the same thing. But black lung is insidious. Even after a miner is retired and out of the dust, the disease can progress. A miner might be able to prove he has the disease, but if he is only partially disabled, he'll lose his case. Five or ten years later, however, he might be totally disabled. In recognition of that fact, black lung regulations allow a miner who has lost a claim to refile, but he has to produce some new evidence showing he's gotten worse. He can't simply relitigate the previous denial.

Though Manning had lost his case less than a year earlier, more than nine years had passed since he had filed the claim, so the initial medical evidence was now years old. More-recent tests showed that Manning's breathing had become significantly worse as his claim dragged on. John submitted this new evidence, arguing that it proved Manning was now totally disabled by black lung.

Soon, the case was before administrative law judge Frederick Neusner, who scheduled a hearing. The prospect of having to stand before a judge and argue Manning's case was more than a little daunting for John. He knew he needed help.

Though the drastic changes to the program in the 1980s had driven away most lawyers who'd been willing to represent miners, a few stalwarts remained. Those who had stuck it out were mostly the true believers, progressive types who saw themselves as champions of the hardscrabble underdogs. They were part of a small, informal network spread out across coal states. Their ranks never grew beyond about twenty or thirty nationwide, and they pretty much all knew one another and shared tips and updates. In West Virginia, one lawyer in particular had a long track record of representing miners and a reputation as someone happy to counsel a newcomer. John made an appointment to see him.

Grant Crandall had first represented miners and widows in federal black lung claims in the summer of 1973, when he was still in law school. An Ohio native, he'd attended the University of California, Los Angeles, and spent his summers in southern West Virginia working with the Black Lung Association, mostly helping with benefits claims. After finishing law school in 1975, he'd moved to West Virginia and set up a small practice.

Crandall's firm subsidized its work on federal black lung cases by taking Social Security disability and workers' compensation cases. "You'd actually go under financially if all you did was federal black lung claims," he said. Black lung–benefits cases, even compared with disability and traditional workers' compensation claims, were uniquely unattractive: The pay was worse; the odds of success were lower; and the costs were higher. Yet black lung remained a priority for Crandall and his firm even as others fled the system.

When John called him in 1991, Crandall was happy to welcome a new recruit. John arrived at Crandall's office in Charleston and found a pile of documents—more regulations—waiting for him. Crandall walked him through the hearing process in what seemed to John like a whirlwind. "It was sort of, 'Here's the diving board. Go jump,'" he recalled.

John did just that, though his first attempt was not as graceful as he'd hoped. Armed with new medical reports, he drove forty-five minutes south to the small town of Mullens in rural Wyoming County, where the hearing was being held. Inside the courtroom, however, he soon discovered he was unprepared for Judge Neusner's questioning. Why, the judge wanted to know, wasn't the previous decision denying Manning's claim res judicata (the Latin-derived legal term for a matter that's already decided and not open for reconsideration)? John stumbled and, in his confusion, failed to get all of Manning's evidence into the record.

In a brief filed a couple of months later, John tried to rehabilitate himself. "Dear Judge Neusner," he wrote, "I apologize for any impropriety on my part at a hearing on the above mentioned case on November 5, 1991, in Mullens, West Virginia, and in this written response. It is not by choice that Mr. Manning and other claimants appear before you without attorneys or with a representative having limited knowledge of the law and legal procedures."

Neusner decided to give John an opportunity to introduce the evidence he'd meant to submit before, but this required further procedural tedium. Nineteen months later, the evidence was in the official record, and a new judge assigned to the case, Stuart Levin, issued his decision: Manning's condition had indeed worsened, he determined. New reports showed that he was totally disabled and identified black lung as the cause. The judge ordered Manning's former employer to pay him benefits.

Despite the victory, John was troubled by the way the system now seemed to function. A few weeks after the judge's award, John took out a memo pad with the New River clinic's letterhead and recorded his thoughts in his usual bullet-point format:

- —Adversarial relationship between individual miners or widows and large corporations is grossly one-sided.
- —Program was intended to be temporary

(a) Compensate disabled miners
<u>and</u>
(b) Clean up the mines

<u>But</u>, companies blatantly violated intent of Congress with wholesale dust sampling fraud
<u>And</u>, still want every advantage against victims of black lung who often cannot even get an attorney.

- —Modest reform legislation, basic fairness issues.

John's analysis of how companies had undermined the dual promises of the 1969 law was similar to that of other miners' advocates. Working on Manning's claim, John had seen for himself the erosion of the first promise—fair compensation for miners. The case proved to be an even better introduction to the world of federal black lung benefits than John realized at the time. The company appealed, and the case dragged on until a court ultimately upheld the award in 1998—eight years after John had taken the case and almost eighteen years after Manning had first filed it.

By then, Manning was dead, and his wife was in a nursing home.

The erosion of the second promise—to clean up the mines and prevent disease—was not something John had seen himself, but as he worked on Manning's claim, it became the talk of the coalfields and the stuff of national newspaper headlines.

7

"The Great Coal Dust Scam"

The Labor Department had promised a bombshell announcement, and on April 4, 1991, a gaggle of Washington reporters gathered at a press conference to hear what it was. For the occasion, Labor Secretary Lynn Martin had brought unusual props, and she now passed them around: small plastic cassettes that housed thin filters made of PVC, each roughly the size and shape of a silver dollar.

The department, Martin said, had uncovered widespread fraud committed by coal companies, and the filters were the proof.

These flimsy, disk-like slivers were linchpins of the system the government had devised to protect miners from black lung. To demonstrate that companies were keeping dust levels in the mine air below the limit Congress had set in 1969, company officials had to collect samples periodically. Federal inspectors also collected samples, but far less frequently. The system relied heavily on coal companies themselves to keep tabs on what miners were breathing.

Miners who had jobs that placed them in areas with the highest levels of dust—usually the people who ran the cutting machines or who, like Gary Fox, drilled bolts into the roof of freshly cut sections—occasionally had to wear a sampling device to measure what they were breathing. It was a cumbersome apparatus—a plastic tube connected a rugged box attached to the miner's belt with a metal-and-plastic contraption that clipped to his shirt collar and extended down past the breast pocket.

The box housed a battery-powered pump that sucked in the mine air through a small slit in the device hanging from the shirt collar. The air flowed to a cylinder, called a cyclone, that spun the air around. Larger dust particles fell to the bottom of the cylinder, and only the smallest dust particles—those capable of penetrating the body's defenses and reaching the lungs—continued their journey through the device. They flowed through the cassette and collected on the filter. When the miner's shift was over, he turned in the whole sampler, and company officials removed the cassette and mailed it to a Labor Department lab. There, technicians opened the cassette, removed and weighed the filter, compared that to the original weight of the clean filter as measured by the manufacturer, and, after making a few calculations, determined whether the mine was in compliance with legal limits.

Overall, the results appeared to be cause for optimism. For years, the average dust levels in samples submitted by coal companies had been well below the limit, and by 1991, only about 11 percent of all individual samples were above it. One of the industry's claims during the debate over the 1969 law—that having to meet this standard would be impossible and might force many mines to close—now looked like a dramatic overstatement. Not only were most companies not struggling to pass the exams; they were making straight As.

The industry also used this impressive compliance record to help write off the concerns of miners and their advocates about the declining rate of approvals for black lung–benefits claims. Of course the rate would go down, they argued; miners who had worked when there were no enforceable dust limits had already been compensated, and thanks to the industry's commitment to compliance, new miners weren't getting the disease. "It should be impossible to get black lung disease under current levels of exposure," Tom Altmeyer, vice president of the National Coal Association, said less than a month before Martin's press conference.

Now, however, Martin's announcement called that apparent progress into question. Flanked by top officials from the Mine Safety and Health Administration and government scientists, Martin directed the reporters' attention to the props. After sampling, a filter ordinarily had a collection of fine dust spread across it. She displayed examples. But an astute technician had noticed something strange, Martin said. Some filters had dust everywhere except in their centers, which was where there should have been the *most* dust.

After a twenty-month investigation, the agency thought it had figured out the cause for those clean spots on the filters' centers, and an official now demonstrated: He blew into the opening of the cassette, which dislodged enough dust from the filter to reduce the weight. It was embarrassingly simple. This rudimentary trick could turn a dust-limit violation into a passing grade.

If the agency was right, then the rosy picture of industry compliance with the law was largely a façade. Companies had a way to avoid the cost of controlling the dust and the risk of being cited, fined, forced to upgrade ventilation and equipment, and possibly even temporarily shut down by inspectors. And it took very little time and no expensive equipment—just a blast of air.

What was most troubling, Martin continued, was what the agency had found after this initial discovery: These "abnormal white centers," as they came to be called, didn't come from just one mine. The more officials looked, the more filters like this they found, and these suspicious samples had arrived from mines owned by an array of companies and spread across most of the country's coal-mining regions. "West Virginia, for example, had two hundred and forty-three separate mines in violation," Martin said. "It seems almost an addiction to cheat at some mines."

In all, the agency found evidence of tampering from almost eight hundred and fifty mines, and it was now issuing a citation for each altered filter, imposing about five million dollars in combined penalties. "It is the largest aggregate fine in the history of the Mine Safety and Health Administration," Martin said. "I am appalled at the flagrant disregard for a law designed to protect coal miners against disabling lung disease."

In keeping with the tradition of coalfield scandals, central Appalachia again was the hotbed of the apparent misconduct. Of the mines that received citations, about 80 percent were in West Virginia, Kentucky, or Virginia. The list of accused violators included more than five hundred companies, though some were subsidiaries of the same corporate parent. Among them were many of the industry's titans. Consolidation Coal Company, one of the largest coal producers in the country, had twenty mines on the list, accounting for almost four hundred citations. Consol promptly vowed to contest every citation, and the company's vice president said, "This contention they're making of an industry-wide conspiracy lacks the credibility that common sense would demand."

Also on the list was Birchfield Mining, Gary Fox's employer, which received nineteen citations. These became part of a lengthy process of trials and appeals that would take years to play out, pitting many of the coal industry's giants against federal regulators.

The citations were only one piece of the Labor Department's strategy to crack down on companies that doctored dust samples, Martin told the reporters. Criminal investigations were ongoing. Three months earlier, the nation's largest coal producer, Peabody Coal, had pleaded guilty to submitting fraudulent samples from mines it owned in West Virginia, Kentucky, Illinois, and Ohio, and it had agreed to pay a five-hundred-thousand-dollar fine, the maximum penalty allowed by law. Martin seemed to be saying that the case wasn't a one-off; it was just the beginning. Indeed, the sample that had first grabbed the attention of an agency lab technician in 1989 and sparked the larger investigation had arrived from a Peabody mine.

Martin's press conference made the evening's national news broadcasts and the pages of newspapers across the country. What became known to many as the "Abnormal White Center Scandal" was under way. Some in the coalfields called it "the Great Coal Dust Scam" and, in a nod to the first shoe to drop, dubbed the curiously altered filters "Peabody Bon-Bons."

The Labor Department's efforts initially appeared to have momentum. Newspaper editorial boards cheered Martin, and she soon announced that the department had revoked the licenses of sixty company officials who had been certified to conduct dust-sampling and increased the fines for a handful of particularly egregious violations.

But even after the heightened fines, the penalties that the Labor Department had the authority to impose were a drop in the bucket to most of the companies. Consol could have paid not only all of its fines but also those of the entire industry using just 2 percent of its $213 million in profits from the previous year.

But it wasn't just about the money. Martin had jabbed a thumb in King Coal's eye; she'd impugned—in very public fashion—the integrity of an American institution. "Our industry has been tried and found guilty without any disclosure of the crime and evidence," National Coal Association president Richard Lawson proclaimed.

The companies on the receiving end of about 70 percent of the citations quickly filed notices contesting them and demanding hearings before an administrative law judge under the Federal Mine Safety and Health Review Commission. Rather than handle each company's case separately, Judge James A. Broderick consolidated them into one.

Broderick decided to split the case into two parts. First, there would be a common-issues trial, which would delve into what caused the abnormal white centers, or AWCs. Then there would be a mine-specific trial at which the judge would apply the findings from the first trial to a test case involving one mine.

Before the first trial even started, however, Broderick mortally wounded the government's case. Everything hinged on what the word *alter* meant. The Labor Department had cited each company for violating a regulation that made it illegal to "alter the weight of any filter cassette." Government lawyers argued that *alter* meant any actions that affected the weight of the sample, regardless of whether they were intentional or accidental. But the judge sided with the companies, finding that, in this context, *alter* clearly meant only intentional conduct. Later, in an opinion blasting Broderick's handling of the case, Marc Lincoln Marks, a member of the commission that heard appeals from administrative law judges, called this "playing Noah Webster."

This seemingly minor semantic point meant that the government had to prove that the existence of an AWC was proof of intentional tampering by companies. Broderick's finding kicked the door wide open for companies and their experts to assert myriad other accidental causes. They were not at a loss for creativity. The common-issues trial and the subsequent mine-specific trial played out in similar fashion, with both becoming spectacles of statistical wizardry and hypothetical explanations that strained credulity. The companies weren't devious, they argued; they were just careless and clumsy. Lawyers and experts for the coalition of coal companies pointed in particular to "rough handling"—dropping or hitting the cassettes, stepping on the hose, or "closing a door or drawer on it, or sitting on it"—as a possible cause for AWCs. Scientists for the coalition testified that they'd conducted experiments that proved the theory valid. To emphasize this point, one offered a courtroom demonstration: He repeatedly dropped a thirty-pound toolbox on a sampling device's hose, producing an AWC on the filter each time.

After listening to the experts, Broderick was convinced that AWCs could be caused by all sorts of apparently common, ham-fisted behavior as well as quirks such as "filter floppiness" or "humidity in the mine environment." It was true, Broderick acknowledged, that they could also have been caused by the exact sort of tampering that the Labor Department asserted, but the judge wasn't satisfied that the department had proved this was "the only reasonable explanation." Marks, in his later opinion, would call this standard of proof—basically, that the government had to rule out all other potential causes—"impossible to shoulder."

The Labor Department tried to introduce testimony from employees of companies that had pleaded guilty in separate criminal cases—workers who had already admitted that the department's explanation for the AWCs was, in fact, exactly what they had done—but Broderick deemed their statements irrelevant and refused to admit them into evidence.

The judge also decided not to give much credence to a mountain of statistical and other circumstantial evidence that the Labor Department submitted. On March 19, 1990, in the midst of its investigation, the Mine Safety and Health Administration had announced that it would no longer accept any sample with an AWC. Suddenly, the number of AWCs arriving at the government lab had dropped precipitously. At the mine selected for the individual test case, Rochester and Pittsburgh Coal Company's Urling no. 1 mine in western Pennsylvania, almost 43 percent of samples submitted before the company learned of the policy change had AWCs; after, the number plummeted to 0.18 percent.

The company's statistical expert, however, reanalyzed the agency's data and testified that the seemingly sharp drop was actually part of a gradual decline that had begun before the announcement of the policy change. Experts also asserted that changes in the manufacturing process by companies that made the filters partially explained the decline. Again, Broderick sided with the defense. Marks, in his later opinion, had his own take on the operators' explanations: "To ask us to believe that is to ask us to believe that elephants fly."

Another key piece of statistical evidence arose in the test case, something that Marks later called "perhaps the single most powerful piece of evidence of deliberate and intentional conduct in the mine-specific trial."

Federal regulators analyze some of the samples taken by inspectors for silica, the particularly harmful mineral dust that often mixes with the coal dust when miners have to cut through certain kinds of rock. Agency technicians evaluate these samples using a different method. Whereas the key metric for regular dust samples is the total weight of all dust, the key metric for silica-dust samples is the proportion of silica. Blowing loose some of the dust, then, would likely affect the former assessment but not the latter. Tampering with silica-dust samples that way could even backfire, as regulators require them to be sufficiently heavy to be considered valid.

Over roughly nineteen months, Rochester and Pittsburgh's Urling mine had submitted seventy-five silica-dust samples, and while more than 40 percent of its regular dust samples contained AWCs, not a single silica-dust sample did. Both types of samples were collected in the same way using the same equipment, so if AWCs had been caused by random accidents, as the defense claimed, both types of samples ought to contain a similar share of them. The wildly different samples that came from the Urling mine were "astonishing evidence" that "strongly suggests" the company tampered with the regular samples, Marks later wrote in his opinion.

But Broderick refused to even consider the silica-dust samples. The method for analyzing these samples in the lab destroyed them, so the defense experts weren't able to examine the filters themselves. Nor had the Labor Department offered testimony by technicians who had conducted the testing. Broderick decided that the silica-dust samples couldn't be used as evidence.

The judge's decision in the common-issues trial made the outcome of the mine-specific trial essentially a foregone conclusion. The Labor Department couldn't exclude every conceivable accidental cause for AWCs in general, and it couldn't do that in the particular case of the Urling mine either. The judge sided with Rochester and Pittsburgh and vacated all of the citations.

The Labor Department appealed, and in November 1995 a panel of commissioners voted two to one to uphold Broderick's decision. One of the commissioners who voted to affirm had been appointed by George H. W. Bush. The other, appointed by Ronald Reagan, had joined the commission after leaving a job as assistant general counsel for an oil and gas company. The dissenter, Marks, had been nominated by Bill Clinton. He'd represented Pennsylvania in Congress as a Republican but he'd also worked for Republicans for Clinton.

Marks concluded his dissent with a pointed rebuke: "Unfortunately, we cannot tell and will never know how many men and women were infected with black lung disease as a result of the illegal actions of the company defendants in these cases. Considering the length of time their illegal actions went on, the number could well be in the many thousands."

A federal appeals court affirmed Broderick's decision in August 1998, and three months later, the Labor Department vacated the thousands of citations against the other companies that had contested them, cases that had been on hold during the seven-year legal fight. Included among them were the nineteen citations issued to Birchfield for samples taken while Gary Fox worked there.

In her original press conference announcing the citations, Martin had insisted that, worrisome as the industry's widespread tampering was, the Labor Department had caught it early on and had put a swift end to it. Reporters had pressed her on how she could be so sure it hadn't been going on undetected for much longer. "We believe that it would have been discovered," she replied. The real story, she added, was that the system to protect miners had worked "exactly as it was supposed to."

But the coal industry hadn't suddenly developed an "addiction to cheat," nor had it suddenly kicked the habit cold turkey. The tampering that produced the AWCs was a nice trick, but it was only one of many methods the industry used to fool the regulators. Even the Peabody case that had sparked the crackdown was about more than AWCs. The bon-bons were the red flag, but deeper investigation found that Peabody was cheating in other ways too. Some of these same tactics came out in the criminal cases that unfolded in parallel with the consolidated civil proceeding.

As just about any coal miner could have told Martin, these practices were hardly sophisticated or new, and her tough talk hadn't put an end to them. For proof, one needed only to look to the underground tunnels where Gary Fox spent, on average, nine or ten hours a day, six days a week, eating dust.

8

The New Coalition

Word began to filter through the coalfields that there was a clinic worker in Scarbro who was gutsy or crazy enough to help miners with federal black lung–benefits claims, even represent them as a lay advocate in some cases. Miners started showing up at the New River Breathing Center with stacks of files they hoped John Cline would review.

As his reputation was spreading, John was becoming an effective legal combatant. In one respect, he already had a leg up on many lawyers: He understood well the tests used to determine disability in claims because they were the same tests he performed every day in the clinic. Building on this base of knowledge, he familiarized himself with the medical literature that frequently arose in claims. And he had another advantage over some representatives, this one geographic: The New River clinic was a short drive from the offices of the godfather of black lung doctors, Don Rasmussen. John sometimes stopped there in the evenings after work, and despite his voluminous caseload, Rasmussen always made time for John, just as he had for many miners' advocates over the years.

When it came to the law, the first and most important thing a miner's representative needed was a command of the regulations. There were multiple iterations of them that might apply, depending on the specifics of the case. John digested hundreds of pages of these intricate rules. "He knew those regs backwards and forwards," recalled Tom Johnson, who was part of the small community of attorneys willing to take federal black lung–benefits cases.

Where John needed more help was with the sorts of things students learned to do in law school—researching case law, writing legal briefs, deposing witnesses. Fortunately for him, the few lawyers who did represent miners in federal claims generally were neither stingy with their advice nor protective of their turf. John got an ad hoc education by calling and exchanging letters with them; he learned the pertinent precedents and skills for each case until he eventually had a broad base of specialized legal knowledge.

He would work out an argument, check and recheck it, refine it, then ask the lawyers if it was right. "Almost always the answer would be, 'Yeah, that looks pretty good to me,'" Johnson said. "He has this part of him that is thinking that he's got to make sure it's a hundred percent right. I'm not sure where that comes from. Probably it's his complete identification with the miners, and he does not want to let them down."

True to his community-organizing impulses, John sought to do more than learn from this small group of miners' representatives; he set out to expand its size. In the past, miners' representatives—attorneys and a few lay advocates—had periodically gathered to share tips and updates, and some of them now thought that it was a good idea to have more of these get-togethers, given the attrition that had come with the changes made during the Reagan years. Working with the National Coalition of Black Lung and Respiratory Disease Clinics, which represented the black lung clinics that received federal funding, they helped organize a conference to bring together lawyers, lay representatives, doctors, and clinic workers in Chicago in November 1994.

Chicago might seem an odd place to hold a gathering of coal miners' advocates, but as it happened, the city was home to an active Black Lung Association chapter. As the coal industry declined in the decades after World War II, thousands of miners had migrated north in search of factory jobs. Many had ended up in Chicago's Uptown neighborhood. Chicago was also home to the small law firm of Johnson, who was one of the lead organizers for the 1994 conference. Johnson had returned to his native city after earning a law degree from Harvard. Unlike many of his classmates, he didn't have his eye on a post at a top firm. "To the extent there was any point in going" to law school, he said, "it was that I was going to try to make a change and do something different, identify with people on the short end of the stick."

After clerking for a federal judge, Johnson went to work for a small legal-services group in Uptown. He helped the neighborhood's residents, most of them poor, challenge evictions, battle unscrupulous lenders, and apply for benefits such as Social Security and welfare.

Given the significant population of former coal miners in the area, some of them clearly suffering the effects of their previous occupation, there was also a need for legal help with black lung–benefits claims. A Brooklyn-born community organizer named Paul Siegel, who had helped form the Chicago Area Black Lung Association a few years earlier, recruited Johnson to the cause. Siegel had become an effective self-taught lay advocate for miners in federal claims. As the claims process grew more complicated and time-consuming, though, Siegel tried to interest lawyers in taking on cases. He found a receptive ear in Johnson.

Siegel delivered a crash course on the subject, and Johnson, who soon started his own firm, learned the rest on the job. He often worked on cases with specialists in occupational medicine at nearby Cook County Hospital, but as more cases became the financial responsibility of coal companies, the sparring over medical intricacies grew more complex. The staff in the hospital's occupational medicine department didn't know the answers to all the questions company-hired physicians were raising. There was, however, a doctor in the pulmonary lab who might find the cases interesting, staffers told Johnson.

That's how Johnson met Bob Cohen. The son of two public-school teachers in Philadelphia, Cohen had earned a medical degree from Northwestern University and joined the bustling public hospital that was a critical resource for the large population of economically disadvantaged Chicagoans. He specialized in pulmonary medicine, treating patients with severe lung disease, many of them afflicted with tuberculosis.

One day, Johnson appeared at his office. "He came to the pulmonary department with a shopping cart full of files and asked me if I would help him look at a case or two cases," Cohen said. "From that point on, it was all downhill, or uphill, or whichever." What Cohen saw in the files fascinated him, and he quickly took a liking to coal miners. Most doctors avoided the time-consuming combat that often came with involvement in a legal claim, but Cohen relished it. "I'm a fighter," he later said. "I like to take on the big guy and help the little guy, and this was just perfect."

By the early 1990s, Cohen had become an indispensable figure in the resurgent black lung movement. "For Chicago, Bob was our Dr. Rasmussen, and we just buried him in cases," Johnson said.

In organizing the 1994 conference, the Chicago crew injected renewed zeal and fresh perspectives. They brought in doctors and scientists whose pioneering research linked breathing coal dust with previously unrecognized varieties of lung disease, and they enlisted lawyers and lay advocates from the small community of miners' representatives to deliver presentations.

This became the new norm. Each year, there would be a national conference, and the West Virginia members supplemented this with their own statewide conference. Attracting new lawyers and doctors to the fold was a tough sell, but they had some success. "We were always out recruiting, but slowly," Johnson said. "We created this little network of people."

This was a critical coalition, a new generation aiming to finish the job that the 1969 insurgents had started. The group was still small and overmatched, to be sure, but its members were dedicated and scrappy. They were fighting a two-front war on behalf of miners, trying to win justice for one miner or widow at a time in benefits claims while also pushing for new laws to level the playing field and clean up the mines so future workers wouldn't be crippled by black lung.

John had a hand in virtually all of the coalition's various endeavors. He evaluated miners and worked with doctors at the New River clinic, represented miners in benefits claims, spoke at conferences, drafted proposed changes to the law, and helped organize Black Lung Association meetings and lobbying efforts. There were almost no lay advocates other than John who formally represented miners and stuck with them when companies appealed, but there were some who helped miners in the early stages or on a more informal basis.

One of them was Debbie Wills, a benefits counselor at a black lung clinic near Charleston who would eventually become the clinic's program coordinator and an outspoken advocate for miners. She met John at one of the annual conferences and started calling him regularly for advice—so often, in fact, that she felt guilty about taking up so much of his time.

"If he told me something," Wills said, "that's exactly what I would do because, as far as I could tell, there was nobody else that knew as much about it as he did or did as much for miners as he did or got as much respect from the leaders of the Black Lung Association as he did. I really tried to figure out how he did stuff and why he did it that way so I could do it too.

"Very honestly," she said, "I wanted to be like him."

The Washington and Lee School of Law in Lexington, Virginia, enlisted John's aid when it established a black lung legal clinic in 1996. Eight or ten students spent an entire academic year helping with a few cases at various stages in the claims process. Under an arrangement with the New River clinic, John screened and referred cases, and he worked with students in some instances.

"These were amazing first clients for our students to have," said Mary Natkin, one of the professors overseeing the program. "They opened their doors. They fed our students."

In the decades since, the legal clinic has represented more than two hundred miners and widows, and it is largely self-sufficient financially, funding its work primarily with fees collected from coal companies after successful claims.

Even in the years before John took on the added role of screening claims for Washington and Lee, his own portfolio of cases was growing, and this became a source of tension at the New River clinic. John did much of his work as a lay advocate at home after hours, but there was no way to keep it from taking at least some time out of his workday. "It cut both ways for me, frankly," said clinic director Craig Robinson. "It was both wonderful that he was helping people in this intense way, but it did mean that he had a fairly limited number of people that he could work with. Sometimes I felt like we may not have been using his skills to touch the most people."

John would later acknowledge as much: "I really was pushing the envelope with Craig to be able to do that."

Ultimately, Robinson allowed John to continue. Despite the complications, he believed what John was doing was important, and he saw how much it meant to him. When a miner who was seriously ill lost a claim, John would fume about the unfairness of the system. "He was on a mission to understand the whole process and all the players," Robinson said.

Another New River staff member who made a similar observation was the nursing supervisor in the part of the clinic that provided primary care, a woman named Tammy Campbell who would become John's wife. Tammy started at the clinic in 1988, shortly after John, and she saw him periodically when he came up from the basement offices of the breathing center to consult with a doctor on the main floor. They spoke a handful of times but never at length. A few years later, Tammy, battling substance abuse, stepped away from the clinic to enter a treatment program. When she returned to her position at New River, she seemed transformed, John thought. Impressed and intrigued, John asked her out on a date, and she agreed.

As the relationship became more serious, Tammy became accustomed to a quirky habit of John's: He always carried around a legal pad, and he would often pull it out, ask Tammy a few questions, and take notes. Did she want kids? Did she like spending time outdoors? What were her thoughts on religion? Tammy understood this was just his way of trying to determine whether they were a good match. Her first marriage, like his, had not worked out. The note-taking didn't bother her except when she was trying to work on something or get out the door and he'd be there with the legal pad, insisting he had just one more question.

(Looking back, John acknowledged it was probably overkill. "That was kind of dumb," he said years later. "I guess I was still thinking I was keeping a checklist for my house building or something.")

It turned out that they were indeed a good match. Tammy had grown up in the town of Terry, a former coal camp just a short drive from where John now lived. Her mother was a cook for the local school, and her father was a coal miner who later left the mines and worked as a carpenter. She had three older brothers, and two of them spent time working in the nearby coal mines. Her oldest brother died after being electrocuted in a mine.

Tammy became aware of black lung as an adolescent. The historic movement in the wake of the 1968 Farmington disaster was the talk of the coalfields, and she saw Rasmussen and the rest of the touring trio of doctors on the news. From watching this insurgency, Tammy gained an understanding of why her father had such horrible coughing fits in the evenings. He eventually filed and won a black lung–benefits claim.

Tammy was the first in her family to attend college. She pieced together courses at a few local institutions and earned a bachelor's degree in nursing. After working at a couple of hospitals and a private doctor's office, she got the job at the New River clinic.

Her view of the world and her values closely matched John's. They began discussing marriage, and in the winter of 1992, John invited Tammy to visit East Aurora so she could see where he grew up. He showed her around town, introduced her to his aunt and uncle, and took her to Niagara Falls.

One evening, when they were sitting in their hotel room, John again pulled out the legal pad and started in with questions. Irritated, Tammy interrupted him.

"Let me ask you just a couple more," John said. When he finished, he lowered the legal pad. "Well, I guess that settles it," he said. "All the things you believe and I believe—I think we could make it. I'd like to get married. How do you feel about it?"

"You know," she replied, "I think I feel the same way."

A few months later, in April 1992, they had a small wedding in the side yard of John's house in Piney View. Two years later, they had a son, Brooks. John's oldest son, Jesse, then twenty-one, lived in the area. A few years earlier, he'd dropped out of college in South Carolina and returned to West Virginia. He declared that he was going to become a whitewater rafting guide—the New River was a natural magnet for tourists seeking outdoor adventures—and he'd done just that. John and Tammy had gone on one of his first runs with him.

Jesse was outgoing and talkative; girls took to him. But he also had a more reserved, contemplative side; he read a lot and occasionally wrote poetry. Jesse was good with Brooks. He'd lie on the couch and hold his younger brother while he slept or coax him into sipping the kale-and-carrot concoction he'd mixed up in his juicer, leaving baby brother with green lips.

John's middle son, Matthew, lived in South Carolina but visited occasionally. Partway through high school, Matthew had left West Virginia and moved to Surfside Beach to live with his mother. After graduating, he found work in painting and construction. Embracing an unfettered lifestyle, he and a few friends sometimes trailed the Grateful Dead as the band toured the country, selling grilled cheese sandwiches and beer to get by. At John's urging, though, Matthew would go on to earn a degree in marine science from Coastal Carolina University. He eventually married, settled down in Johns Island, South Carolina, and had three children. Returning to the craft he had learned from his father, Matthew became a skilled carpenter.

Even as John's family responsibilities increased, he often spent evenings working on his steadily growing load of black lung–benefits cases. More and more files took up temporary residence in the Piney View house. John would hammer out legal briefs on a boxy Tandy computer. "He was up the whole night working on it, until he got it exactly the way he wanted it," Tammy recalled. He sometimes read her the briefs when he'd finished or told her about his frustrations with a system he viewed as unfair to miners. "I would think, *For sure, something has to be done*," she said.

The work meant her husband was often preoccupied in his free time, but she supported him. She saw his conviction and knew from her own life how much the black lung–benefits program could mean to working-class families. When she was fifteen, her father had died suddenly of a heart attack, and the modest monthly black lung–benefits payments that she and her mother continued to receive had helped sustain the family.

She saw that this same small measure of financial security and dignity was what John was trying to win for other miners and their families.

John also shared with her a pattern he believed was emerging in his cases—something more than the structural unfairness of the system; something that went beyond even the bare-knuckles tactics that, while arguably callous, were considered acceptable in the legal arena; something that he believed, even in a world of moral ambiguity, was simply wrong. It was 1994 when John caught his first glimpse of this pattern. That summer, a miner showed up at the New River clinic with a medical report that raised John's suspicions and set in motion a two-decade struggle with some of the region's most powerful forces.

9

"Outlaw"

Underground, Gary Fox on occasion would chew out laggards or bristle at orders he found unreasonable, but that's where he left those workplace complaints—underground. They didn't invade his conversations at home with Mary, with one exception: He did not like Don Blankenship.

Gary was far from alone in his assessment of the man who became his new boss in 1993. Blankenship had made plenty of enemies during his rapid ascent to the top of A. T. Massey Coal Company, and he was just getting started. Massey was one of the nation's leading coal producers when Blankenship took over as chairman and CEO in 1992, but he wasn't content with prosperity. He craved dominance.

His rise came as shifting economic forces were transforming the coal industry, and he seized the opportunity to have a hand in crafting the new reality, one that would shape the lives of Gary and his fellow miners.

Blankenship was a true son of West Virginia, a badge he displayed frequently with pride. He was born in 1950 in a small eastern Kentucky town, but before long, his mother separated from her husband and took her son across the border to Delorme, West Virginia, where she raised him by herself in a small home with an outhouse. She opened a store and toiled constantly to make a go of it. When he was old enough, Blankenship worked the cash register, developing his natural knack for numbers.

He excelled in school and enrolled at Marshall University in Huntington, near the Ohio border. To earn tuition money, he got a job as a coal miner, and he graduated with a degree in accounting in 1972, finishing in just three years. For about a decade, he worked as an accountant for food companies, spending time in Georgia, Illinois, and Colorado. He got married and had two children, but he later got divorced.

In 1982, Blankenship started at a Massey subsidiary, Rawl Sales and Processing Company, near Matewan, the West Virginia town where he'd attended high school. Within a couple of years, he was running the Rawl mining operations. This placed him at the center of a bitter fight between Massey and the United Mine Workers that left him with an undying hatred for the union.

Amid a downturn in the industry, his boss, CEO E. Morgan Massey, looked for ways to cut costs, and he set out to break the union. The first clash had come shortly before Blankenship assumed control of Rawl Sales, and it unfolded at the mine site operated by Massey subsidiary Elk Run Coal Company, where Gary would later work. For decades, central Appalachia had been a UMW stronghold—one that miners had shed blood to establish during the Mine Wars of the early twentieth century—but Massey opened the new Elk Run complex as nonunion. The UMW tried to organize the site, and tense standoffs on picket lines had escalated into violence. The company had convinced a federal judge to issue an injunction against the UMW, and the union's efforts had ended in defeat.

In 1984, a larger contract dispute erupted, and the UMW staged strikes at multiple Massey sites, including Rawl Sales. Blankenship brought in strikebreakers, and violence again broke out. Someone fired shots into Blankenship's office, and a bullet shattered the screen of his Zenith television (he kept the TV for years as a memento). He denounced the UMW in the media, saying that the union had become "too powerful" and that miners "don't have to be slaves of the union any longer." After fifteen months of turmoil that included picketing and the shooting death of a nonunion coal-truck driver, the two sides reached an agreement that ended the strike. The result was widely viewed as a win for Massey.

These bloody conflicts are often regarded as a turning point for the UMW, the beginning of the union's decline in the region.

Blankenship became E. Morgan Massey's handpicked successor, and he assumed the company's top post in 1992. Despite the fact that he now ran one of the nation's largest coal companies, he remained in the home where he'd lived for years, a relic that had been built decades earlier for a mine superintendent in the small community of Sprigg, West Virginia, not far from where he'd grown up. He rarely visited Massey headquarters in Richmond, Virginia. Instead, he ran the company from a double-wide trailer set up just across the border in Kentucky.

Blankenship staked the company's future on central Appalachia even as other producers abandoned the region or had to shutter their operations there in response to the changing market. Cheaper fuel sources, especially natural gas, were displacing coal. Within the industry, companies were investing more heavily in coal deposits out west, particularly in the Powder River Basin reserves stretching across parts of Wyoming and Montana. There, the coal was closer to the surface, making it cheaper and easier to mine, and it contained less sulfur, making it attractive to electric utilities that had to comply with new environmental regulations. Productivity increased, but employment plunged. Central Appalachia took the hardest hit, losing mining jobs by the thousands.

Central Appalachia did retain one distinct advantage, a sort of geological blessing: metallurgical—or "met"—coal. The region had rich deposits of this variety of the material, which had the qualities—high energy content, few impurities—necessary for use in the steel-production process. These reserves were at the core of Blankenship's bet on the region. He acquired large stores of the high-quality coal that would give Massey the flexibility to provide what global markets demanded, whether it was low-sulfur fuel for U.S. power plants or the grist for steelmaking in rapidly industrializing nations. The bet paid off, bringing soaring revenues.

Blankenship possessed qualities well suited to the task of building an empire in a tumultuous and fiercely competitive industry. He was smart, meticulous, relentless. One of his most notorious exploits: A

West Virginia jury hit Massey with a fifty-million-dollar judgment, siding with the owner of a small coal mine who accused Massey of driving his company into bankruptcy, leaving the local executive financially ruined and more than one hundred union miners unemployed. The heavily indebted owner said Blankenship had warned him: "You don't want to sue Massey. We spend a million dollars a month on attorneys." As the appeals dragged on, Blankenship spent more than three million dollars during an election for a spot on the West Virginia Supreme Court of Appeals, funding vicious ads attacking the liberal judge he hoped to unseat. It worked, and the court, with its balance now tipped in Massey's favor, vacated the award.

"What you have to accept in a capitalist society generally," Blankenship had proclaimed during the earlier union fight, "is that it's like a jungle where a jungle is the survival of the fittest." He chafed at what he viewed as unnecessary constraints on personal freedom. Unions and government regulators were favorite targets of his ire, along with "greeniacs" and the "liberal media."

Blankenship was remaking not just the industry but the entire landscape of central Appalachia. He embraced a controversial mining practice known as mountaintop removal. The technique was both a lucrative innovation and a horrifyingly efficient force of destruction—blow off the mountaintop, haul away the coal. Companies were technically required to "reclaim" the moonscapes that resulted, but rebuilding a centuries-old Appalachian forest was beyond the capacity of even the most well-intentioned operators. The people living in the hollows below felt the brunt of the blasting. Dust coated their homes, and waste ended up in their creeks. Without vegetation to help absorb rainfall, flooding intensified. Residents complained of contaminated water and clusters of strange illnesses.

Massey's operations generated vast amounts of waste that had to go somewhere. Lakes of sludge held back by dams loomed over communities and, in one instance, an elementary school. These were repositories of a material produced when coal was washed in a bath of chemicals to remove impurities—slurry, a black liquid containing toxic substances and heavy metals. One notorious spill from a Massey impoundment in eastern Kentucky contaminated the water supply of thousands.

As Blankenship's clout grew, he became more politically active. Though he was physically imposing—tall and heavyset, with black hair and mustache—he wasn't a charismatic speaker; he spoke in a soft monotone. But the words that came out were Molotov cocktails. He called climate change a "hoax" and produced a documentary titled *Regcession* that purported to explain how the Environmental Protection Agency was "destroying America." In West Virginia, he ensured it was politically toxic for any public official to suggest that coal might not sustain the state's people indefinitely.

His message resonated among many in the coalfields who depended on coal for their livelihoods. He was a straight-talker who insisted he was saying what the politically correct didn't have the guts to utter. He understood the people living in mountain hollows; he was one of them. (Though he sometimes stayed in a four-story mountaintop villa in the coalfields, traveled in a helicopter, and vacationed in Monaco; his annual compensation eventually surpassed seventeen million dollars.) He donated money for new baseball fields, for medical care for the poor at community clinics, for scholarships at local schools. And he employed more people than perhaps anyone else in southern West Virginia.

While other companies were closing mines and laying off workers, Massey was hiring. That cold economic reality led Gary Fox and hundreds like him to Don Blankenship's payroll.

By 1993, miners were no longer being Mabenized; they were being laid off. Maben Energy CEO Paul Kizer had survived high-profile scandals, but he couldn't escape the creditors. He had sold some of his mines, including the Birchfield mine where Gary worked, and he'd declared bankruptcy.

One day, Gary and the two men who rode to work with him arrived at the Birchfield mine and found the gates locked. A supervisor informed them that they no longer had jobs. So they turned around, drove an hour and a half back across Bolt Mountain, and started looking for work.

A few months later, Gary started as a roof bolter for Massey subsidiary Elk Run. Beside the Big Coal River near the small town of Sylvester in Boone County was a chrome-black industrial metropolis erected by the company. The complex extended down a mountain hollow, and Elk Run had bored into the hillsides along the way, opening a handful of underground mines. For some reason—Massey miners weren't quite sure why—most mines were named after chess pieces: White Knight, Black King, Bishop. Belt lines extended their tentacles from the hills, conveying coal to stockpiles and a plant; there it was ground up, sorted, sent to a railyard, and loaded on railcars that carried away hundreds of tons of coal each day.

Other laid-off Birchfield miners, including Gary's two carpool buddies, also ended up at Massey mines. Some joined Gary at Elk Run, though they rarely saw one another. The company shuffled workers from mine to mine within the complex as need dictated.

For many of the miners, taking these new jobs required swallowing hard and accepting the Massey way. Gary and some of his coworkers had always worked at union mines. Massey officials stated the company's dim view of the United Mine Workers in no uncertain terms, Gary's coworkers recalled. "They made it clear they didn't want no union," said Frank Ratcliff, who had befriended Gary at Birchfield and also went to Elk Run. "They run you through the grind before they hire you. You knew going in not to cause waves or they'd get rid of you."

"I didn't like it, but you didn't have much choice," said Raymond Keffer, who joined Massey's Marfork operation after being laid off at Birchfield. "You needed a job, so you put up with it."

Union mines were hamstrung by needless strictures and bloated with liabilities such as health care obligations, Blankenship sometimes said. Low labor costs gave Massey an edge on the competition. A self-described bean counter, Blankenship micromanaged his operations to a remarkable degree given the extent of the company's holdings. He constantly monitored production and demanded explanations for even brief shutdowns or delays. When truck drivers at one operation wanted a $1.91 an hour raise or a manager wanted to pay a contractor $750 to freeze-proof equipment before winter, Blankenship reviewed the requests himself.

No personnel decision was too small, it seemed. In one case, a mine at the Elk Run complex was having problems with the belt lines carrying coal outside, leading to repeated delays, so Blankenship sent an engineer to take a look. Elk Run's vice president, Stan Edwards, later said during a deposition that he thought he'd resolved the issue by giving the mine's two foremen a week to fix the problem, which they'd done. "Well, that didn't suit the chairman," Edwards said, referring to Blankenship. Edwards was directed to fire one of the foremen: "The man that had had brain cancer and only had one eye."

Edwards concluded the tale by citing an acronym some Massey managers internalized: DADS, for "Do as Don says." They sometimes found cans of Dad's Root Beer sitting on their desks as a reminder.

What Don often said was a familiar command: "Run coal." He coaxed record-breaking production totals from his workforce using a mix of carrots and sticks. Crews got bonuses for exceeding production goals. "A lot of times our bonus was more than our paycheck," recalled Harold Ford, who worked as a section foreman at Elk Run. Delays and missed goals might trigger a handwritten note or dictated memo from Blankenship. One read: "I'm looking to make an example out of somebody and I don't mean embarrassment." Another: "You have a kid to feed. Do your job."

The problem was that these production goals were almost impossible to meet by operating within the law. Health and safety regulations mandated various tasks throughout a shift, and each one took time—time that wasn't spent mining coal. "You can't mine that much legally," one former section foreman at Elk Run said during a deposition. "You have to do something illegal."

Blankenship dispatched memos offering direction on what tasks to prioritize. One memo, which he dictated for the superintendents of all of Massey's underground mines, had the subject line *RUNNING COAL* and said: "If any of you have been asked by your group presidents, your supervisors, engineers or anyone else to do anything other than run coal (i.e.—build overcasts, do construction jobs, or whatever), you need to ignore them and run coal. This memo is necessary only because we seem not to understand that the coal pays the bills."*

This was the Massey way, and it was very different from what Blankenship said publicly. When he took over as CEO, Blankenship implemented a system he called "S-1, P-2, M-3," which he said represented his priorities: safety first, production second, measurement third. The company had an exhaustive list of safety rules that exceeded regulatory requirements, Blankenship said publicly. These rules looked nice on paper, some former Massey miners said, but they often didn't translate to the workplace. "It was a big joke," Ford said.

"It was kinda turned around into S-two, P-one," said Carter Stump, who worked as a superintendent at Elk Run.

Blankenship touted the company's impressively low injury rates, but these, too, were sometimes exercises in semantics. The rates were based on the number of days an injured employee was not able to work, but this failed to detect what some called "the walking wounded"—managers had injured miners report for work, then assigned them light tasks or just had them sit outside the mine for a shift. *No days missed,* the official log would indicate.

Under Blankenship, Massey amassed a prodigious number of safety and health violations, but it didn't take an accountant of Blankenship's ability to realize that the value of the added production attained by disregarding the rules exceeded the meager fines regulators could impose. (Even today, a serious violation that puts miners at risk of grave injury carries a maximum penalty of about seventy thousand dollars, and the fine is usually far less.) In a memo to Blankenship, a top Massey lawyer

* One week after sending that memo, Blankenship sent a follow-up stating that the previous memo was not intended "to imply that safety and S-1 are secondary. I would question the membership of anyone who thought that I consider safety to be a secondary responsibility. The point is that each of you is responsible for coal producing sections, and our goal is to keep them running coal. If you have construction jobs at your mine that need to be done to keep it safe or productive, make every effort to do those jobs without taking members and equipment from the coal producing sections that pay the bills."

relayed the concerns of a former federal mine inspector hired by the company to advise on health and safety issues: "The attitude at many Massey operations is 'if you can get the footage, we can pay the fines.'"

There were ways to minimize even these relatively paltry penalties. When an inspector showed up, "they done everything by the book," one former section foreman at Elk Run said. "When they wasn't there, they'd just run outlaw."

There was, of course, a reason that safety regulations existed and a cost to ignoring them. Massey's record of accidents culminated in the 2010 Upper Big Branch blast that killed twenty-nine miners, the worst mine disaster in forty years.

For those killed by fires and explosions, there would be memorials and annual commemorations. There would be investigations, trials, media coverage, public outrage.

But there was a largely unseen toll, and though a full accounting of it is impossible, it is almost certainly far greater. Many of the same practices that maimed and killed also filled mine tunnels with clouds of dust that gradually accumulated in miners' lungs. Time and medical uncertainty obscured the link between cause and effect. Any legal claims arose in unwatched administrative forums. Deaths came slowly, one by one, unnoticed by all but the family members who bore witness to the suffering that preceded.

Miners tended to worry most about immediate threats—explosions, fires, roof collapses. It was not that they didn't realize the danger from the dust; it was just that they spent most of their time and mental energy trying to run good coal without getting blown up or crushed. Those who avoided these violent deaths—men like Gary Fox—too often found a slower, more agonizing death awaiting.

On the occasions when Gary did bring up Don Blankenship, Mary saw that flash of anger he usually reserved for coworkers he thought were slacking off. Gary didn't go into specifics, just said Blankenship wasn't running things the right away, that he ought to be fired.

Gary had never minded long hours. He was happy for the overtime pay, and he wasn't a union zealot. Still, there was something to be said for having a backstop. At the Birchfield mine, there had been at least the pretense of union protection. At Massey, you had to protect yourself.

"I'll never in my life forget that first day at Elk Run," Gary's coworker Frank Ratcliff said. "That was the first nonunion mine I ever worked at. The first time I walked in there, I felt naked."

At Birchfield, if Gary was really sick, he would take the day off. Not at Massey. Flu, fever of 103—it didn't matter; he was bolting top. "You didn't have a choice," Mary said. "It was kind of like put to you, 'Do it, or . . .'—you know."

Gary had good reason for putting up with it. When he started at Elk Run, his daughter, Terri, was sixteen. Soon, she'd finish high school, go to college, graduate debt-free, get a good job—that was Gary's dream. It had been since May 1972, when he and Mary were wed. He'd been ready to have a child right away. Once, he'd gone with Mary to the hospital when she had a pain in her side, and before the doctor could utter "walking pneumonia," Gary asked if she was pregnant.

Over the years, they'd become a trio—Gary, Mary, and Terri—and wherever one was, the other two were bound to be close by. If Mary decided to run to the grocery, if Terri needed new shoes at the mall, all three went. Their inseparability became a running joke among family and friends.

That is, they were inseparable on Sundays, the one day Gary was consistently off work. The other six days, Mary and Terri rarely saw him. For much of his time at Elk Run, Gary worked the evening shift. He'd start his hour-long commute in the early afternoon, before Terri was home from school. When he returned in the middle of the night, often around three a.m. or later, Terri was asleep, and when she went to school the next morning, he was asleep.

Thus, Sunday was a treasure, often a day of spontaneous adventures. The family would just get in the car and go. They saw the state's natural beauty, went to amusement parks, drove down to Gatlinburg, Tennessee, just for a good meal. Gary's appetite was legendary. He'd polish off a steak dinner and announce he could go for some pancakes. Mary figured it was the constant hard labor that kept him so thin.

Gary bought Terri a used Honda Civic, but he didn't let her drive it until she knew how to change the tires and the oil. He didn't want her to have to depend on anyone but herself, which was why he and Mary also drilled into her the importance of an education.

She didn't need much convincing. She was "a little bit of a nerd," as she later put it—Advanced Placement classes, National Honor Society. She excelled in science, especially biology. In 1995, she enrolled at West Virginia University.

Gary was bursting with pride. Everyone within earshot at Elk Run knew his daughter was going to college, and at the state's flagship university no less. Yet now that she was three hours away in Morgantown, he began to realize what he'd given up so she could be there: Attending her soccer games. Watching her pose for pictures before a high-school dance. Just discussing the day's events at the dinner table.

If a soccer game fell on a Sunday, he'd go. When Terri broke her ankle during one game, Gary ran onto the field and, over her protests, carried her off. There was no questioning his intense commitment to his family, but the fact remained that he had missed much of his daughter's childhood. "I think he felt guilty about being away and working so much," Terri said. "He always wanted us to have a better life, and he made sure of that."

Mary, too, missed out on time with him. The man she'd married, her best friend, was rarely around. "Did I ever resent it? Yeah," she said. "Did I hate him for it? No, because I knew he was doing it for me." She added, "It's not that he picked his job over that. He knew that he had to be there, or get fired."

Shortly after Terri went off to college, Mary tried to convince Gary to leave the mines. For years, when Gary was at work and the phone rang late at night, Mary's and Terri's minds would go racing through disastrous possible scenarios. It had always been a wrong number, but they'd still had trouble getting back to sleep. Then, one night, it wasn't a wrong number. Gary had been taken by ambulance to the hospital, someone from the company told Mary. That was all. He couldn't say what happened or how badly Gary was hurt. Mary rushed to Raleigh General Hospital, where she found Gary in a bed, still caked in coal dust. Falling rock had landed on his shoulder, but it wasn't serious.

Her fears, once hypothetical, suddenly seemed real. She asked him to quit, but he couldn't, or wouldn't. Just a few more years and Terri would be finished with college.

Mary and Terri knew what black lung was; for people living in southern West Virginia, it was hard not to. But it didn't occupy their minds the way roof falls and explosions did. Gary did have coughing spells from time to time, but they hadn't thought much of them. He didn't seem to be short of breath.

Still, Gary had been keeping track of how his lungs were working. In 1989, he'd told a doctor that he was having some trouble getting his breath when he exerted himself, and tests showed his ability to move air in and out of his lungs was mildly impaired. In 1996, another round of testing revealed worsening obstruction of his airways. Though still relatively modest and well above the standards for total disability under the federal black lung–benefits program, the impairment was enough to qualify as a partial disability under the state workers' compensation system.

These reports were among the files John Cline reviewed years later. Tracking a miner's breathing-test scores over time was part of his job at the New River clinic, so he recognized the story that these reports were telling about Gary's lungs.

"If he was monitored closely," John later said, "you could have predicted what happened to him."

10

Cotton

Charles Caldwell came by the New River Breathing Center in the summer of 1994 with a few files he hoped John Cline would review. Caldwell was a patient at the clinic, and Dr. Dan Doyle had been treating him for black lung. The previous year, John had helped him file a federal black lung–benefits claim, and Caldwell had selected Dr. Don Rasmussen from the list of Labor Department–approved doctors to conduct the medical exam, paid for by the government, that would be used to determine whether he qualified for benefits.

Rasmussen had run Caldwell through the normal battery of tests at his clinic in Beckley and determined that he had advanced black lung. He was sixty-three years old and had trouble climbing more than two or three flights of stairs. His worsening breathing problems had forced him into early retirement a few months earlier, though he had already put in more than his share of work over the years.

Born in 1930, Caldwell grew up in Kanawha County. His father, a coal miner, died of a heart attack when Caldwell was seventeen years old, so Caldwell dropped out of high school and went to work as a section hand for a railroad company, digging ditches and laying track, to provide for his mother and younger brother.

When he was twenty, he married a local girl named Patsy who was eighteen and had already lost both her parents. She always called him by the nickname he'd acquired as a child, "Cotton," though she was never sure of the moniker's origins—maybe the light wisps of hair he'd had at the time. They had a son and a daughter.

Caldwell went underground in 1957 and worked for about three and a half years hand-loading coal at a mine in Fayetteville. When there was a downturn in the industry, he found work on road-construction crews as a driller or a mechanic, sometimes having to travel as far as Minneapolis. In 1973, he returned to mining, and he worked the next twenty years at surface mines, most of that time as a mechanic at Hobet Mining.

Though he worked aboveground, he still got a heavy dose of dust. "I wish you'd seen his clothes when he came home," Patsy said. "It's a wonder I didn't have black lung." Six days a week, he made the ninety-minute drive from their home near Fayetteville to the Hobet mine in Madison.

By 1993, he'd developed a frequent cough, and he got so short of breath doing the lifting and walking he had to do on the job that he thought he had no choice but to retire. Rasmussen's testing showed "marked impairment in respiratory function"—enough to render him totally disabled—and a radiologist who examined Caldwell's chest X-ray saw evidence of the advanced stage of disease, complicated pneumoconiosis,* which, under the system's rules, is supposed to lead to an automatic award of benefits for miners.

This provision within the regulations is a recognition of the severity of this stage of disease. Those stricken with it face a high likelihood of premature death, but only after almost indescribable suffering. A lung

* In the black lung–benefits system, *pneumoconiosis* has a meaning similar to the lay term *black lung;* it encompasses classic coal workers' pneumoconiosis as well as other conditions, such as emphysema and chronic bronchitis, that might result from inhaling coal-mine dust. Because the dust in a coal mine frequently contains both coal and silica, the benefits system also considers silicosis and mixed-dust disease as pneumoconiosis. In the general medical sense, though, pneumoconiosis includes a variety of diseases caused by dust inhalation—asbestosis, silicosis, and siderosis (welder's lung, caused by breathing iron dust), among others. The word *pneumoconiosis* is a combination of etymological roots meaning "lung" and "dust."

transplant is often the miner's last hope, and even this brings its own hardships and may be only a brief reprieve.

Doyle had sent Caldwell for X-rays and a CT scan, and a radiologist had also interpreted those films as likely complicated pneumoconiosis.

In most cases, this is the body of evidence that a miner puts forward to prove his claim: the records from his treating physician and the breathing-test scores and report by the doctor he chose to conduct his Labor Department–paid exam. Reports by doctors on the approved list varied in depth and quality. John and the other staffers at the breathing center almost always suggested going to Rasmussen. He knew how to write a report that applied the medical evidence to the specific legal standards of the benefits system—a skill that some physicians who examined miners did not possess.

Rasmussen was a rare type of expert witness. He had helped countless miners win benefits over the years, but he had maintained a reputation of almost unimpeachable integrity. Judges didn't always credit his reports over others, but basically no one viewed him as the sort of biased hired gun that lawyers often deride as "whores." One doctor who has long been an expert of choice for the coal industry said of Rasmussen: "He's an amazing guy. We don't always agree, but I have a lot of respect for him."

Lawyers representing coal companies tried to attack his opinions, but as one attorney who handled many miners' claims recalled, "They never laid a glove on him in all of the depositions I did with him, and that's because he was so honest and would never go beyond what he could say scientifically."

When deciding whether to take a miner's case, John and some of the lawyers who represented miners often treated Rasmussen's reports as a sort of due diligence. It wasn't uncommon for Rasmussen to conclude that a miner simply didn't qualify, but when he determined a miner was totally disabled by black lung, it was probably a solid case.

Rasmussen's report on Caldwell along with Doyle's treatment records provided a sturdy foundation, enough to convince the Labor Department district director to issue an initial determination that Caldwell qualified for benefits.

This, however, was just the beginning. As often happened when the miner won this first stage, lawyers for the coal company contested the award, and the case proceeded to an administrative law judge. That was when defense lawyers usually cranked up their formidable evidence-generating machinery. Miners had Rasmussen and a few other doctors whose reports would hold up before a judge, but company lawyers had an entire stable. They would send the miner's X-rays and CT scans to certain favored radiologists, specialists who often insisted the lesions seen on the films weren't evidence of black lung. In the relatively rare case in which a miner underwent a biopsy, defense lawyers sent the slides of lung tissue to their own pathologists, who tended to minimize or explain away any dust and scarring. The attorneys would then give these expert reports to company-hired pulmonologists, who incorporated all the test results into reports that ticked through each of the elements a miner had to prove in order to win, frequently undercutting one or more of them.

Miners' lawyers had their own shorthand for the usual defense strategy: ABBL, for "anything but black lung." For example: The lesions on the X-rays were probably tuberculosis scars, or maybe they were evidence of sarcoidosis or histoplasmosis or interstitial pneumonitis or any of the other lung ailments that could look similar to black lung on film. And the miner's breathing problems were really attributable to one of these other diseases. Or to obesity. Or to the undisputed champion among the scapegoats: cigarette smoking.

The first stage in building an edifice of negative medical evidence was an exam conducted by a doctor of the defense's choosing. A miner who filed a claim was required to submit to another round of testing, this time by a physician from the defense stable.

Hobet sent Caldwell to Dr. George Zaldivar, a specialist in pulmonary medicine who held multiple prominent posts at the Charleston Area Medical Center, the largest hospital in the state's capital. He was a favorite choice for many coal companies, and miners' advocates came to know him well. When the Washington and Lee legal clinic was up and running, Professor Mary Natkin and her colleagues encountered Zaldivar in so many cases that they could predict what his opinion would be. They held mock depositions in which students would question "Dr. Zaldivar" (a clinic staff member who had his arguments down pat).

The same year that Zaldivar examined Caldwell, an academic journal published an article he had written with two other doctors, both of whom had been producing research that provided ammunition for the coal industry for years. They had reviewed the records of more than six hundred federal black lung–benefits claims and concluded that the emerging evidence linking coal dust with emphysema and chronic bronchitis was mostly unfounded—welcome news for coal companies, whose liability would balloon if the growing body of literature establishing the link took hold. The article could help companies continue to blame certain miners' ailments on cigarette smoking alone.

In March 1994, a month before this journal article hit the presses, Caldwell made the one-hour drive to Charleston and underwent an examination by Zaldivar. A few weeks later, Hobet's lawyer had sent the doctor's report to Caldwell and the Labor Department.

It was this report that Caldwell now showed John in the breathing center office. John flipped through it—cover letter, Zaldivar's write-up of Caldwell's medical history and the physical examination, the results from the day's lung-function testing, and that was it. Something was off. John had seen enough reports by Zaldivar to know that they usually included an X-ray reading and a narrative that summarized his conclusions. The report on Caldwell didn't contain either.

John figured Caldwell must have left part of the report at home. Was this everything the company's lawyers had sent? John asked him. Caldwell insisted it was.

His suspicions aroused, John agreed to act as Caldwell's lay representative. He decided to call Hobet's lawyer to figure out what had happened. He found the phone number below the letterhead of the firm he had begun seeing in a number of cases, a powerhouse that had stood by the coal industry's side for more than a hundred and fifty years and become the firm of choice for many of its biggest players in black lung–benefits cases: Jackson Kelly.

11

The Firm

The law firm that is known today as Jackson Kelly has been litigating, lobbying, and legislating in the coalfields for longer than West Virginia has existed as a state. Lawyers from the firm played key roles in drafting the state's constitution and shaping its contours in courtrooms, and they have been there, working largely behind the scenes, in the midst of many fights that defined the state's tumultuous and sometimes bloody history.

The firm traces its roots to a small practice formed in 1822 by Benjamin H. Smith in what was then Charleston, Virginia. After studying law under a man who became a prominent Whig politician, Smith married into a family of standing in the region's burgeoning salt industry. He soon focused on a subject that would be his life's work, one that has largely determined the social and economic landscape of West Virginia: land—more specifically, who had rightful ownership of it.

The largely unsettled region that was then western Virginia was a sort of Wild West of land speculation. Entrepreneurs, many of them located outside the state, coveted the area's coal, timber, and other natural resources, and they bought up parcels at fire-sale prices. Given the sloppy or nonexistent records and the difficulty of surveying the mountainous terrain, disputes were inevitable. The situation called for a good lawyer, and Smith made it his mission to establish the rules for determining who owned what. This commercial certainty, he believed, was the necessary foundation for any prosperous economy.

On top of his prodigious caseload, Smith, a member of the Virginia legislature, took the lead in writing land laws. When West Virginia seceded from Virginia and entered the Union as the thirty-fifth state in 1863, he was a delegate to the state's constitutional convention, playing a central role in drafting the provisions related to land titles. "The state's economic development, especially its natural resources, has been largely predicated on the basis of property law as defined, interpreted, litigated, and implemented by Smith, his contemporaries, and the major West Virginia law firms which followed," a businessman and state legislator later wrote.

By 1865, Smith had joined with another prominent attorney, Edward B. Knight, to form the firm Smith and Knight. When the state held a second constitutional convention in 1872 to make revisions, Knight served as a delegate and was, like Smith before him, a principal crafter of the provisions affecting land ownership. "The land-title provisions, written as usual by a select committee of land lawyers, escaped debate entirely," historian John Alexander Williams wrote.

The product of this gathering earned the nickname "the Lawyers' Constitution," and a critical assessment published in the *New York Times* called it "a contrivance gotten up to make litigation the principal business in West Virginia to the great impoverishment of suitors and the enrichment of the swarms of one-horse political lawyers that now feed upon the body politic."

Though the intent of the drafters was unclear, Williams observed, the effect of what they wrote was "the eventual transfer of titles from small owners to mining and lumber corporations in later years," which took place "under the supervision of 'distinguished land attorneys.'" In other words, these early land laws and the court decisions interpreting them—defined in large part by these lawyer-statesmen—are a good starting point for those seeking to understand how a state so rich in natural resources could have so many citizens trapped in poverty.

Knight's son continued the family tradition, and when his father retired in 1892, he and two attorneys of similar stock formed the firm Brown, Jackson, and Knight. James F. Brown had served in the state legislature, and he held leadership posts with several coal companies and avidly promoted oil and gas development. His father had served with Smith as a delegate to the first constitutional convention and won election as one of the first justices of the West Virginia Supreme Court of Appeals, the state's highest court. Malcolm Jackson, also a prominent Charleston attorney, served in the state legislature as well.

The rapid displacement of the mountaineer lifestyle by booming industrialism was a boon to the lawyers at Brown, Jackson, and Knight. The firm not only helped the coal industry establish control of large swaths of the coalfields but also defended the industry against those who challenged that control.

The formation in 1890 of the United Mine Workers in Columbus, Ohio, inaugurated an era of iconic labor battles. By the turn of the century, the union had won major concessions in coal states such as Ohio, Illinois, Indiana, and Pennsylvania, and it boasted about a quarter of a million members. But much of West Virginia, especially the isolated counties in the south, remained beyond the UMW's reach, and when the organizing efforts did arrive, bloodshed often wasn't far behind.

The Mine Wars that resulted have become West Virginia lore. Miners on Paint Creek in Kanawha County went on strike in 1912, demanding that coal companies enshrine basic rights in a union contract. Fellow miners in nearby Cabin Creek soon joined them. Coal bosses responded by evicting families from their company housing, and these outcasts retreated to disease-ridden tent villages. Mary Harris "Mother" Jones, the fiery labor activist whose image John Cline would later frame and place on his office wall, returned often to West Virginia to aid the union's organizing efforts, and she was by the miners' side as the bullets flew on Paint Creek.

On the other side of this fight, standing with the coal bosses, were the lawyers at Brown, Jackson, and Knight. They hired a stenographer to follow Mother Jones around the coalfields—an attempt to gather evidence that she was inciting violence—and recruited the first armed guards, men who essentially became a private police force around the mines and company towns. Before long, the skirmishes between miners and the guards became violent. Gunfire rang out in the mountains, and men on both sides fell. The governor declared martial law and sent in the National Guard. Eventually, a newly elected governor brought the fighting to an end by imposing a compromise settlement. Lawyers at the firm helped some of the companies navigate the multiple government investigations in the aftermath of the bloody strike.

A measure of the elite status the firm had attained came in 1929 when it moved into new offices. It was among the first tenants of the twenty-story tower that served as the headquarters for the city's oldest bank (of which one of the firm's partners had been vice president) and housed more than twenty coal companies, as well as the industry's trade association. Brown, Jackson, and Knight took over the entire sixteenth floor of this epicenter of power and wealth in the state capital.

In 1941, the firm added to its stable of lawyers former West Virginia governor Homer A. Holt, whose time in office had been marked by a combative relationship with labor interests. Two years earlier, he had deployed his clout in an attempt to stanch bad publicity over one of the least known but most devastating industrial disasters in U.S. history. In 1930, Union Carbide and Carbon Corporation hired workers to dig a tunnel, to be called the Hawks Nest Tunnel, through Gauley Mountain to divert water from the New River to a hydroelectric generating station that would supply electricity to nearby plants. These laborers drilled through rock that contained high levels of silica, filling the tunnel's air with the finely ground dust that has long been known to obliterate lung tissue when inhaled. Some workers developed silicosis so severe that they died in about a year; others lingered on in the nearby village of Gauley Bridge, which became known as "the town of the living dead." Though the true death toll is unknown, a congressional inquiry placed it at 476; a later study estimated it at 764.

Holt reviewed a draft of a guidebook to West Virginia's culture and history that had been assembled by writers participating in a federal program, and he took exception to a section on the Hawks Nest disaster and other labor issues. He contended the book "may be distinctly discreditable to the State and its people." The writers protested what they deemed a "white-wash," but by the time the manuscript reached publication, its criticisms had been toned down.

After his term as governor ended, Holt worked for six years at Brown, Jackson, and Knight, then moved to New York to become chief counsel for none other than Union Carbide. In 1953, he returned to Charleston and rejoined the firm, which soon changed its name to Jackson, Kelly, Holt, and Moxley.

In 1969, when John Cline marched with striking miners and the original Black Lung Association advocates to the state capitol building to demand that the legislature pass a bill compensating victims of black lung, lawyers from the firm had already been inside for weeks, lobbying on behalf of the coal industry to weaken or kill the bill. The "prestigious Charleston law firm"—by then named Jackson, Kelly, Holt, and O'Farrell—had numerous connections to people in key legislative posts and "was considered the most politically influential" in the state, journalist Brit Hume noted in his 1971 book about a revolt within the union. "Its members had been operating in the hallways and back rooms of the Capitol for years. Some had served in the legislature. They were well-known and capable. They knew all the pitfalls and shortcuts involved in getting things done."

In 1976, the United States celebrated its two hundredth birthday, and the firm celebrated its one hundred and fifty-fourth. The same year, historian John Alexander Williams, in the first edition of his definitive book on West Virginia, took stock of the firm's lofty status and the role it had played in shaping the course of history in the state: It was "West Virginia's oldest and most prestigious law firm," he wrote.

> A lawyer could be born into a firm like this; he could work his way into it, combining strength of intellect with the proper amounts of deference and tact toward his superiors; or he could be selected for preferment from among the scores of ambitious politicians who crowded into the capital year by year. However he made the grade,

the man who made it was not likely to quibble about the defects in West Virginia's economy, for that economy assuredly worked for him. West Virginia thus remained the way it was because the most powerful West Virginians liked it that way.

In the 1980s, the firm set its sights even higher, seeking to parlay its dominance in West Virginia into prominence on the national stage. It established an office across the border in Kentucky in 1985 and would eventually set up shop in Washington, DC; Denver; Pittsburgh; Akron, Ohio; and Evansville, Indiana. Its client list included major companies and trade associations in industries such as coal, natural gas, oil, chemicals, and more.

Jackson Kelly attorneys submitted comments to federal agencies on behalf of clients, trying to shape the drafting of new rules on everything from air and water pollution standards to mine safety and health. The firm guided clients through the complex application process to build new pipelines, and it participated in investigations when federal mine safety regulators showed up to look into a fatal accident. When clients didn't like new regulations, the firm sometimes sued the government in an attempt to invalidate them, and when the government or aggrieved workers or citizens accused clients of wrongdoing, the firm defended them. And those were just a few of its many services.

Still, the firm's bailiwick was representing the industry that dominated its home state. "We understand the coal industry from the inside," Jackson Kelly's website declares, noting the firm has served the industry since the mid-1800s. "From the Powder River Basin to the coalfields of Central Appalachia, our attorneys are not afraid to get their hands dirty and are frequently found at our clients' surface mines, prep plants, underground mines, and mine offices investigating accidents, providing training, or preparing for trial.... Not only is Jackson Kelly part of the history of the industry, it is part of coal's future."

At the same time as the firm began this expansion in the 1980s, changes in the federal black lung–benefits program created a growing demand for lawyers to represent coal companies. As more and more claims became the financial responsibility of companies rather than the government trust fund, these companies needed lawyers to help defeat miners' claims. It was in this seemingly obscure area of administrative law that the firm found a particular niche. Other firms had a few attorneys who handled federal black lung–benefits cases; Jackson Kelly built an entire team of lawyers dedicated primarily to defending these claims.

During a 1988 congressional committee hearing in Beckley, a United Mine Workers official complained about the plummeting rate of benefit approvals and the growing length and complexity of claims; he pointed to the vast disparity in resources and legal representation as a major reason. Whereas only a few lawyers were willing to handle a miner's claim, he said, "Jackson and Kelly, a large law firm in Charleston, which represents coal companies in black lung cases, has approximately eleven attorneys working the federal black lung circuit."

At the time John Cline started taking cases as a lay representative, one of the firm's leading black lung lawyers was William S. Mattingly, who would later be formally designated head of the black lung unit. After earning a law degree from the University of Cincinnati in 1985, Mattingly served as a clerk for one of the Labor Department's administrative law judges, then joined Jackson Kelly's Morgantown office. He quickly rose to prominence in the world of federal black lung benefits. He gave presentations and wrote updates on case law for legal journals, and in addition to his caseload, he weighed in on proposed changes to black lung regulations on behalf of coal companies. He eventually became an adjunct lecturer at West Virginia University's law school teaching appellate advocacy, and he was a perennial shoo-in for a spot on the Best Lawyers in America list in the category "Workers' Compensation Law—Employers."

Mattingly offered his take on the history and evolution of the federal black lung–benefits program in a 2003 article for the *West Virginia Law Review*. The 1969 law creating the program "was conceived in response to political pressures," he wrote. There had indeed been growing recognition of the dangers of breathing coal dust, he said, but the 1969 law was largely a result of the "legislative rush to respond" to the 1968 Farmington explosion that killed seventy-eight. The program Congress created was "flawed from its inception."

Echoing the coal industry's arguments, Mattingly wrote, "The black lung program, in the guise of a disability program, was more legitimately viewed as an entitlement or pension program for miners with more than ten years of coal mine work." The plummeting approval rate that had followed the enactment of tougher standards under the Reagan administration was, contrary to what miners' representatives believed, a natural and just development based on a few factors: The amended rules had corrected "a flawed award system"; the measures to control dust exposure had reduced the incidence of disease; employment in underground mines had declined because of mechanization and a shift to more surface mining; and an ever-growing number of cases were being fought not by government lawyers but by the coal companies' attorneys, who, for whatever reason, "have been much more successful in defending claims."

Mattingly acknowledged that miners pursuing claims "must confront the vastly superior economic resources of their adversaries" and said the increasingly complicated system had come to resemble "a three-ring circus." But he contended that this circus risked inflicting injustices not just on miners but also on his clients, the coal companies: "Trying to control all of the action under the three-ring big top is due process. At its core, due process guarantees a meaningful hearing after proper notice before personal liberties or property are taken, providing equity and fair play."

In other words, though the chaotic system might disadvantage miners in some respects, it also threatened to saddle coal companies with liabilities they shouldn't have to shoulder, such as paying people who weren't truly disabled or whose illnesses were caused by years of smoking. Each claim, Mattingly wrote, carried an average liability of about $175,000 for the company.

Miners and clinic workers came to know Mattingly and his colleagues at Jackson Kelly well. At the black lung clinic near Charleston where Debbie Wills worked, miners showed up with papers bearing the firm's letterhead and with the firm's name dripping disdainfully from their tongues. "Miners are very familiar with Jackson and Kelly—that they were company lawyers, that they would do anything to prevent them from getting their benefits," Wills said.

At the New River black lung clinic, Susie Criss had her share of run-ins with Jackson Kelly as well. She had grown up in nearby Fayetteville, the sister of two coal miners and a federal mine inspector. She'd started at the clinic as a receptionist, and John later trained her as a benefits counselor. She helped miners manage the steady stream of paper coming at them from Jackson Kelly and came to see how the firm operated. "They are very good at their job," she later said. "They cause a lot of grief for coal miners who feel like they're being singled out, and I have to explain to them constantly: 'This is not personal. They're representing the company.' They're a fierce adversary."*

When Charles Caldwell showed John his case file in the summer of 1994, John was just beginning to learn Jackson Kelly's modus operandi, but other miners' representatives had already seen enough to convince them to steer clear of any showdown with the firm. Earlier in the year, a U.S. Senate committee had held a hearing on proposed reforms to the benefits program, and afterward, a lay representative from Tennessee had submitted a statement for the record. In it, the miners' advocate described the consequences of the increasingly unbalanced legal matchup in federal

---

* I reached out to the Jackson Kelly attorneys mentioned throughout this book by e-mail, certified letter, and, when possible, phone. I provided a list of specific points and general themes that I hoped to discuss and requested an interview. With one exception—Mary Rich Maloy, who agreed to discuss only her work outside of Jackson Kelly and whose comments are reflected in a footnote in chapter 12—the attorneys either refused to talk with me or did not respond to my requests.

The firm's general counsel, Michael Victorson, sent me this e-mail:

"The firm received your recent inquiries for interviews on topics covering many years of the firm's history for your soon-to-be-released book. Given the unambiguous slant in your previously published articles, the summary of your book as already posted on Amazon, and the nature of the questions posed in your emails, we do not believe there is an opportunity for an objective, fair discussion. Jackson Kelly has proudly represented thousands of clients and been engaged in a broad spectrum of important legal issues. Our clients know us for advocacy, excellence, and value."

In reply, I wrote: "I understand your position on my previous work, but I do think it is possible for us to have a fair, objective discussion. I remain very interested in doing so." I did not receive a response.

black lung–benefits claims. He helped miners at the initial stage before a Labor Department official, he said, but if the miners won, the company almost always appealed. He tried to find lawyers to represent these miners, but "this has become impossible."

He explained why:

> In every case that I have been associated with where the coal company is represented by the Law firm of Jackson and Kelly, the coal miner looses [*sic*] any and all attempts of every avenue they might pursue in getting their claims through. It has even come to the point in this area to where you can call an Attorney and ask their assistance for a miner and the first question out of their mouth is who is representing the coal company, and when you tell them that Jackson and Kelly Law firm is on the opposing team then the Attorney will not take the case because they already know they will loose [*sic*] the matter before the administrative Law Judges there in Washington.

Whether because of courage or ignorance or both, John Cline didn't have such qualms.

12

Discovery

John Cline was frustrated. He'd been calling the Jackson Kelly office in Charleston and leaving messages with the assistant for lawyer Mary Rich Maloy for weeks but gotten no response.

He was trying to find out why there seemed to be parts missing from the report Maloy had sent to Charles Caldwell containing the results of his medical examination by Dr. George Zaldivar. John figured he and Maloy could just talk and figure out what had gone wrong. But it was now September 1994—more than two months after he'd agreed to represent Caldwell—and his attempts at outreach seemed to be getting him nowhere.

John thought someone within the small group of lawyers who represented miners might be able to help him get answers, and one in particular came to mind: Bob Cohen, who had a small practice in the northern West Virginia town of Fairmont. This was not the same Bob Cohen who had become an essential medical expert in Chicago. For clarity's sake within the small community of black lung experts and advocates, the two were known as "Lawyer Bob" and "Dr. Bob."

John called up Lawyer Bob and described Caldwell's case. What, he asked, should he try next?

"My God," Cohen replied, "you've got to put it in writing."

It was the kind of advice that seemed obvious to a seasoned lawyer:

Apply pressure by creating a paper trail. But John was of the mind-set that people ought to sit down and talk through their differences together. When he and his father had run their building company, they hadn't even had formal contracts with most home buyers; people just trusted one another and generally did the right thing.

John sent a letter to Maloy outlining his attempts to get information about Zaldivar's report, and he copied the administrative law judge hearing the case, Frederick Neusner. One week later, Neusner issued an order giving Maloy a choice: Either turn over whatever missing pieces of the report there might be or risk having any opinions that referenced Zaldivar's exam excluded from evidence.

Shortly before the deadline Neusner set, Maloy sent John a three-page narrative report and two X-ray readings, all on the letterhead of Dr. George L. Zaldivar. This was exactly what John had suspected was missing. Zaldivar had graded the X-ray from his exam and the one from Dr. Don Rasmussen's earlier exam as "complicated, category B"—the same severe stage of black lung that the radiologist who reviewed the film for the Labor Department had seen. In the narrative report, Zaldivar concluded that Caldwell's pulmonary impairment was caused by the disease.

In other words, Jackson Kelly had sent Caldwell to a doctor of the firm's choosing—one whom they used frequently and who had a reputation of rarely finding black lung—and even he had determined that Caldwell had the advanced form of the disease that automatically qualified him for benefits. The incomplete report that Maloy had submitted to Caldwell and for the official Labor Department record made Zaldivar's findings seem ambiguous, and Maloy had used the disclosed portions to help build a defense. But in reality, Zaldivar had told her in no uncertain terms—in a report addressed directly to her—that Caldwell had a meritorious claim. Nonetheless, rather than advise Hobet Mining to pay the claim, Maloy had submitted the partial report and continued to contest Caldwell's case.

John wondered what other undisclosed documents the firm might have. Figuring that out, however, would be more challenging. Jackson Kelly had refined a legal argument that it believed justified the withholding of much of the evidence the firm collected. Zaldivar's report was one of the rare exceptions; the document fell into a category of evidence that the firm felt it did have to disclose (though only if the other side asked, which many lawyers and virtually all unrepresented miners wouldn't think to do). If asked for pretty much any other evidence it was holding back, the firm would dig in and fight.

John knew he needed help, so he set out to learn from Lawyer Bob, who knew Jackson Kelly's defense strategy well. Tangling with the firm was perhaps the clearest expression of the mind-set animating Cohen's career, an approach he characterized as "fighting one bastard or another." Like John, he was a New York native, and he'd been surprised at being welcomed in West Virginia in 1974, given his self-described biographical details: "I'd been a conscientious objector to the war. I had a beard, and I was Jewish. And I was from the North and had left-wing politics."

Working as a lawyer for mostly poor, rural clients in West Virginia hardly seemed a likely career for him based on his childhood circumstances. Born in 1946, he grew up as an only child in Scarsdale, a wealthy New York City suburb where his parents had moved so he could attend the area's highly rated schools. Cohen thought the town "reeked of money," as he later put it. "I felt alienated from the materialistic culture, and I sensed that there was some other world out there."

As a teenager, he was captivated by the unfolding civil rights movement. While enrolled at Brown University, he participated in a program that allowed him to spend a semester at Tougaloo College, a historically black institution near Jackson, Mississippi. He became involved with the Mississippi Freedom Democratic Party as it challenged the state's traditional Democratic Party, which excluded African-Americans. The movement's leaders became his lifelong heroes.

After graduating with a degree in political science in 1968, Cohen went to work for a welfare rights organization in Rhode Island. In 1970, he decided to try law school, enrolling at Boston University.

He hated it almost from the start. He came to believe the dominant model for legal education was based on a pernicious lie: that zealous advocacy for clients, no matter who they were or what their agenda, was a lawyer's only duty; that, in pursuit of this zealous advocacy, a lawyer ought to do whatever the law allowed; and that whatever outcome followed from this was just and justifiable. As Cohen saw it, the purpose of law school seemed to be to take smart young people, break them down, and build them back up as amoral apostles of the law.

He would come to regard Jackson Kelly's handling of black lung–benefits cases as perhaps the ultimate embodiment of how the doctrine of zealous advocacy could be used to justify actions that had devastating consequences for average people.

Cohen spent one summer during law school working for a legal services organization in Detroit and fell in love with a coworker named Kathy, whom he later married. That fall, an acquaintance connected him with rank-and-file miners who were rebelling against what they saw as the corrupt United Mine Workers hierarchy. Working as a poll watcher during a contentious union election, he quickly took a liking to coal miners.

After Cohen finished law school, he and Kathy moved to West Virginia, settling into a small house in a rural stretch of Marion County in the northern part of the state. Kathy worked as a kindergarten teacher, and Bob became a staff lawyer for an organization that provided the area's low-income residents with an array of legal services—workers' compensation, Social Security disability, accidents, wills, divorces.

One day, a handful of men came up to him after a meeting and asked him to help with their black lung–benefits claims. He had no experience with the benefits system, but he took their cases and won. He started to take on more miners' claims.

Meanwhile, Kathy earned a law degree herself, and the couple joined with another attorney and opened a small practice in Fairmont. Bob spent a good portion of his time on black lung–benefits cases, and that's when his years of scrapping with Jackson Kelly began.

The idea of a small-town legal practitioner taking on the state's largest and most powerful firm appealed to his underdog mentality and presented an ideal situation for the deployment of what he later called the "Ewok approach"—a reference to the small, furry creatures who helped the good guys take down the evil empire in the third film of the original

Star Wars trilogy. He explained: "They would go scurrying around and pull these monstrous, evil machines down by tying ropes around their legs and pulling on them. That's what I did. I was an Ewok. I was fast. I was mobile. I was smart. I would take advantage of things. The people that were up against me were big and clumsy. I had to figure out how to bring them down."

Part of this approach was doing something that virtually no other lawyers did in black lung–benefits cases: pursuing discovery. This process—essentially, each side requesting documents and information from the other side—was a staple of most civil litigation, but it was virtually nonexistent in black lung–benefits cases. Lawyers for coal companies often sent formal written questions, including document requests, to miners, but most of the time, there wasn't much for them to disclose. Unlike the defense lawyers, miners usually didn't have the money to pay for additional expert reports and specialized testing.

Miners and their representatives, however, rarely filed requests for discovery. There were a few reasons for this. The first and most obvious was that many miners didn't have lawyers and didn't know they could ask for documents, let alone how they might file such a request. Even when miners did find lawyers, most of them either didn't think to file for discovery or didn't want to bother with the process. Attorneys tended to approach black lung–benefits cases much the same way as they did traditional workers' compensation and Social Security disability claims—areas of law where discovery usually played a relatively minimal role. The black lung–benefits system also created financial disincentives for lawyers to press for discovery. The claims were already more drawn out and less remunerative than other workers' compensation or disability claims, and fighting over production of documents only exacerbated that problem.

Cohen understood why other lawyers didn't pursue discovery as he did. "It was exotic; it was difficult," he said. "You do it, and you're getting into a nasty fight, no question about it."

But Cohen relished the fight, even if it did mean his cases became especially complicated and time-consuming. Jackson Kelly was the tall, lumbering machine, and he was scampering about below, tying ropes around its legs.

He didn't personally dislike the lawyers who made up the black lung unit. Bill Mattingly, the group's leader, seemed like a decent guy—he was friendly; he attended the local Methodist church; he donated time and money to local service organizations.

The unit's other members evinced similar bona fides as pillars of their local communities. They were parents who volunteered at their children's schools and coached their sports teams, loyal alumni who guest-lectured at their alma maters and mentored young lawyers, neighbors who enjoyed gardening and served on homeowners' associations, nationally recognized attorneys who donated their expertise to worthy causes, such as representing victims of domestic violence. Years later, after retiring, Mary Rich Maloy volunteered with a local legal-aid society and the American Civil Liberties Union and won an award for her many hours of pro bono work.*

Nonetheless, Cohen thought the unit's lawyers—or at least some of them—had adopted the law-school line and drunk the zealous-advocacy Kool-Aid. The black lung–benefits system magnified the injustices that could result. The underlying premise of the adversarial legal system was that each side had a zealous advocate and that, from their combat in the legal arena, a just outcome would emerge. In the black lung–benefits system, however, the coal company's zealous advocate often found little opposition in the arena, just a retired miner, his widow, or a lay representative. And few people were watching the spectacle. These cases were

* Maloy told me by phone that she gave her time to these organizations because she believed in their mission, which included representing mainly poor and vulnerable clients, such as women seeking protective orders against abusive partners and prisoners challenging living conditions. I suggested that this type of work seemed quite different from her years representing coal companies in black lung–benefits cases and asked how the two fit together. "I've done a lot of different things," she said, "and I think in all situations a lawyer represents their clients the best they can. I think that's what I've done in all cases." She declined to discuss her work at Jackson Kelly. "I know you're bent on this issue, and I disagree," she said, referring to articles I'd written about the firm's conduct. "I don't think I can change your mind, and I don't think you can change mine."

not dramatic courtroom showdowns that captured public attention; they were technical administrative proceedings that most people didn't even know existed. There was little outside scrutiny that might cause a lawyer to rethink controversial tactics.

In time, John would arrive at a similar evaluation of Jackson Kelly's black lung unit, regarding it as a sort of machine. Often, multiple lawyers from the firm teamed up on a claim. John came to believe the firm had an overarching strategy for fighting claims that largely transcended individual attorneys. In the notes John would stuff in case files over the coming years, Jackson Kelly seemed to assume the status of an autonomous entity, JK: JK filed a motion, JK refused to disclose a report, and so on.

As both John and the firm prepared for the next round in Charles Caldwell's claim, John was becoming familiar with the JK strategy, which Cohen later would dub "pyramiding": the firm's attorneys solicited reports from various doctors but disclosed only the favorable ones to their consulting pulmonologists, the opposing side, and the judge.

Lawyers are allowed to consult various experts, and if they arrive at different conclusions, lawyers can choose to submit the reports that support their case. But John and Bob came to believe that Jackson Kelly's strategy amounted to much more than that. By withholding valuable pieces of information and instead submitting pieces that were inaccurate or misleading, the firm was manipulating the evidence to achieve the desired outcome. The top levels of the pyramid—the consulting pulmonologists' reports and testimony that incorporated the various types of evidence and were critical to the case—were not the result of an assessment of all the evidence; they were merely a product of what the lawyers had decided to disclose.

In Caldwell's case, and in many others, some of the key pieces at the foundation of the pyramid were the readings of X-rays and CT scans. This was a common battleground in benefits cases. Miners sometimes submitted radiology reports indicating they had black lung, but for each of these positive readings, there were often multiple negative readings submitted by the company lawyers, enabling the defense to argue that miners did not have the disease. When deciding which reports to credit, judges frequently compared the radiologists' résumés.

The doctors who had treated Caldwell or evaluated him for the Labor Department as part of his claim had interpreted his films as showing complicated black lung. But they were coalfield physicians; though they had extensive experience with black lung, they didn't possess elite institutional affiliations or lengthy curricula vitae studded with prestigious awards and lists of published articles and formal presentations. The doctors who had interpreted Caldwell's X-rays as negative for black lung did possess those trophies.

To obtain the negative readings on which to build its case against Caldwell, Jackson Kelly had gone to the same place it had so many times before: a small unit of impeccably credentialed radiologists at one of the world's most prestigious medical institutions.

13

"Superior Credentials"

John Cline had begun to notice a trend in the X-ray readings that coal company lawyers obtained, and the ones Jackson Kelly had submitted in Charles Caldwell's case were prime examples. His shorthand for the recurrent issue appeared in letters to colleagues and handwritten notes to himself: *TB vs. CWP*. The lesions on the X-ray weren't black lung (coal workers' pneumoconiosis, or CWP), the defense went; they were scars from healed tuberculosis. Or maybe a similar bacterial or fungal infection. Anything but black lung.

It was hardly surprising that lawyers had found doctors whose reports reliably supported their cases. The jarring part was where they had found those doctors: the Johns Hopkins Medical Institutions. They were not hacks for hire; they were longtime radiologists and professors at the world-renowned medical center. They read X-rays and CT scans at a hospital perennially ranked as one of America's best, and their extensive curricula vitae were filled with publications, presentations, and honorary posts.

The reports in Caldwell's case, like so many others John had seen, bore the hallowed institution's letterhead. Coal companies paid a premium for this prestigious imprimatur, and it often proved to be worth the added cost. In the black lung–benefits system, judges usually relied heavily on physicians' credentials to resolve disputes, and virtually none could match the Hopkins radiologists'.

They could evaluate films relatively quickly and churn out multiple reports each hour. Their reports formed much of the foundation for countless Jackson Kelly evidentiary pyramids. These doctors' clout combined with their prodigious output meant this small unit had a far-reaching impact; over the years, it touched the lives of thousands of miners seeking benefits.

Some doctors at university-affiliated medical centers performed this sort of litigation consulting on their own time, but Hopkins had decided years earlier to make it an official part of its work as an institution, though this remained little known outside of black lung circles. It had established within the radiology department the Pneumoconiosis Section, the members of which spent roughly half of their time reading X-rays and CT scans for outside clients. Some of the films they read were from patients at a local TB clinic or workers at large steel, oil, and rock-products companies that had contracted with the hospital to monitor the health of their employees. But the doctors also spent much of this time reading films that came from law firms defending against black lung–benefits claims.

The leader of the Pneumoconiosis Section, Dr. Paul Wheeler, had been part of the original team when it was formed in the 1970s. He had undergraduate and medical degrees from Harvard University and a long list of publications, presentations, and plaudits to his name. He had served as an army medical officer in Vietnam, and at Hopkins he had helped develop a computer system that enabled radiologists to convey their interpretations to treating physicians more quickly. Slightly built, he kept his hair close-cropped, and his speech was sharp.

Radiologists had come and gone over the years, but Wheeler was the stalwart. His productivity was prolific. Companies sometimes paid for additional readings by one or two of his colleagues in the unit, and these reports tended to echo Wheeler's very particular opinions. "The newer ones were trained by me, so they have my philosophy," he later said.

Wheeler's reports had an almost robotic consistency, and his depositions sounded like the same tape-recorded conversation played again and again. He had a very specific and narrow view of what black lung looked

Gary Fox in the underground mine where he worked in 1994. *(Courtesy of the Fox family and Shifoto Studios, Coal City, WV)*

Gary and Mary Fox. *(Courtesy of the Fox family)*

Gary Fox and his daughter, Terri, both of them avid West Virginia University Mountaineer sports fans. *(Courtesy of the Fox family)*

Gary and Terri, taken around the time Gary retired from the mines. *(Courtesy of the Fox family)*

Miners driving metal bolts into the mine roof to pin unstable layers of rock together and render a cave-in less likely, Clearfield County, PA, 1976. Roof bolter (Gary Fox's job for most of his career) is one of the most dangerous jobs in an underground coal mine. *(earldotter.com)*

Shuttle car operator in a low coal seam, Logan County, WV, 1976. To save time, money, and wear on equipment when extracting a thin seam of coal, companies sometimes decide not to cut away the surrounding rock. This can leave miners working long shifts in tunnels as low as three feet high. *(earldotter.com)*

The New River Company's tipple in the coal camp of Skelton in 1948. Coal dust often blew over from this facility, which was used to load the fuel onto railcars, and coated the nearby home where Gary Fox grew up. *(New River Company, 1976)*

No. 1—Mt. Hope (1937-70)	No. 2—Scarbro (1937 To Date)	No. 3—Carlisle (1937-61)	No. 4—Whipple (1937-54)
No. 5—Cranberry (1937-58)	No. 6—Summerlee (1937-58)	No. 7—Lochgelly (1937-64)	No. 8—Sprague (1937-54)
No. 9—Skelton (1937-75)	No. 10—Wickham (1937-49)	No. 11—Prudence (1937-43)	No. 11—Stanaford (1944-61)
No. 12—Harvey (1937-54)	No. 13—Glen Jean (1937-58)	No. 14—Price Hill (1940-54)	No. 15—Kilsyth (1940-49)
No. 16—Oswald (1940-53)	K—Kilsyth (1949-58)	M—Mabscott (1949-54)	ORCO Reverse

Miners were paid in these metal coins, known as scrip, that were redeemable only at the company store. Inflated prices kept coal camp residents in perpetual poverty. *(New River Company, 1976)*

Samuel Dixon left his native Yorkshire, England, in 1877 and founded the New River Company in southern West Virginia. His domain came to include the coal camps of Skelton, where Gary Fox grew up, and Scarbro, the town that would become home to the medical clinic where John Cline worked. *(New River Company, 1976)*

Miners demanding a state law to compensate black lung victims march down a main thoroughfare in Charleston, WV, on their way to the capitol in February 1969. More than forty thousand miners staged a wildcat strike that crippled the state's largest industry. The rank-and-file uprising spurred passage of not only state legislation but also the landmark Federal Coal Mine Health and Safety Act of 1969. *(Douglas Yarrow)*

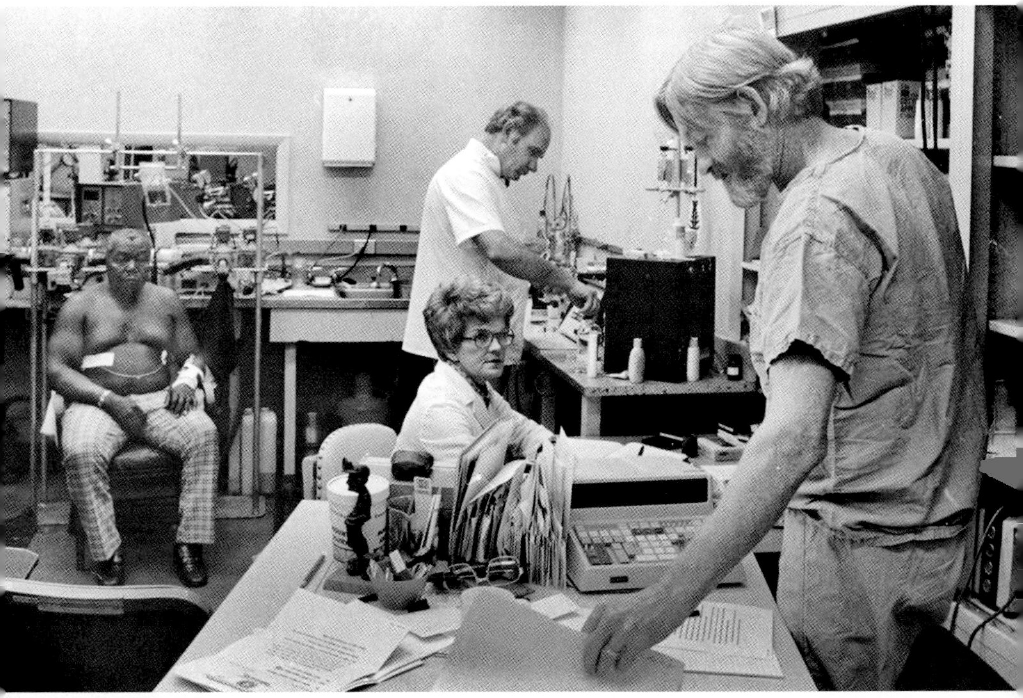

Dr. Donald Rasmussen examining a miner at a clinic in Beckley, WV, 1976. He became a hero to miners for his pioneering research documenting the effects of black lung in the 1960s, and he continued to examine miners for decades. *(earldotter.com)*

Dr. Rasmussen conferring with a miner before lung function tests, Beckley, WV, 2007. Roughly forty years after he provided the scientific backbone for the grassroots movement that led to the Federal Coal Mine Health and Safety Act of 1969, Rasmussen was still treating miners. *(earldotter.com)*

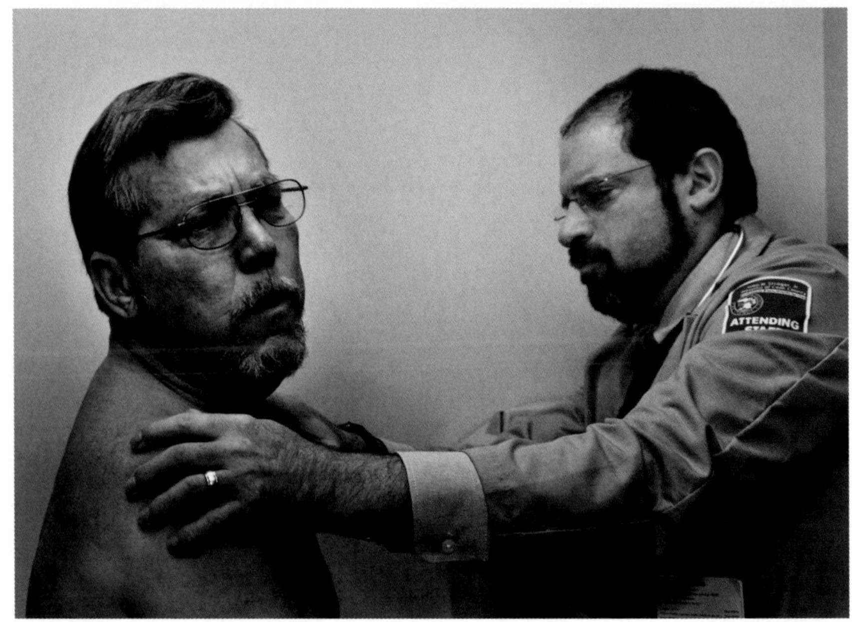

Dr. Robert Cohen examining a miner before a lung function test, Stroger Hospital Black Lung Clinic, Chicago, 2011. Chicago has a relatively large community of former residents of Appalachia who migrated there in search of work amid a coal industry downturn. Beginning in the 1990s, Cohen helped revive organizations that advocated for miners. *(earldotter.com)*

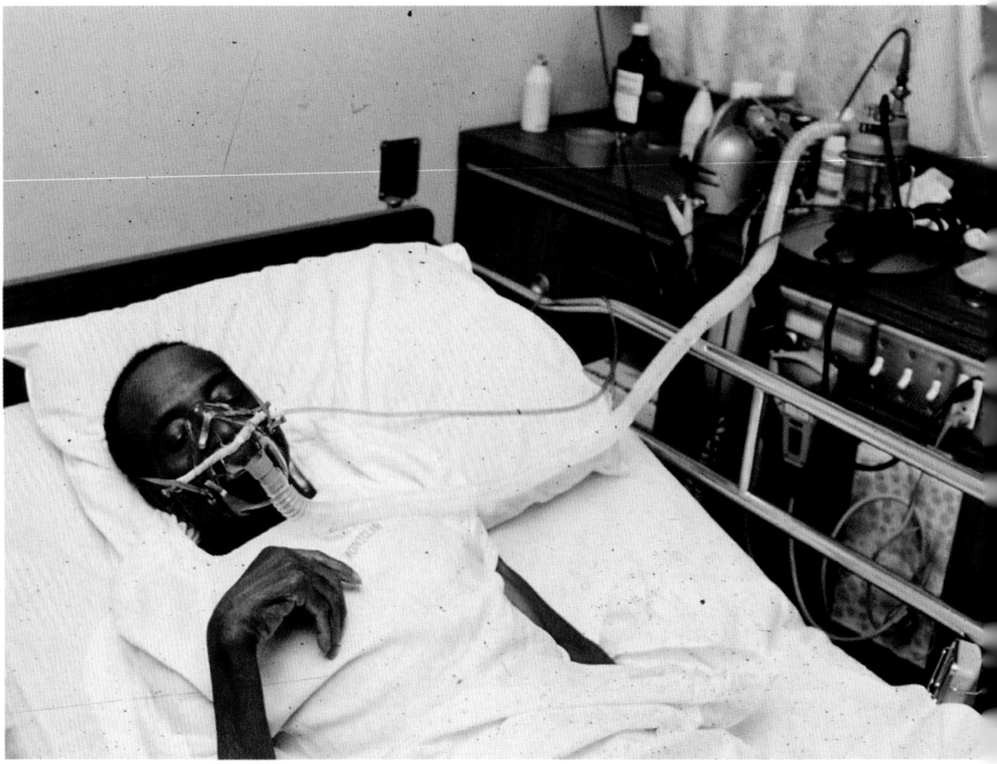

A former coal miner stricken with black lung in 1972. *(Courtesy National Archives, photo no. 26-G-3422)*

like on X-rays and CT scans, and he voiced it over and over, supporting it with the same similes, metaphors, and anecdotes. He wanted to see a "shotgun pattern" of rounded lesions in the central part of the mid- and upper lung regions. Their distribution should be "quite symmetrical"; that was because the airways deposited dust on the lungs "like a drop of water falling on the center of a sponge." Large spots on the outer areas or the highest reaches of the lungs were "out of the strike zone." So were any markings on the pleura, which was like a "liner" or "Saran Wrap surrounding the lung." The presence of calcified granulomas—likely the hardened remnants of a past inflammation—indicated that another disease was to blame. For a large mass on the film to represent the advanced stage of black lung, it should be accompanied by a background of smaller nodules of sufficient size—"It's easier to build a house out of cinder blocks than it is out of pebbles."

It all sounded logical, and Wheeler stated it authoritatively. But it didn't jibe with much of the medical literature or the views of many leading physicians and medical organizations.

For example, numerous published studies dating as far back as 1974, some of them led by researchers at the National Institute for Occupational Safety and Health (NIOSH), found that lesions frequently did occur in portions of the lung that Wheeler considered "out of the strike zone" and that these markings didn't have to be rounded, as Wheeler demanded, but could vary in shape.

A study published in a prominent radiology journal—of which one of Wheeler's colleagues in the Johns Hopkins radiology department had served as editor for twelve years—compared X-rays and CT scans with later autopsy findings for the same patients and found that advanced disease sometimes did extend to the "Saran Wrap." (Years later, John brought up this study, which had been conducted in Japan, during a deposition, and Wheeler replied: "I am an American. I'm over here. I understand my science over here.")

The directors of both the NIOSH Respiratory Health Division and the American College of Radiology's task force on pneumoconiosis—the organizations that developed the standards for detecting pneumoconiosis on X-rays and have taught them for decades—disputed Wheeler's views. The appearance of the disease can vary, both said when I spoke to them years later, and some of Wheeler's criteria were incorrect or overly restrictive. "When you take this very strict view where you put in all these rules, none of which are a hundred percent, what will happen is you'll wind up excluding people that have the disease," said the NIOSH official, Dr. David Weissman.

Wheeler acknowledged in depositions that he didn't attend seminars or conferences or keep up with the medical literature or the latest research and consensus opinions regarding black lung. "I don't think I need medical literature," he said in one deposition.

The strongly held views Wheeler repeated again and again seemed to have crystallized early in his life, largely as a result of the strong influence of two men: his father and his mentor. When challenged during depositions, Wheeler frequently invoked what he seemed to regard as the ultimate authority: his mentor, Dr. Russell Morgan, who had run the Hopkins radiology department for years and served a brief stint as dean of the medical school. Wheeler came to Hopkins as a resident in radiology in 1963, and he developed a deep admiration for Morgan, whom he considered a "certified genius."

In response to the 1969 law that created the black lung–benefits system and the government-run X-ray surveillance program, NIOSH began putting together a program to certify doctors as experts in detecting pneumoconiosis; they would be known as "B readers." Because of Morgan's expertise in imaging and statistics, NIOSH asked him to devise an exam that physicians would have to pass to earn this distinction. Morgan selected a handful of doctors to help him develop the B-reader exam, and Wheeler was one of them. Beginning in 1974, Morgan trained him to read X-rays for pneumoconiosis, and the standards that he learned then were the ones he would routinely recite for decades and that he would impart to the other members of the Pneumoconiosis Section.

Ironically, the method that Wheeler used for evaluating X-rays violated the rules of the very system underpinning the exam he'd helped

Morgan develop, though he didn't see it that way. The regulations governing the benefits system specify that X-ray readers should use a system established in 1950 by the International Labour Organization and widely used throughout the world. This system was what NIOSH and the American College of Radiology taught in seminars, and the B-reader exam tested proficiency in applying this system. When grading X-rays, physicians are supposed to compare them to a set of standard films depicting various abnormalities and mark the appropriate boxes on the form corresponding to any spots (or opacities) they see. The system uses a twelve-point scale to grade the concentration of small opacities. Large opacities are classified as category A, B, or, the most advanced, C, which describes a lesion stretching across an area equivalent to the entire right upper lung zone. If a miner proved that a large opacity of any category had been caused by coal-dust exposure, he was considered to have complicated pneumoconiosis, which automatically resulted in an award of benefits.

X-rays and CT scans are particularly important in the cases of miners who have evidence of complicated pneumoconiosis. These miners might not meet the threshold for total disability based solely on breathing tests, but their impairment may worsen even if they never breathe another speck of coal dust. Their cases, then, often hinge on what radiologists mark on this form, meaning that Wheeler's readings mattered most in the cases of miners with the worst stage of the disease.

There is an inherent subjectivity to grading chest films, and some variation among radiologists is normal, even expected. But Wheeler's readings were often wildly different from those of the other doctors who had weighed in on a particular case. When Wheeler evaluated X-rays, he often saw small and large opacities—representing nodules and masses on the lungs—but did not mark the corresponding boxes on the form because he believed the lesions were more likely the result of a disease other than pneumoconiosis. He explained why he thought this in the section of the form reserved for comments. In his telling, this was the medically responsible thing to do. He sometimes quoted a section from the International Labour Organization guidelines that stated that there was no particular pattern on film unique to pneumoconiosis and that other ailments could produce scarring that looked similar.

He neglected to quote the next paragraph of the guidelines, which stated that X-ray readers should nonetheless mark the appropriate boxes for whatever lesions they saw, then use the comments box to explain why they felt one disease or another was more likely.

This was a critical point within the black lung–benefits system. Grading the nodules and masses did not necessarily equate to a finding of pneumoconiosis, but failing to grade them was a definitively negative reading. The first approach was the beginning of a discussion; the second precluded discussion entirely.

Wheeler occasionally derided the system he was supposed to be using; it had flaws, he said. He wasn't convinced that all of the standard films—the exemplars of various appearances of pneumoconiosis—actually represented the disease. Some didn't fit within the standards he'd learned from Morgan, and they hadn't been proven by pathology, he claimed.

Yet somehow he maintained his certification as a B reader. Every four years, a physician has to take an exam to demonstrate continued proficiency in applying the system, and Wheeler always passed. If he had filled out the forms during these exams the way he filled them out when evaluating films for coal companies' lawyers, he almost certainly would have failed. In other words, he seemed to read X-rays one way during the test and another way in benefits cases.

NIOSH's Weissman later made clear that Wheeler's approach was simply wrong. "You're supposed to grade what's there," he said. "You're not supposed to alter what the grade is based on what you think the underlying cause is. That's not using the system properly."

NIOSH, however, had limited authority. Its police power extended only to the B-reader exam and the X-ray surveillance program that tracked prevalence of the disease, not to the benefits system.

One of Wheeler's longtime colleagues in the Pneumoconiosis Section, Dr. William Scott, later reiterated Weissman's point when I reached him by phone. Of Wheeler and his approach—not grading lesions he suspected did not represent pneumoconiosis—Scott said: "He was a good friend of mine. He taught me a lot about radiology. He was a good teacher. But as far as my readings—my readings are mine, and his are his."

When I told Scott that I had seen many cases in which he had followed Wheeler's approach, he responded incredulously: "You would always grade the opacities," he said. As I began to describe specific examples, he hung up the phone.*

Wheeler did have an opinion as to what was really causing the nodules and masses he saw in miners' lungs. Reading the comments he wrote in case after case, one might conclude that there was an epidemic among coal miners of undiagnosed TB or histoplasmosis.

The latter is an infection usually caused by breathing fungal spores contained in bird and bat droppings. The most common way to be exposed to these spores is by rooting around in contaminated soil. By Centers for Disease Control and Prevention estimates, between 60 and 90 percent of people living near the Ohio and Mississippi River Valleys have been exposed, but most don't become ill from it. Those who do usually have flu-like symptoms for a few weeks, sometimes longer. Histoplasmosis can be life-threatening, but mostly for people with weakened immune systems.

When John Cline started handling cases in the early 1990s, TB was Wheeler's bogeyman of choice, but over time, histoplasmosis took on that role. That was because, Wheeler later said, he kept encountering cases in which he suspected TB based on X-rays but skin tests for the disease came back negative. From this he concluded that another disease that could cause a similar pattern on film must be responsible, he said, "and histoplasmosis is far more common than TB in this country." (Later, he increasingly added to the mix mycobacterium avium complex, or, as he described it, "bird TB.") He often referred to the likelihood that a

* In the case of Charles Caldwell, for example, Scott read three X-rays. In the comments section on each form, he wrote that he saw multiple masses but thought they likely represented a granulomatous infection such as TB. He did not grade any of the lesions he saw. Other physicians reviewing the same films graded them as consistent with complicated pneumoconiosis.

miner had "granulomatous disease," a category including these alternative suggestions.

Wheeler's preoccupation with TB and histoplasmosis seemed to derive, at least in part, from personal experiences, which he sometimes related in depositions. During his one-year residency in pathology, he had performed an autopsy on a woman who had been treated for breast cancer. Masses had appeared on a follow-up X-ray of her lungs, and a radiologist logically had assumed her cancer had metastasized. But when Wheeler examined her lungs, he found not cancer but undiagnosed TB. He occasionally told the story of a colleague who had contracted histoplasmosis after spending "one night in an abandoned chicken coop." His own father, he said, had suffered from both ailments. His TB had "self-cured," Wheeler said, but his histoplasmosis had become so severe that Wheeler's mother was "scared stiff" that he might die. He eventually recovered.

It might seem that it shouldn't be that difficult to determine if a miner has TB or histoplasmosis. Skin or blood tests can detect exposure to TB bacteria, and blood or urine tests can detect the fungus that causes histoplasmosis. But Wheeler had counterarguments: The tests were imperfect and could produce false negatives, or the miner might have had the disease years earlier without realizing it and recovered, leaving behind the nodules and masses but not active infection that the tests could pick up.

When I later ran this scenario by doctors specializing in pulmonology, they said that, although undetected disease could account for a few abnormalities on chest films, these lesions should be easily distinguishable from pneumoconiosis.

Weissman said Wheeler's contentions about histoplasmosis were incorrect. It's true that, in areas where the infection is endemic, it wouldn't be uncommon for a person's chest X-ray to bear markings of the lungs' reaction to the fungus. But these would appear as a few nodules, not the extensive patterns Wheeler was evaluating. Rarely would it closely mimic pneumoconiosis. "That shouldn't be hard, in general, to make that differentiation," Weissman said.

Another reason for Wheeler's belief that the patterns were not black lung appeared to be his underlying assumption that the disease had become rare. He seemed to think that after the enactment of the 1969 law, regulators had waved a magic wand and eradicated the disease. "High dust exposures that would be capable of causing coal workers' pneumoconiosis has [*sic*] been illegal for decades," he said in one case. Two more Wheelerisms appeared frequently in the comments section of his reports: the miner was "unlikely" to have advanced black lung because he was "quite young," and complicated black lung was found mostly in "drillers working unprotected during and before World War II." During one deposition, he acknowledged he hadn't talked with miners about their working conditions in many years, and when asked if he'd ever been in a mine, he replied: "Aside from the exhibit in the Chicago Museum of Science and Industry, no. But I understand that much of it's now mechanized with cabins that are air conditioned and a lot of water that's sprayed on the faceplate, unlike the conditions in the past."

Any coal miner would have had a good laugh at that, but Wheeler had a less amusing corollary: If miners working after the 1969 law took effect contracted complicated black lung, it might well be because of their own reckless behavior; in essence, getting sick was their own fault. "Maybe he doesn't follow the rules," Wheeler said during one deposition. "Maybe he drills or something like that specifically without a face mask."

In another deposition, when John said that many of his clients were continuing to describe massive dust exposure, Wheeler responded: "I would think that the workers themselves would be blowing the whistle on their companies. Maybe not. Maybe they are just perfectly content to undergo lethal exposures, but I sort of doubt that's in their DNA."

On a few occasions, miners' lawyers confronted Wheeler during depositions with the evidence NIOSH researchers had been gathering about a resurgence of complicated pneumoconiosis among miners who had worked their entire careers after dust limits had taken effect and who were, in Wheeler's view, "quite young." Wheeler dismissed the NIOSH data, saying that the cases the agency was documenting were "not proven." This was perhaps the ultimate Wheelerism, an expression of a core belief that had formed early in his life.

At the suggestion of his father, a bacteriologist, he had done a one-year residency in pathology at Western Reserve in Cleveland. "My father felt that if I was to be a radiologist I should have my feet firmly planted in reality, which is pathology," Wheeler said. This hardened into a belief that Wheeler included in virtually every report and wielded as a talisman when challenged in depositions: Pathology was reality; radiology was just a shadowy approximation. He wasn't saying radiology was worthless; rather, that it was only a starting point. When it came to many types of lung disease, including black lung, the lesions seen on film weren't proof; they just indicated that certain illnesses were more or less likely. The proof was in the lung tissue itself.

Getting that tissue in a biopsy—a surgical procedure that involved cutting out a small piece of the lung—was critical to getting an "exact diagnosis" and providing proper treatment, Wheeler insisted.

Most black lung claimants, however, had not undergone biopsies. Wheeler seemed to believe there were two possible reasons for this. One: Some coalfield physicians were incompetent or sloppy. When John pressed him during a deposition, Wheeler repeatedly expressed puzzlement as to why John's client hadn't just gotten a biopsy. That was the standard of care in the United States, he said—"Except apparently in your part of the country."

In a letter that the coal company's lawyers submitted to support their defense in one benefits case, Wheeler wrote, "I question the decision to place more weight on" a positive X-ray reading by "an employee of Mountain View Hospital in Norton, Virginia...while discounting reports by two senior Johns Hopkins radiologists." He filled about a page of the two-and-a-half-page letter with a recitation of his qualifications and Hopkins's accomplishments, then continued: "I know evaluating reports involves understanding quality of information, the experience of the individuals reporting and the standings of Institutions." He and his colleague, who had also read one of the miner's X-rays, "are clinical radiologists at one of the two or three best known hospitals in the world."

Wheeler's criticism of Appalachian physicians often went hand in hand with the second possible reason he seemed to think could explain why miners didn't get biopsies: They knew they had something other than black lung, but they were trying to put one over on the government and the coal companies so they could get a handout. He lectured John during one deposition: "As long as we're dealing with X-ray patterns, we're going to have constant uncertainty, and in this day and age, there is not need for that uncertainty unless somebody feels that getting a biopsy would disprove their case."

In fact, there are good reasons why a miner would elect not to undergo a biopsy. First, it's not necessary in most cases. The standard tests used to diagnose black lung as outlined in the federal regulations and guidelines from groups such as the American College of Chest Physicians are an X-ray, pulmonary function study, and physical examination, together with an occupational history. Second, it's potentially dangerous. Though the risk is relatively low for a healthy person, it increases significantly for people with damaged lungs. Possible complications include infections, collapsed lungs, and, in a worst-case scenario, death. These dangers were documented in the medical literature and borne out by the experiences of doctors who actually treated coal miners.

Dr. Jack Parker, the longtime chief of Pulmonary and Critical Care Medicine at the West Virginia University School of Medicine, former head of the NIOSH office tracking coal miners' health, and regular instructor for NIOSH and American College of Radiology educational courses, later told me that he often advised miners who have serious lung disease not to risk a biopsy.

John pointed out these risks to Wheeler during one deposition. When Wheeler chastised John's client for not getting a biopsy, which he insisted was "quite safe," John said, "I have represented a number of miners that have had that procedure and have a collapsed lung as a result of it."

"That's temporary," Wheeler shot back.

It might seem easy to regard Wheeler as little more than a hired gun, but the reality appeared more complicated. Coal companies did pay a hefty fee for the services he and his colleagues in the Pneumoconiosis Section rendered. As of 2013, Wheeler said during depositions at the time, an X-ray reading cost five hundred dollars if there were markings that had to be interpreted, and an evaluation of a CT scan cost seven hundred and fifty dollars. By contrast, miners' lawyers usually paid somewhere around eighty dollars for an X-ray reading and two hundred and fifty for a CT scan evaluation.

But the money didn't go directly to the doctors at Hopkins; it went to the radiology department. Wheeler said repeatedly during depositions and interviews that he was a salaried employee whose pay wasn't tied to reading films for companies. Another member of the Pneumoconiosis Section testified in one case that the doctors got bonuses if the unit was "productive." Wheeler insisted he didn't know how his salary or bonuses were calculated. Hopkins public relations officials later refused to clarify.

Wheeler said in a 2013 deposition that Hopkins charged six hundred dollars an hour for his testimony. For short depositions, he said, the money went to a scholarship fund at the medical school. For those that took longer, half went to the fund and half went to the radiology department. He said he liked this setup: "I personally feel for all the sins in my life, if I'm ever asked to explain my justification for living in front of a higher order, that scholarship which will, as long as Hopkins is here, be training a decent scientist in what I call the enthralling calling. I think that will stand me in good stead."

Whatever Wheeler's motives, he spoke with a sense of righteous indignation. He was a bulwark of medical integrity, a defender of the moral high ground against negligent or lazy physicians who placed their patients at risk and greedy trial lawyers and their malingering clients who sought to extort American businesses. Maybe it was disingenuous; maybe it was a rationalization he'd convinced himself was true; maybe he genuinely believed it.

Wheeler's criticisms were not completely unfounded when he began reading X-rays for the Pneumoconiosis Section in the 1970s. But, as with his views about the disease's appearance on film, his views on the legal system seemed to be frozen in time, and he either refused to recognize or remained willfully ignorant of the major changes in years since. His perception of the newly created federal black lung–benefits system was intertwined with his involvement in asbestos-related civil lawsuits, a field that had recently burst on the scene and become a cash cow for some plaintiffs' attorneys. Wheeler's duties included reading the X-rays of workers potentially exposed to asbestos, sometimes for companies that faced lawsuits from hundreds or even thousands of former employees and their families.

In 1984, Wheeler told a U.S. Senate subcommittee that "a tidal wave of asbestos claims threatens the judicial system, industry, insurance firms, and eventually the Government." He expressed frustration at a system that allowed workers with relatively minor X-ray abnormalities and little or no evidence of disability to prevail in court. His criticism did have some basis in fact. Some big-time plaintiffs' firms did troll for clients, employ radiologists of questionable quality, and pressure companies into settlements. But throughout his testimony, he interspersed references to the black lung–benefits program, as if there were little difference between the two legal regimens. He told the senators that there was "clear evidence of fraud in the black lung evaluation process."

The General Accounting Office had found evidence of some questionable black lung–benefits claims being awarded under the liberalized standards that took effect in 1978, but within a few years, much stricter rules were in place that caused approval rates to plummet to 4 percent. Unlike in asbestos lawsuits, lawyers in black lung–benefits claims couldn't file class actions, negotiate settlements, or make lucrative contingency-fee arrangements. A minimally positive X-ray was often of little use to a black-lung claimant, and winning a claim was no windfall. During the Senate hearing, Wheeler gave as an example one X-ray he'd seen that could be misread as showing asbestosis and said that the patient nonetheless "has an X-ray which, in the right hands, would be worth probably half a million dollars."

A miner who won a black lung–benefits claim would have to live for an additional 132 years to collect that same amount in monthly payments. If Wheeler was aware of the stark differences between asbestos lawsuits and black lung–benefits claims, he didn't mention it, and he voiced similar outrage over what he saw as unjustified payments in both systems.*

Altogether, Wheeler held a number of views that were decidedly outside the mainstream, he could come off as outright hostile to miners and the entire idea of the black lung–benefits system, and he arguably was violating federal rules by not following the International Labour Organization system when grading films. Yet none of this seemed to have diminished his usefulness to coal companies and their lawyers.

Some miners' advocates worked diligently to challenge Wheeler, with some success, but they were constrained by certain peculiarities of the black lung–benefits system. They didn't have the time or resources to compile a comprehensive accounting of Wheeler's record that could establish bias or reveal whether his opinions had been proven wrong in other cases; they were scrambling to fight individual claims.

Convincing a judge to discredit Wheeler was a tall order. Some judges practically gushed about Wheeler in decisions that were peppered with phrases such as "outstanding qualifications," "superior knowledge of radiology," and "highly distinguished record." They gave companies a return on their investment in pricier X-ray readings, often referencing "the prestigious Johns Hopkins University Medical Institute." There were numerous variations on a familiar refrain: "Given Dr. Wheeler's superior credentials, I defer to his interpretation of the X-ray and find it to be negative."

For judges who were inclined to look skeptically at Wheeler's readings, there were challenges. The decision likely would be longer and more difficult to write. They weren't allowed to consider their experiences with Wheeler in other cases, and there had to be something in the record—a statement in a report or a comment during a deposition, for example—that they could use to pick apart his reasoning. Often, this evidence just didn't exist. Pressure to clear a backlog of cases only exacerbated these difficulties. Retired judge Edward Terhune Miller said

* I repeatedly attempted to reach Dr. Paul Wheeler by phone and certified letter to discuss the material about him in this book, but I received no reply.

that trying to reach a just decision while operating within the system's rules sometimes proved frustrating. He felt the law compelled him on occasion to credit physicians he personally distrusted and issue denials he personally thought unjust. "It's very difficult not to recognize in the course of your larger experience that there are certain doctors who simply never find pneumoconiosis," he said. "Judges deal with these things the best they can."

If the record did contain some of Wheeler's more problematic views, judges sometimes tried to work with those. In a few cases, judges seized on Wheeler's statements about miners being "quite young" and surmised that his readings were clouded by an assumption that it was impossible to get complicated black lung under current conditions. In others, judges found his views on the necessity of biopsies contrary to the law's requirements and his speculation about TB and histoplasmosis unfounded.

But there were perils to this approach. Company lawyers routinely appealed awarded claims to the Benefits Review Board, which some administrative law judges considered overly conservative—quick to vacate and remand an award or to affirm a denial.

For example, in one case in which every radiologist except Wheeler and a Hopkins colleague saw complicated black lung—the same one in which Wheeler declared, "I don't think I need medical literature"—a judge discounted Wheeler's opinion and awarded benefits. The board twice vacated the award and remanded the case, finding in both instances that the judge's characterization of Wheeler's credentials had been inadequate. After the case had bounced up and down for more than three years, the third award stuck.

As difficult as it was for a miner with a lawyer or lay representative to overcome Wheeler's readings, doing so was an almost hopeless proposition for a miner forced to represent himself. Wheeler's reports could go unchallenged, and if defense lawyers chose to depose him, they could elicit statements that seemed like pronouncements from on high while steering him clear of expressing any troublesome beliefs that might undermine his credibility.

This recurring scene would play out again in the fall of 2000. Wheeler would sit with a court reporter in the halls of the Johns Hopkins radiology department and, under the deft guidance of Jackson Kelly attorney Bill Mattingly, recite his usual criteria yet again to explain away the masses he'd seen on the chest films of yet another southern West Virginia coal miner, Gary Fox.

14

"No Bad Samples"

The federal government's pursuit of coal-industry cheaters, sparked by the Great Coal Dust Scam, eventually led criminal investigators to Massey Energy Company's Elk Run complex, where Gary Fox worked. As the Labor Department's ill-fated attempt to fine companies that had submitted peculiarly altered dust filters—the Peabody bon-bons—played out, the Justice Department pursued a parallel track of criminal cases. Prosecutors brought charges against individuals they alleged were responsible for exposing miners to dangerously high concentrations of dust.

One of those people, apparently, was Thomas L. Duncan. Until recently, he had been a section foreman at Elk Run's White Knight mine. Duncan's crime, a misdemeanor, was "willful violation of a mandatory health or safety standard." It was his job to oversee the collection of dust samples that the company submitted to regulators. He was supposed to ensure that miners working in the high-exposure jobs at the coalface wore the sampling pumps, but he admitted that he had actually taken the pumps and placed them in the power center a few hundred feet back, a spot where there was relatively little dust.

Soon after being charged, he pleaded guilty and agreed to cooperate with prosecutors and testify if necessary, but the Justice Department's pursuit of industry criminals seemed to have stopped just above the bottom rung of the corporate ladder.

For Duncan, the case was devastating. Massey fired him, and he had to surrender his certification to conduct dust-sampling, which essentially barred him, at least during his one-year probation term, from getting a job at another mine. He lived in the small coalfield town of Logan, and seven family members depended on him almost entirely for support. Money was so tight that the judge allowed him to pay the six-hundred-dollar fine in installments over six months.

At his sentencing hearing, Duncan got the opportunity to speak. "I've just got to say that what I done, I know it was wrong," the forty-seven-year-old now-former coal miner said. "I just want to apologize for it. The Court don't know really how sorry I am for doing it in the first place, but I was under a lot of stress. I just want to say I'm sorry for it."

It was obvious to Duncan's lawyer that his client was little more than a fall guy. Massey's relentless production targets combined with the relative scarcity of mining jobs in the area at the time put intense pressure on Duncan, he said. "Tom Duncan's a hard-working, decent man," he said, addressing the judge. "He intended not to hurt anyone; I hope it's obvious to the Court. He worked in the same conditions as the people that these rules are designed to protect because he's one of the people that these rules are designed to protect." Every day, he had walked the same tunnels and breathed the same air as the men he supervised.

The government's prosecutions of Duncan and other coal-company employees, contractors, and companies themselves over the decade following the 1991 Peabody case yielded numbers that seemed impressive: more than 180 convictions or guilty pleas, two million dollars in fines, and even brief jail terms in some cases. It was far less clear whether punishing people like Duncan significantly curbed cheating. The penalized workers had not invented the techniques for which they were convicted, and miners continued to describe witnessing or participating in them years later.

The type of sampling fraud to which Duncan had admitted was one of the oldest tricks in the book, and it had been widespread throughout the coal industry for years. There were variations: Pumps ended up in the intake air not far from the mine's entrance, in dinner buckets, anywhere but in the dusty atmosphere where miners actually worked. Section foremen like Duncan knew they were expected to get "good" samples. If they didn't, the company line went, regulators would issue violations and require improved ventilation, then profits would vanish, and there'd be layoffs.

There were a few tried-and-true ways to get a good sample. Some involved plain fraud; others involved calculated exploitation of legal loopholes. In a regulatory system that relied heavily on companies to police themselves, cheating was almost inevitable, and over the years, United Mine Workers officials complained—repeatedly, but with few results—about "the fox guarding the chicken coop."

The evidence of fraud and gamesmanship was almost as old as the 1969 law that created the sampling system. Beginning in 1970, companies had to start controlling dust levels and submitting samples to prove compliance, and the permanent standard for the concentration of dust allowed in the air miners breathed—2.0 milligrams per cubic meter of air—took effect at the end of 1972. Though the industry had called this standard virtually impossible to meet, the Interior Department reported in 1973 that companies apparently had done the impossible: 94 percent of underground mine sections had achieved compliance.

This should have been good news for miners like Gary Fox, who went underground the following year. But a 1975 General Accounting Office report found that the Interior Department's rosy assessment was, at best, dubious. There were "many weaknesses in the dust-sampling program which affected the accuracy and validity of the results" and made it "virtually impossible to determine how many mine sections are in compliance," the report said. A primary flaw was the control of much of the sampling process by the companies themselves. The GAO noted that it couldn't prove that operators were intentionally cheating but said there was significant evidence that their samples frequently didn't reflect the air miners were actually breathing. Auditors had focused on the department's Mount Hope office, which policed much of southern West Virginia, and when they examined the samples submitted by one hundred and twenty-five mines, they found that about 18 percent of samples supposedly collected from miners working in the dustiest areas indicated concentrations of just 0.1 milligram per cubic meter. Skeptical GAO officials took their own samples in outside areas at mine sites and found that even the offices contained enough dust to register at least 0.2.

Yet companies continued to submit implausibly low samples and face few consequences. During the time Gary Fox worked underground at Elk Run, the company submitted dozens of samples registering 0.1. Miners who worked at Elk Run later gave their assessment of these results:

"Ain't no way."

"Even in a closet, you'll get a .1."

"It doesn't take a rocket scientist to figure it out.... When you consistently get .1, .1, .1, .1, somebody is screwing with the pump."

The first scandal broke in 1975, and the company accused of screwing with its pumps was Consolidation Coal Company, the nation's second-largest coal producer and the owner of the Itmann mine where Gary had started the previous year. A former employee had informed federal investigators that the company was faking samples at a handful of mines in southeastern Ohio; this led to indictments against managers and the company itself. Employees testified that they had thrown out "heavy" samples and "made our own dust"—sweeping a little onto the filter with a broom, for example, rather than testing the actual mine air. Two lower-level employees were convicted, and two more pleaded guilty.

The more senior officials and the company went to trial. The judge dismissed 142 of the 172 counts in the indictment and refused to allow prosecutors to utter the words *black lung,* reasoning that doing so would only "arouse the jury's passion." Consol argued there was no evidence that higher-ranking officials knew about whatever shenanigans their subordinates might have pulled, and the trial ended in acquittals for all of the remaining defendants.

This 1979 verdict was "a devastating defeat for the federal government," the *Washington Post* reported, and a Justice Department attorney conceded, "I think it may have an adverse effect on future cases of this kind."

In 1980, the government tried again. Union miners at an operation in southwestern Virginia owned by a subsidiary of Pittston Coal Company, the nation's largest independent producer, told a federal investigator that company officials routinely placed sampling pumps in nondusty areas. A grand jury returned an eighty-nine-count indictment against the company and eleven of its officials. But the company's lawyers buried prosecutors in dense motions, and, apparently fearing that the judge was about to side with the defense and issue a decision that would make enforcing the rules much more difficult in the future, the U.S. Attorney's Office struck a deal with the defendants. All charges against company officials, who had faced a total of more than three hundred and fifty years in jail, were dropped, and the company pleaded guilty and agreed to pay a $100,000 penalty, far less than the maximum potential fine of $850,000. "We were not interested in having a bad precedent here," the lead prosecutor said. Union miners called the deal a "sellout," and a local president said, "I'd never heard of a man's lungs being bargained off like a used car."

From the end of the Pittston case in 1980 until 1991, when the Peabody bon-bons triggered a spate of prosecutions, the government won convictions in just four dust-related cases, exacting a total of about $145,000 in fines. No company officials went to jail; the most any of them got was three years of probation.

Meanwhile, it was clear the cheating hadn't stopped. In 1978, miners publicly revealed the contents of the industry's bag of tricks at a series of hearings held by the Mine Safety and Health Administration (MSHA): Company officials put sampling pumps in fresh air or even outside the mine, employees said. They trashed samples that looked too dusty. They had miners sign the data cards stating that they had worn the pumps during the shift in advance, and if miners refused to sign, they were sent home and docked a day's pay.

Miners continued to describe the use of many of those same tactics in the years that followed. It was an open secret within the industry and among people who regularly worked with miners, such as lawyers and employees at black lung clinics. "I heard those stories hundreds of times, maybe thousands of times," said Debbie Wills, who started as a benefits counselor at a clinic near Charleston in 1989. Lawyer Grant Crandall had been hearing the stories since he started representing miners in 1975. "It wasn't like it was hidden," he said.

Brenda Halsey Marion, the nurse at the New River clinic who hired John Cline, said miners had been describing rampant fraud to her for decades. In the 1980s, "that was the story," she said. "In the 2000s, that's the story. It's been pretty consistent."

Statistical analyses offered further evidence. In one, published in 1984 in the *American Journal of Industrial Medicine,* researchers compared data on company and inspector samples taken at the same mines and found that suspiciously low dust concentrations were far more common among company-submitted samples. The results, they wrote, "suggest widespread underreporting." A 1990 study published in the *American Industrial Hygiene Association Journal* reached similar conclusions: "Taken along with the previous analyses and expert judgment on low samples in operator data, there is strong evidence that these low samples do not well represent actual exposure conditions."

Miners' advocates grew frustrated with the lack of government action to address the obvious problem. "There was already a record of tampering in 1978," Davitt McAteer, who ran a public-interest law firm and would become the nation's top mine safety and health regulator under President Bill Clinton, said. The government "was given an abundance of anecdotal evidence that indicated that the tampering was sufficiently widespread that the agency should have been concerned about it."

But the same political forces that led to the drastic decline in black lung–benefits awards also helped stall reforms to the dust-sampling program. In the wake of the 1975 Consol scandal, momentum for reform had built. MSHA proposed a rule that would require companies to allow a representative of the miners' choosing to observe every phase of the sampling process. Government researchers enlisted contractors to design a tamper-resistant cassette and prioritized development of a sampling device that would give real-time readouts of dust levels and be more difficult to manipulate.

These efforts stalled during the 1980s. There was a "perceived decrease in urgency," and the government had "other funding priorities," a GAO report found. Money for work on the tamper-resistant cassettes and better sampling pumps dried up. In 1985, the Reagan administration shelved the proposed rule that would have allowed miners' representatives to keep tabs on company sampling.

The flaws in the system designed to prevent black lung had been evident since the year Gary Fox first went underground, but when he went to work at Elk Run, almost two decades later, little had changed.

In depositions and interviews, miners who worked at Elk Run at the time said fraud was commonplace. Continuous miner operator Ricky Tucker testified that, during company sampling, bosses would approach him with a demand: "I need your dust pump. I've got to take it back there in the fresh air and make sure we don't get no bad samples." The company line remained the same: If they got a bad sample and regulators required costly fixes, "they'd have to shut the mines down."

Carter Stump, who worked alongside Gary for some of his time at Elk Run, said that supervisors usually took the sampling pumps off miners and put them in clean air. "The boss took them in, and they always stayed in the intake," he said. "You didn't wear 'em."

Over the years, though, some things had changed. The industry had evolved, and chief among those who had adapted and thrived, at least in central Appalachia, was Don Blankenship. The shifting landscape over which he presided came with heightened dangers for miners.

"The cheating never changed; what changed was the mining," Charleston lawyer Tim Bailey said.

Companies had already gouged out many of Appalachia's thickest coal seams, leaving thinner bands surrounded by more rock largely untouched. Now, thanks to rising demand and advances in mining technology, it made economic sense to go after these thin seams. For miners, however, that often meant cutting more rock, which added a heavy dose of silica to the mixture of dust they breathed. It was a recipe for nastier, more rapidly progressing disease. Stump recalled some days at Elk Run's White Knight mine when the product his crew sent out on the belt line contained more rock than coal.

Tools and techniques had also evolved. Massive machines called longwalls could shave off panels of coal hundreds of feet wide, but they were expensive and required certain geological conditions, so many miners—including Gary Fox—kept working in sections using continuous mining machines. A more efficient twist on the traditional method for deploying these machines grew in popularity among mine operators, ramping up both productivity and the amount of dust in the air. Massey had embraced the method, and Gary spent the bulk of his time underground at Elk Run working on these "super-sections." The key point on which they differed with traditional continuous miner sections was ventilation, the lifeblood of underground miners. Fans forced air into the mine, and as it moved through tunnels, it swept away both the gases that could fuel explosions or suffocate miners and the dust that could end up in miners' lungs.

In a traditional section, the air entered on one side, swept across the coalface, and flowed back out on the other side. That meant only one continuous miner could be running at a time; otherwise, miners would be working downwind, directly in the path of the dust clouds leaving the mine. In a super-section, however, the air entered in the middle of the section, reached the coalface, then split and flowed in opposite directions. Its path resembled a capital T, and miners often called it "fishtail air." It carried the gases and dust out of the mine on both sides of the section.

The advantage of this was that two continuous miners could be running at the same time, and workers would be downwind for only a few brief periods per shift, which federal rules allowed. That, at least, was the theory. The reality, according to miners who worked at Elk Run, was very different.

Running a super-section required sketching out a detailed plan and sticking to it. A section might be made up of ten entries, each about twenty feet wide and separated by the unmined coal pillars that helped support the roof. One crew might handle entries one through five, and the other crew would take six through ten. They had to cut the entries in a specific order, called a cycle. A company's plan had to win the approval of regulators, who imposed strict limits on how often miners could work downwind.

But the temptation not to follow this plan—to mine off-cycle—was overwhelming. It took time to move the mining machine and its attendant equipment from one entry to the next. A crew could extract more coal in less time by focusing on just two or three entries. The result looked a little like an out-of-whack Tetris board where the player had become fixated on a small portion of the screen. This disrupted ventilation and left miners working downwind far more often than allowed.

At Elk Run, mining off-cycle was "pretty much expected," Stump said. Miners worked downwind "all the time," said Harold Ford, a former section foreman for Elk Run. Other miners said the practice was common at Elk Run and other Massey mines.

The rhythm of the mining process meant that those who got the worst of it were the roof bolters like Gary Fox. Ford, who ran a bolt machine for years before becoming a foreman, recalled, "Many times I bolted, couldn't hardly see." Gary himself would later tell John Cline that he'd often had to bolt in the dust-filled air downwind of the mining machine.

Other time-saving techniques of the unlawful variety further increased the amount of dust miners breathed. Keeping up proper ventilation required constant maintenance throughout the shift—tasks that took time away from running coal. Fresh air had to be guided to the working face; otherwise, it would just shoot off in all directions through the grid of tunnels and dissipate before it could do its job. To direct the airflow, miners hung curtains from the roof, built wall-like stoppings, and cut openings called overcasts. As they moved forward, cutting deeper into the seam, they continually had to advance the curtain, build more stoppings, and cut more overcasts. At Elk Run, miners said, those things often were not done. The curtains could be a nuisance. Crews had to maneuver around them, and equipment sometimes snagged and ripped them down. It was common to see the curtains taken down and placed out of the way, miners recalled.

Many miners didn't like breaking the rules, but they felt they didn't have much choice. The Massey way combined lofty production demands with bare-bones staffing. The standard crew at many other companies included a miner helper who was responsible for tasks such as hanging curtain. Massey crews often didn't have a miner helper, leaving the other miners scrambling to fill in, according to former Massey workers. On super-sections, many companies used two roof-bolt crews. Massey sometimes used just one. That meant bolters like Gary were constantly racing back and forth across the mine in an attempt to keep up. "We worked them like convicts," one former Elk Run superintendent and vice president said during a deposition. "We never had enough people to operate that mine in a safe manner."

Despite being perpetually shorthanded, crews were expected to meet ambitious production targets. Company higher-ups insisted these were merely goals, but many miners perceived them as quotas. "If you hung curtain and run legally by law, you would not run the footage that we needed," Stump said. Other Elk Run miners had similar recollections. "You can't do it legally," one former section foreman said during a deposition. "Anybody says they can, I'll tell them they're crazy."

Looking back, Stump said his time as a superintendent weighed on him. "Yes, it bothered me," he said. "I can say I did try to run halfway legal.... All they looked at was footage. They didn't care how you got it. You have the pressure on you all the time."

This pressure flowed down from the top of the Massey hierarchy. In his infamous *RUNNING COAL* memo, Blankenship had told superintendents not to take time away from mining to perform certain tasks meant to keep the mine in safe working order. In a handwritten message to a manager, he instructed, "Do not cut any overcasts." And in another memo to a manager: "We'll worry about ventilation or other issues at an appropriate time. Now is not the time."

Blankenship recorded some of his phone conversations, and in one that later became public, the caller said regulators seemed to think that Massey often wasn't performing the tasks needed to keep up ventilation. Blankenship replied, "Yeah, I'm sure that there's a certain amount of things that section bosses do and so forth, but the truth of the matter is black lung is not an issue in this industry that's worth the effort they put into it."

In 2008, Massey hired a former MSHA official to advise on issues including ventilation, and the official flagged dust-sampling as a primary problem in a memo the following year. The company's approach "shows weakness, lack of concern, high non-compliance records and lack of experience," he wrote. A few months later, clearly frustrated, he wrote in an e-mail to another Massey official, "There [*sic*] still cheating on the Respirable Dust Sampling."

In a memo to Blankenship in 2009, one of Massey's top lawyers related a host of problems that the former MSHA official had described. "Massey is plainly cheating on dust sampling at some of its operations," she wrote, recounting one of his concerns. Some mine foremen, she continued, "have admitted that they cheat on dust sample day. They feel that in doing so, they are carrying out what they were told to do."

Despite the problems with self-policing, there was a stopgap built into the sampling program, at least in theory. Federal inspectors visited each mine four times a year, and they could collect their own dust samples.

That's where the loopholes came in—legal cheating. During these quarterly spot checks, some miners said, inspectors were treated to a "dog and pony show." The last-minute preparations before the rise of the curtain weren't legal, but the inspectors didn't see this part. Inspecting a coal mine wasn't like inspecting a restaurant and catching the staff picking cockroaches out of the salad bar. It took a while for officials to get to the actual mine face. In the meantime, the silent alarm had been sounding, triggering frenzied acts of ass-covering.

Sometimes, former Elk Run miners said, the bosses would call the underground phone and announce: "We've got company." Other times, the belt line would suddenly stop running, and when the underground crew called outside to see what the problem was, officials would relay news of the visitors. Depending on the mine, the crew had at least thirty minutes, maybe much longer, to clean up, hang the curtains, and do whatever else was needed to make the section look like a model of compliance.

With the inspector watching, foremen couldn't just put the sampling pumps in the fresh air, so they had to find other ways to ensure a "good sample." These were the rare days when the ventilation plan the company had put on paper actually corresponded to reality, miners said. They hung curtains and mined on-cycle. For good measure, they might redirect air from other sections of the mine to the area where the inspector was sampling.

The bosses knew the rules well and used them to their advantage. Under the regulations, MSHA would consider a sample valid if the pump ran for eight hours—not the nine, ten, or more that miners often worked. The rules also allowed companies to dramatically cut production on sampling days—50 percent of the normal haul when collecting company samples and 60 percent of normal when inspectors were sampling. Stan Edwards, who worked as a superintendent at Elk Run, said during a deposition that he usually calculated the exact amount of coal they'd have to mine to reach that production minimum. "I'd do the math and everything for them. And then I'd let my foreman know this is what we have to run for the pumps to be good," he said. "Otherwise, if you don't get that, then they're going to come another day and you're going to lose another shift fooling with them."

The way regulators assessed compliance also helped. Inspectors usually put sampling pumps on five miners working different jobs on the same shift. Only the average of all five samples had to be below the limit. That meant that the low results from miners who worked in less dusty areas could cancel out the high results from miners who worked in dustier spots. In other words, even if the miners with the most high-risk jobs—roof bolters like Gary Fox, for example—were exposed to concentrations of dust that were well above what the government had deemed safe, their sections might still be considered compliant.

If samples nonetheless came back too high and inspectors issued a violation, the company was allowed to take five samples of its own, and if these came back under the limit, MSHA considered the hazard abated.

There was another wrinkle in the dust-sampling program that could be problematic for companies. MSHA tested some samples for silica, and if this mineral constituted more than 5 percent of the total mixture of dust, the agency would set a reduced overall dust standard that the company had to meet. The agency applied a formula to determine what that new standard would be—the greater the proportion of silica, the lower the standard.

During the time Gary worked underground at Elk Run, the company's mines were frequently on reduced standards. Some samples contained more than 20 percent silica. That translated to a dust limit of 0.5 milligram per cubic meter or less, a mark that was extremely difficult to hit, even if the company scaled back production and enhanced ventilation.

There was a way, however, to make reduced standards disappear. Years after Gary stopped working, federal investigators discovered it at one Massey mine. The company had played a shell game with its mining machines. A reduced dust standard was attached to mining equipment, not the section of the mine where it was operating. Replace that equipment and the standard went back to normal. By shuffling around machines, Massey could avoid having to meet a reduced standard even though the same section repeatedly tested out of compliance. "That was a trick that everybody used," former Elk Run superintendent Edwards said during a deposition.*

As violations piled up, some in the coalfields came to regard Massey as an outlier. "Other companies did stuff wrong too, but I think Massey

* Don Blankenship declined my request for an extensive interview. I spoke with him briefly by phone, and he said: "No one ever did more for mine safety than Don Blankenship did, and people that worked at Massey know that. The press has lied about that for thirty years." I asked about accounts by former Massey employees of widespread dust-sampling fraud, and he replied: "I don't know anything about it. I do know that the sampling system has been flawed from the beginning." In his experience, he said, some miners didn't like wearing sampling pumps, and it was difficult to ensure that they didn't take them off. "It's on the individual," he said. He declined to answer follow-up questions.

Some Massey managers, in depositions or interviews, also denied allegations of dust-sampling fraud or wrote them off as rumors, defending the company and portraying Blankenship as the victim of political attacks by regulators, the United Mine Workers union, and the media. Frank Foster, who worked as safety director for the Elk Run Coal Company from 1994 to 1997 and went on to become safety director for all of Massey, said during a 2013 deposition, "Don Blankenship is one of the most safest CEOs I've ever worked for.... Any safety enhancement we wanted to do in Massey, Mr. Blankenship always—he may ask me questions, why we wanted it, and he may even go so far as how much it was going to cost the company. But he has never turned me down on any type safety enhancement that I ever requested."

was the black hat," said Bill Harvit, a lawyer who, like Blankenship, had grown up in Mingo County and worked summers in the mines to pay for school. "The level of disregard for worker safety was never as bad as it was at Massey."

But some practices, including many used to game the dust samples, were common throughout the industry. Some of Gary Fox's coworkers at the Birchfield mine and others owned by Maben Energy described taking down ventilation curtains and seeing sampling pumps placed in clean air, for example.* Getting a call when an inspector showed up was routine. "Everywhere I've ever worked, as long as I can remember, it's always been that way," Ford said.

Researchers at the National Institute for Occupational Safety and Health, trying to understand the causes of black lung's resurgence, later studied a cluster of disease, interviewing miners who had developed complicated pneumoconiosis about their work experience. "Very consistent themes emerged," the authors wrote in 2018. Among them: mining off-cycle, failing to maintain ventilation, cutting through copious amounts of rock, and placing sampling pumps in clean air.

Thus, some came to think of Massey as simply the biggest and meanest in a crew of bullies, the industry's dark side writ large.

"During the relevant time period when these guys are getting exposed, who's the big boy? Massey. Plus, you get that corporate attitude of mine, mine, mine, don't keep up ventilation," said lawyer Tim Bailey, who grew up in southern West Virginia and is the descendant of miners. "And now, bingo, you've got all these guys who are coming down not just with simple pneumoconiosis but complicated pneumoconiosis."

---

* The president of Birchfield Mining when Gary Fox worked there, Dale Birchfield, said: "Curtain was there. I'm not going to say that the men always hung it. But it was there for them. Some felt like hanging it, some probably thought it was in the way.... Back then, we probably didn't fully understand the importance." He said that he, like many others in the mining industry, had since learned much more about prevention of black lung. Looking back, he said, he should have placed more emphasis on keeping the curtain up. Asked about the accounts of former Birchfield miners who described dust-sampling pumps being hung in fresh air, he said: "I hope that didn't happen. Sure, they're cumbersome and get in the way, but it was necessary."

That was true of some of the men who worked alongside Gary Fox. Stump said he'd never given much thought to all the times he'd seen sampling pumps hanging in the intake air—"until my breathing got worse." He was diagnosed with complicated pneumoconiosis and won a federal benefits claim. Reflecting on his time underground, he said he would have done things differently, or at least tried. "It would have been done right," he said between bouts of coughing. "I would have used the curtain. Now, walking from that building up here, I don't have no wind, can't breathe. But you didn't think about it then. You was thinking about making money for the company."

Some miners who ended up with black lung voiced anger against their former employers, but many more struggled to explain a more complicated tangle of feelings about their current plight—a mixture of fatalism, regret, and self-reproach: Maybe they wouldn't be in this state now if they hadn't been so young and foolhardy, hadn't felt invincible, hadn't been blinded by the bonuses and the gifts and the free trips. Maybe they could have spoken up or quit. A sense of personal responsibility remained even though they knew that they hadn't really been presented with a choice. Mixed in was a belief, encouraged by some company bosses, that preventive measures were futile; as long as people mined coal, some of them would get black lung.

In truth, it was not inevitable that miners would end up with black lung, and the bulk of the responsibility for ensuring they didn't rested with company officials and regulators. By the 1990s, it was clear that what needed to be done was not to teach rank-and-file miners personal responsibility but to fix the laws that allowed them to be victimized twice: stricken with a preventable disease, then cheated out of benefits.

For years, that had been the goal of the informal coalition of miners' advocates—the clinic workers, lay representatives, lawyers, doctors, union officials, miners in Black Lung Association chapters. They had developed proposals and advocated for them, with few results. They'd become keenly aware that the policies that shaped miners' lives were dictated not just by science and logic and notions of fairness and justice but by the ideology of whoever was calling the shots in Washington. Since 1980, reform had basically been off the table.

But by 1993, changing political winds offered some cause for optimism. The advocates recognized a rare moment of potential, and they were determined to seize the opportunity.

15

Window

On a warm September day, John Cline and thirty other members of the Black Lung Association chapters from Fayette and Kanawha Counties piled onto a bus bound for Washington, DC, where they would rendezvous with chapter members from Kentucky, Virginia, Indiana, and Illinois. It wasn't the first time they'd made this five-hour trek northeast, but if things went well, it might be one of the last.

They'd gotten the hang of lobbying on a budget. They'd rent a bus—the "Hope Coach," some members called it—and meet at a strip mall near Beckley, armed with thermoses of coffee and a change of clothes. They'd stop for the night just outside DC in Manassas, Virginia, where they often slept four to a room in the Howard Johnson or the Super 8. In the morning, they'd descend on the capital, and miners would walk the halls of congressional office buildings, some wearing Black Lung Association hats and T-shirts, some dragging oxygen tanks, brushing past professional lobbyists who charged more for a few days of their services than the national association had in its bank account, which was roughly four thousand dollars. To fund these treks, the association held cookouts and concerts, received occasional small donations, collected dues, and passed the hat during meetings.

For a few years now, they'd been making these trips. John and others in the informal coalition of miners' advocates had identified what they regarded as the main flaws in the black lung–benefits system and developed proposals to fix them. Coal-state lawmakers had incorporated them into legislation, but these bills remained in congressional purgatory session after session.

In 1993, as John and the other members boarded the Hope Coach, they had reason to think the outcome might be different. Bill Clinton was in the White House, and Democrats controlled both houses of Congress.

After a day of walking through the halls of Congress to speak with members and staffers, the miners and their allies watched a subcommittee's markup of the bill, then held a rally on the Capitol Building lawn, where three representatives from coal states spoke.

"So," the president of the Fayette County chapter later reported in an update for members, "the quest continues and the 'Hope Coach' will hit the road again until Congress can approach the black lung issue with some sense of empathy and understanding of how this disease can destroy a miner and his family. Let us pray the road will not need to be traveled too much longer."

The author of this report was a retired miner who had quickly become a leading voice in the national black lung movement and a close friend of John's. Mike South had grown up in the small coalfield town of Beards Fork, about thirty-five miles north of Beckley. His father, a longtime coal miner, died of black lung. South enlisted in the air force after finishing high school and served in Vietnam as a mechanic. Back home, he rekindled a relationship with a high-school classmate named Kathryn, and the two soon married. He went underground, but after he spent about a decade mining, his lungs were so shot that Dr. Don Rasmussen informed him that if he didn't quit, he wouldn't live to see his two daughters grow up. It was only then that doctors discovered South had a rare condition called alpha-1 antitrypsin deficiency that predisposed him to lung disease. By the time he was in his early forties, he was as impaired as miners who were decades older and had spent much of their lives underground. In 1989, he won his federal black lung–benefits claim.

South began attending Fayette County Black Lung Association meetings, and he and John took to each other. South was both much younger and much sicker than many members, but he seemed not to harbor an ounce of self-pity. He dealt with his situation by deploying his sharp sense of humor, almost mocking death, and he combined that wit with his innate artistic talent in his advocacy. He was a gifted oil painter, but he'd been forced to stop when the fumes became more than his lungs could handle. So he'd switched to pen-and-ink, and his drawings graced flyers, T-shirts, and the walls of the New River clinic. The association sold his art to raise money. Each bore the signature he used: a compass rose with the cardinal direction *S* circled.

He was not physically imposing or given to fiery rhetoric, yet the president of the United Mine Workers later called South "this nation's most widely known black lung activist." He had a beard and a full head of auburn hair, and he would appear at gatherings dragging an oxygen tank behind him, his labored breathing impossible not to notice, and deliver a message of simple eloquence, an appeal to basic notions of fairness and justice. As president of the Fayette County chapter of the Black Lung Association, he lent his moral authority to the national push for legislative reform.

A few months after the 1993 lobbying trip to DC, the president of the National Black Lung Association, Allen Hess, sent an update to members: "We are closer to having our bill passed than we have ever been. . . . We are expecting to have a happy 1994!"

The legislation Congress was considering, called the Black Lung Benefits Restoration Act, contained more modest reforms than the association had sought a few years earlier. Repeated defeats and political realities had led the association to drop some of its more ambitious proposals. The current bill would, among other things, limit the amount of medical evidence each side could submit in claims and make it easier for widows to win.

Some members of Congress, primarily Republicans, worried the legislation would create a "coal miners' lottery" or an "extra pension" and that it would impose "new and excessive costs" on the industry. Nonetheless, with members of the Black Lung Association watching from the gallery, the House passed the bill in May 1994. Five months later, when the

Senate adjourned, the bill was still stuck in committee, dying there just as it had in past years. Hess sent a solemn update to members: "This is a major disappointment for all of us."

By the end of the year, the national leadership had settled on a new approach and had a new president: Mike South. He recognized the changed political landscape and what it meant for the reform efforts, and he offered a succinct diagnosis in trademark fashion: "3 things are needed to pass a bill through Congress," he wrote. He listed them, with each accompanied by a drawing: "1) 2 bat wings 2) 3 frog hairs 3) the 'aye' of Newt!!"

The nation had gone to the polls one month after the reform bill died in the Senate, and Newt Gingrich's "Republican Revolution" had swept the Democrats from power, giving the party control of both the House and Senate for the first time in forty years.

South and other leaders realized that legislative reform was dead for the moment, but that was not their only option. The most sweeping changes would have to come from Congress, but some reforms were within the authority already granted to the Labor Department under existing law. The group shifted its focus to regulatory reform, where the odds of success seemed better.

President Clinton had installed officials more sympathetic to miners than their predecessors in key Labor Department posts. The new head of the Mine Safety and Health Administration, for example, was Davitt McAteer, who had previously run a public-interest law firm in West Virginia and had long been an outspoken advocate for miners. South and others began meeting and corresponding with this new set of bureaucrats to discuss what might be possible.

The change in administration had come at a propitious time. Throughout U.S. history, reforms to protect miners were almost always enacted only after a disaster, such as the Farmington explosion that sparked the 1969 law. The Great Coal Dust Scam had a similar effect, supplying momentum for changes to both the rules designed to prevent black lung and those governing the benefits program. The Labor Department had created a task force to examine the dust-sampling program, and inspectors had blitzed hundreds of mines, collecting samples and assessing company dust-control practices. In 1992, the task force had issued a report recommending, among other things, rule changes to cut down on legal cheating: Stop letting companies mine half as much when sampling, and issue violations if any single sample—not just the average of five—exceeded the limit. It also recommended reviving the research on tamper-resistant cassettes and sampling pumps that could provide continuous, real-time measurements.

Researchers at the National Institute for Occupational Safety and Health issued similar recommendations in a 1995 document that provided ample scientific support for reform of both the sampling and benefits programs. The 360-page tome incorporated the latest epidemiological and medical studies and determined that the dust limit established by the 1969 law, 2.0 milligrams per cubic meter of air, was not strict enough; the agency recommended halving that limit.

The report also sought to end the decades-long debate over whether coal dust alone could cause obstructive lung diseases such as emphysema and chronic bronchitis—a critical issue that often arose in benefits claims. The coal industry's position had long been that only smoking caused these illnesses. The NIOSH document cited a large body of evidence to refute this. Nonsmoking miners also developed obstructive lung disease, and researchers had repeatedly documented the relationship between increasing coal dust exposure and worsening impairment.

Still more support for reform came the following year in a report from an advisory committee that Labor Secretary Robert Reich had tasked with issuing recommendations that would "bring this disease to an end." The committee, which included labor and industry representatives as well as "neutrals," had held meetings in West Virginia, Kentucky, Virginia, Pennsylvania, and Utah, and miners had traveled from across the coalfields to speak, as had industry officials and researchers. John Cline and Mike South had spoken during the session in Charleston.

The committee's report, released in November 1996, concluded that the entire system meant to prevent black lung needed an overhaul. Like the earlier Labor Department task force, the committee recommended switching to single-shift sampling and expediting research on real-time, tamper-resistant sampling devices. It also suggested that regulators consider lowering the dust limit. The report expressed serious concern over the ongoing cheating, both legal and illegal, that miners had described during the hearings. The current system of company self-policing "has been severely compromised," the committee wrote. "One of MSHA's highest priorities should be to take full responsibility for all compliance sampling."

This process of hearings, reports, and recommendations had consumed the entirety of Clinton's first term, and what remained to be done would be "a massive undertaking," as a Labor Department lawyer put it in a 1996 internal memo. Some of the proposed changes fell under the department's administrative authority, meaning it could simply change its policy. Others, however, required going through the formal rule-making process, which had become tedious and time-consuming in large part because of new requirements imposed by legislation, executive orders, and federal court rulings. The department planned a two-pronged approach, with MSHA addressing prevention and the agency that managed the workers' compensation program addressing the benefits system.

In January 1997, the latter agency proposed a rule that incorporated many of the Black Lung Association's suggestions. Recognizing the "vastly superior economic resources" of companies that allowed them "to generate medical evidence in such volume that it overwhelms the evidence supporting entitlement that claimants can procure," the agency said it planned to limit the number of reports each side could submit and require judges to give added weight to the opinions of miners' treating physicians. Other proposed changes would make it easier for miners who had initially won and started receiving benefits payments but lost on appeal to avoid having to pay back what they'd been given.

Comments poured into the Labor Department, which was obligated to consider all of them. Members of the Black Lung Association coordinated with attorneys, lay representatives, and other advocates across the country to compile a submission supporting various pieces of the rule, and the

United Mine Workers and NIOSH sent in favorable comments of their own. Meanwhile, the industry worked to undermine the regulation, and some of its allies in Congress began to gin up opposition.

When members of the Black Lung Association convened a year later, in 1998, the rule hadn't been finalized, and time was increasingly becoming a concern. "There is still a lot to do and there is only a short window of time," the association secretary recorded in the meeting minutes. In the final days of the Clinton administration, miners' advocates tried to spur on the Labor Department. They wrote department leaders, enlisted the aid of sympathetic members of Congress, traveled to DC yet again.

On December 20, 2000—one month before Clinton would leave office—the department issued the final rule implementing the reforms to the benefits program. In a document that spanned 123 pages of the *Federal Register,* the department meticulously refuted each of the industry's primary objections. Most of the key provisions in the proposal remained intact.

One part of the rule that had drawn particular opposition was the acceptance of NIOSH's findings that coal dust could cause conditions such as chronic bronchitis and emphysema. Now, with the final rule in effect, miners' representatives didn't have to argue on a case-by-case basis that this causal relationship was possible; company lawyers could no longer claim that smoking alone caused these ailments. Miners' representatives did, however, have to show a connection between the disease and each particular claimant's dust exposure.

Meanwhile, in its parallel efforts on prevention, the Labor Department had taken a calculated risk that proved costly. MSHA officials had decided that the agency wasn't legally required to go through the lengthy rule-making process to begin using single samples taken by inspectors to determine compliance, and it announced this policy change in a February 1998 notice. The National Mining Association sued, and a federal court decided in its favor. MSHA had to start over, this time following the procedures for issuing a formal rule.

In July 2000, the agency did so, proposing two prevention-related rules. Together, they would create a new regime in which inspectors took over all sampling, companies had to operate closer to normal production levels when they were there, and violations would be issued for any single sample over the limit. But regulators had just months to plow through a rule-making process that often took years.

The following January, as a new president took office, the proposed rules on prevention remained unfinished. George W. Bush owed much of the credit for his narrow victory to West Virginia and the coal industry. The state had not gone to a non-incumbent Republican since 1928, but with the help of a sustained and well-funded push by the coal industry, Bush had won West Virginia's five electoral votes—the exact margin separating him and Democratic candidate Al Gore in the final tally. "State political veterans and top White House staffers concur that it was basically a coal-fired victory," the *Wall Street Journal* reported.

The investment paid off for the coal industry, as Bush soon began easing environmental restrictions. Less noticed was a finding published in the *Federal Register* three years into his term: MSHA, now under the leadership of a former mining executive, had yet to finalize the prevention-related rules proposed in 2000, and now it had decided that it ought to wait until more research could be done on continuous-sampling devices. Just as in the Reagan administration, the agency had quietly taken reform off the table and placed it on a shelf to gather dust.

It was just politics, a stark lesson for those members of the black lung coalition who hadn't already learned it. Progress on policy goals depended as much on political pressure as on sound arguments. "Occasionally, the pressure swings in our direction," said Grant Crandall, who had left private practice for a post as general counsel of the United Mine Workers in 1996. "Most of the time, it's a pretty uphill battle. You just keep after it as best you can, and occasionally, you get a little window of opportunity and you try to grab it and get as much done as you can."

The Clinton administration had been that window, and in retrospect, it looked like a missed opportunity. "There were a lot of things that were left hanging there," Crandall said.

Miners and their advocates had won some important changes to the benefits system, but many of the problems they'd initially set out to fix remained. The efforts on prevention had fared worse. A decade after the Great Coal Dust Scam had made national news, the rules hadn't changed.

For now, all they could do was regroup and hope they would get another chance.

16

Crooked Rafter

John Cline had put the lawyers at Jackson Kelly on their heels by sniffing out and then forcing them to hand over the missing pieces of Dr. George Zaldivar's report on Charles Caldwell, but the firm had a strategy for continuing the fight. At the heart of this defense were Dr. Paul Wheeler and his colleagues at Johns Hopkins.

Zaldivar's determination that Caldwell had complicated pneumoconiosis, supported in part by his own reading of two X-rays, was damaging, but there was a way to deal with such inconvenient information: Trump Zaldivar's readings with those by esteemed radiologists, then give those readings to other pulmonologists to incorporate in a report rebutting Zaldivar's overall conclusions.

That's exactly what Jackson Kelly lawyer Mary Rich Maloy did. Having previously obtained negative readings of Caldwell's X-rays and CT scan from Wheeler, she bolstered these with additional rereadings from two of his colleagues. Wheeler had reprised his greatest hits: Though there were apparent masses on the films, they were out of the strike zone; they weren't symmetrical; and there was no shotgun pattern of smaller cinder blocks to build the masses. They were "compatible with healed granulomatous disease," and the location "favors TB over histoplasmosis." His colleagues echoed those views.

The chest films were particularly important in Caldwell's case. To win, a miner had to prove four things by a preponderance of the evidence: that he had pneumoconiosis, that it was caused by his coal-mine employment, that he was so disabled that he could not perform his usual work in the mines, and that pneumoconiosis caused or substantially contributed to this total disability.

The burden was on the miner to build this four-legged chair. If the coal companies' lawyers could chop off any one leg, the whole thing would collapse. Some miners had multiple potential paths to victory, but in claims such as Caldwell's, the first leg was the make-or-break point for the whole case. Under the regulations, a finding of complicated pneumoconiosis automatically locked in all four legs. Defense lawyers commonly used Wheeler like a buzz saw to remove the first.

These crucial radiology reports then became grist for analyses from pulmonologists that could seal the defense's case. At the top of the evidentiary pyramid, these experts could survey everything below and incorporate it all into an overarching medical opinion that judges often relied on heavily when deciding cases.

Just as readings from radiologists at Johns Hopkins frequently formed the base of the pyramid, the upper reaches were often the province of a relatively small group of pulmonologists frequently consulted by coal companies' lawyers. Prominent among them was Dr. Gregory Fino. He, and other physicians like him, sometimes provided multiple reports in a single case. If some new piece of information damaged the company's defense, the lawyers could return to Fino; he might explain it away or revise his opinion. This could help defense attorneys deal with setbacks such as the disclosure of Zaldivar's full report.

Fino played this role in Caldwell's case. He had previously written a report finding that Caldwell didn't have black lung, but he'd been given only the partial Zaldivar report. Now Maloy submitted a new report from Fino. This time, he'd been given not only the previously missing parts of Zaldivar's report but also the additional radiology reports from Johns Hopkins. In light of all of this, Fino wrote, "I disagree with Dr. Zaldivar." His conclusion remained unchanged: Caldwell was disabled and had extensive scarring in his lungs, but the cause was not his years of mining. Fino couldn't say with certainty what the true cause was, but he could say that it was *not* black lung.

Judges occasionally looked askance at this. In a few cases, judges discounted Fino's opinions after finding some of his statements demonstrated "evidence of bias," "bias against the claimant," or "bias... in failing to attribute changes to coal dust exposure." In one case in which Fino had changed his opinion, a judge wrote, "Trying to reach an accurate estimate is one thing; flip flopping and embracing a discarded theory when confronted with inconsistent data is another thing, suggesting a possible ideological bent."

Nonetheless, Fino often proved to be an effective resource for coal companies and their lawyers. Whereas Wheeler didn't keep up with medical literature on black lung and refused to stray from the views he'd formed decades earlier, Fino engaged with the evolving science in his reports and depositions. When the Labor Department proposed the rules reforming the benefits system in 1999, Fino had cowritten a lengthy report for the National Mining Association opposing the changes and challenging in particular the department's conclusion that coal dust could cause disabling obstructive diseases such as emphysema and chronic bronchitis. When the regulation was nonetheless finalized, he had an apparent change of heart. Pressed during a deposition about views he'd expressed just two years earlier, he said: "I don't believe in that anymore. I think I did at one time." This change, he said, was based on the medical literature, not on the fact that, had he held to his past views, the judge could have discounted his opinion.

Fino also performed medical exams for companies. He'd drive from Pittsburgh, where he had a private practice, to a location in central Appalachia where coal companies had directed miners to go, and he'd stay for a day or two, during which he'd examine maybe twenty or twenty-five miners. In a typical year, he'd make this trek half a dozen times. For a few years, he regularly held these bulk evaluations in eastern Kentucky, but that ended in 1999 when, at the request of the Kentucky Board of Medical Licensure, he agreed to stop practicing in the state.

The board had investigated him after three miners whom he'd

examined in a motel room filed complaints saying the conditions were unsanitary. Fino's normal procedure at the time was to rent a few rooms in which he'd conduct physical examinations and pulmonary-function tests. He'd then send miners to a nearby medical center for X-rays and blood-gas testing.

The board's investigation concluded that, because the exam and the breathing tests weren't invasive, it wasn't unsanitary to perform them in motel rooms. But there was a separate problem: Fino didn't have a license to practice in Kentucky when he conducted them. He'd had one previously but failed to renew it. He said it was a technical mix-up and not anything intentional, and he'd since applied for a new one. The board, however, told him that if he didn't withdraw his application for a license, it would start formal disciplinary proceedings.

Like Wheeler, Fino would weigh in on Gary Fox's claim at the request of Jackson Kelly.*

There were other pulmonologists like Fino, and companies often solicited opinions from a few of them in each case. Their reports came in from medical centers and university hospitals across Kentucky, West Virginia, and Virginia and into Ohio, Missouri, and Colorado. Some of these doctors also traveled periodically to the coalfields for bulk evaluations.

Not all of the doctors who submitted reports favorable to miners had spotless records either. A few had apparently never met a miner who didn't have black lung. There were certain radiologists whom John Cline didn't use because he thought they tended to overdiagnose the disease. One pathologist who sometimes weighed in on behalf of miners lost his

* When I met Fino in 2013, he had begun conducting exams in a hospital in the small southwestern Virginia town of Grundy. His red Porsche Cayenne SUV stood out in a parking lot full of muddy trucks and well-worn sedans with bumper stickers that said FRIENDS OF COAL. The miners who filled the hospital's waiting room, some in wheelchairs or dragging oxygen tanks, wore gruff expressions. Some said they had heard about this "company doctor." After a day of exams, Fino defended his record, telling me: "I'll say it like it is. Eight to ten percent of all exams I do, I find them to be disabled due to black lung." When he got back to Pittsburgh, he checked his records and told me that, during the past eighteen months, he'd examined 305 miners at the request of coal companies and found total disability due at least in part to black lung in about 20 percent of them. If the records before judges didn't reflect that, he said, maybe the company's lawyers had decided not to submit his reports.

license to practice in Kentucky, though it wasn't related to his work in black lung cases; he was busted for doling out pain pills to people he'd never seen.

The ranks of claimant-side experts for hire, however, were relatively thin, largely because of simple market forces: Coal companies had money to pay experts; miners usually didn't. The doctors hired by coal companies typically charged in the ballpark of five hundred dollars an hour for reviewing records, writing reports, and testifying at depositions, and they might charge anywhere from a thousand dollars to two thousand or more to conduct an exam. A year of this part-time work could easily bring in hundreds of thousands of dollars.

Usually, miners relied primarily on the report from the exam paid for by the Labor Department. If they wanted additional reports, miners' representatives in West Virginia frequently turned to Dr. Don Rasmussen, who charged a flat fee of four hundred dollars for a report that could take hours to write. This was still a substantial sum for many miners, who were often afraid to spend the benefits payments that came from the Labor Department trust fund while coal companies appealed initial awards, worried they'd have to pay all of it back if they ultimately lost. Companies spent more on one CT scan reading by Wheeler than a miner received in benefits for a whole month.

This uneven contest of experts now began to play out in Caldwell's case. Maloy submitted reports from additional doctors in the usual stable. John tried to counter by getting reports from Bob Cohen (Dr. Bob), who had reviewed the evidence, and Dan Doyle, who had treated Caldwell at the New River clinic.

But John had another card to play. After consulting with the other Bob Cohen (Lawyer Bob), he filed a request for discovery. It was just a one-page letter, not the fine-tuned motion Cohen used, but it had much the same effect.

Maloy responded with the same legal argument that Cohen had been trying to pierce. Jackson Kelly acknowledged that, if the opposing side asked, the defense had to turn over the report written by the doctor it had chosen to examine the miner. Citing federal rules, the firm divided all other experts into two categories: testifying and non-testifying. The first included all doctors whose reports the firm had decided to submit for the record. The second included any other doctors the firm had consulted but whose reports it had decided not to submit.

Jackson Kelly's position was that the firm was obligated to turn over any other reports from testifying experts—though, again, only if the opposing side asked. In other words, if an expert had written multiple reports and the defense had submitted any one of them, the rest were fair game. Some of the documents John had sought in his recent discovery request fell into this category, so Maloy gave him an answer: There weren't any other reports from those three doctors.

The rest of the documents John wanted fell into the other category, and it was here where Jackson Kelly drew the battle line at which John and the firm's lawyers would fight for the next two decades. Any reports that might exist from non-testifying experts were privileged, the firm argued. It refused to turn them over.

Under Jackson Kelly's argument, these reports were *work product,* a sanctified term among lawyers. The doctrine protects attorneys from having to disclose materials they gather in anticipation of litigation, allowing them to get the candid advice and differing views they need to develop their case. But the protection is not absolute; the opposing side can force disclosure if it demonstrates a "substantial need" for the documents and an inability "without undue hardship to obtain the substantial equivalent of the materials by other means."

Administrative law judge Frederick Neusner found that John cleared that bar, and he ordered Jackson Kelly to turn over any other readings of the chest films that it might have. Maloy registered her continuing objection but complied. There was only one report, but it was exactly the one John had suspected might exist.

The author of the report was Dr. Jerome F. Wiot, a renowned radiologist at the University of Cincinnati. Jackson Kelly frequently used his readings, so John had been suspicious when there was nothing from him in Caldwell's case.

Wiot was perhaps the one person alive whose opinion could

trump Wheeler's. Wheeler himself had called Wiot "certainly the senior pneumoconiosis expert in this country right now, particularly on the East Coast or in the eastern United States." Jackson Kelly had argued in a recent case that "Dr. Wiot is clearly recognized as the leading radiologist in the interpretation of x-ray films for pneumoconiosis." Retired judge Edward Terhune Miller recalled, "The lawyers always presented him as the ultimate authority."

Wiot had been on the American College of Radiology's task force on pneumoconiosis since it began its seminal work on the disease at the federal government's request after the passage of the 1969 law. He had since held top leadership roles on the task force, and he had headed its educational program for years. When the government asked the American College of Pathologists to develop standards for diagnosing pneumoconiosis, the group needed a radiologist to help correlate X-ray and pathology findings, and it selected Wiot.

Wiot was notoriously conservative in his readings, but unlike Wheeler, he did find advanced black lung on occasion. That was exactly what he'd done in Caldwell's case. He had reviewed the X-rays and CT scan and found the same advanced stage of black lung that Caldwell's doctors had seen. "In summary, this patient's chest x-rays as well as the CT scan are consistent with complicated coal worker's pneumoconiosis," he'd written to the lawyers at Jackson Kelly.

A picture of Jackson Kelly's strategy was coming into focus. The doctor the firm had chosen to examine Caldwell had concluded he had complicated disease, and then their first-choice radiologist—Wiot, whose report was dated a month earlier than Wheeler's—had told them the same thing. Rather than rely on the experts it had many times before, the firm had withheld their opinions and appealed the initial award of benefits. Then it received reports from Wheeler, Fino, and others and used those instead.

From one perspective, there was nothing wrong with this. It was what lawyers did. They were supposed to be zealous advocates for their clients, to consult various experts and prepare a case that put the client's best foot forward. They weren't doctors; they didn't know who was right and who was wrong. They just assembled the strongest defense they could within the bounds of the law.

John and Lawyer Bob had a very different perspective: Jackson Kelly was playing a cynical game with miners' lives. The black lung–benefits system was not just another area of civil litigation. It was a remedial administrative program, set up by Congress with the specific purpose of compensating ailing miners. The system had its own rules, as the government recognized the distinct challenges in determining the presence and severity of the disease, the long history of denial and neglect the program was designed to correct, and the immense disadvantages miners faced in the legal arena with coal companies. It was not supposed to be a sort of legal free-for-all.

Even setting aside the particulars of the system, John and Bob didn't think the documents Jackson Kelly claimed it could withhold were covered under the work-product doctrine. The protection was commonly associated with things like lawyers' notes and correspondence with potential expert witnesses. What Jackson Kelly had were formal reports that were no different than the ones the firm had chosen to submit; the lawyers just didn't like what these said.

Whether Jackson Kelly's approach violated the law or not, John and Bob thought, it was unethical and wrong. They found Jackson Kelly's purported agnosticism disingenuous and its incomplete disclosures to the judge, the miner, and its own experts manipulative. If the firm wanted to argue that Wiot wasn't credible, fine. But the pulmonologists evaluating the full body of evidence—or what they likely believed was the full body of evidence—ought to have a chance to weigh his opinion for themselves. What would Fino have concluded if he'd been given Wiot's report?

Tellingly, Jackson Kelly apparently didn't want to find out. John had submitted Wiot's report for the official record, and if Maloy went back to Fino or any other pulmonologists for a report that addressed this newly disclosed evidence, she didn't submit it. Instead, as the date of the hearing approached, Maloy informed Judge Neusner that Hobet Mining had agreed to pay Caldwell's claim.

Charles Caldwell and his wife, Patsy, were relieved that the three-year ordeal was over. At times, Charles had gotten so disgusted with the process that he'd considered stopping, but Patsy had given him pep talks: "We done started this, and we're gonna finish it."

They had, but Charles didn't have much time to enjoy the victory. Not long after Jackson Kelly conceded the case, his health began deteriorating rapidly. He was coughing and hacking up black phlegm. His breathing grew shorter and more labored, and his doctor put him on oxygen. He was in and out of the hospital, and Patsy took care of him at home between stays.

He'd always liked to fish, and he'd bought a small boat after he retired. Now, he didn't have the breath to use it. One day in August 1999, his longtime fishing buddy Gene came over and sat next to the hospital bed Patsy had set up in the living room. Charles lifted his oxygen mask so he could talk—he sure would love some fresh fish. Gene left and came back later with his catch, already prepared for Charles. "Gene's wife, she fixed that fish, fried it for him, made biscuits, and brought it to him," Patsy recalled. "That was a Saturday, and he died on Monday morning."

Caldwell's case had taught John lessons he'd never forget: Be persistent, put everything in writing, and, most important, look for what's missing. "After this case," he said, "I was always thinking about what wasn't there."

He came to believe that his background—not being a trained lawyer—was in some ways beneficial. He hadn't been drilled in the sanctity of the work-product doctrine or the inviolability of certain procedural norms. It wasn't that he wanted to throw the law books out the window; it was just that he believed above all in commonsense notions of fair play, which didn't always map neatly onto legal doctrines.

He realized he didn't think about cases the way most lawyers did. He looked at them through the eyes of a carpenter, and he regarded that as an advantage. Later, while sipping coffee at a Bob Evans restaurant in Beckley with Arland Griffith and Daniel Richmond, who had been mainstays on Cline Building Corporation jobs and with whom he was still close, John explained his thinking.

"Arland and Daniel could come up to a house, and if you had even a rafter crooked, they could see it," John said. "It's sort of like we were talking about—see what's missing." Richmond nodded, dipping his hat. "They had instincts and vision," John continued. "Part of it's experience, but part of it was the way they thought."

John shared those traits, and he began applying them to black lung–benefits cases. The natural tendency was to look at the barrage of evidence coming from the coal company's lawyers and figure out how to counter it with evidence of your own. Partly out of temperament and partly out of necessity—countering the entire evidentiary onslaught wasn't feasible or affordable—John tried to step back and look at the bigger picture, see how everything fit together. He started creating simple charts. Each individual X-ray got a row; each doctor who had read at least one of them got a column—where were the blank cells? He spent hours mapping out what evidence was there, then thinking about what was not.

John's next major run-in with Jackson Kelly taught him a less encouraging lesson: Prying loose whatever the firm might be withholding depended not just on facts and arguments but also on luck—specifically, the judge to whom a case was assigned. To John's frustration over the years, not all judges would take the approach Neusner had in Caldwell's case.

That included administrative law judge George Fath, who oversaw the case of Calvin Cline (no relation to John). Cline, whom John agreed to represent in 1995, had worked more than thirty years underground, much of that time in "low coal," crawling through tunnels only two or three feet high. He'd retired fifteen years earlier because of worsening breathing problems, and doctors had diagnosed him with black lung. Since then, he'd filed two claims. Both times, a Labor Department claims examiner had awarded benefits but had been overruled by an administrative law judge. The second reversal—a denial by Fath—had come shortly before John took the case.

John promptly filed a motion for reconsideration and asked Fath to order Jackson Kelly to turn over any withheld materials. He again was suspicious of Wiot's conspicuous absence, and he said so in his motion. In her response, Jackson Kelly lawyer Ann Rembrandt didn't deny that the firm had consulted Wiot but said that whether or not it had done so was "irrelevant." John could not clear the high bar required for overriding the work-product doctrine, she argued.

Fath sided with Jackson Kelly and denied John's motion. John appealed to the Benefits Review Board. The board usually decided cases based on the written record, but Rembrandt requested an oral argument, citing the complexity and importance of the case. If the board sided with John, it could set a precedent that endangered the firm's litigation strategy. The board scheduled a hearing date.

John felt unprepared for that level of legal sparring, so he enlisted the help of Bob Cohen, who filed a supplemental brief and argued before the board.

Labor Department lawyers also came to John's aid. They filed a brief that bolstered the arguments John and Bob were making: The black lung–benefits system was never intended to be like high-stakes civil litigation. It had its own rules, which were established to accomplish the main goal Congress had outlined: "truth-seeking." Given this and the stark resource disparities between miners and coal companies, allowing lawyers to shop for evidence and withhold what they chose presented a particular risk of producing an unjust outcome.

"Clearly," the department's lawyers wrote, "the intent of the Act and regulation is to require submission of all available evidence, favorable and unfavorable, as early as possible in the proceeding so that a claim may be resolved quickly and correctly." They asked the board to send the case back to the judge with orders to compel the disclosure of all reports from both sides.

In its decision, the board staked out a middle ground that neither side found satisfactory. The three-judge panel acknowledged the particular rules that applied in benefits claims but found that they didn't require automatic disclosure of all evidence. The decision created the basic standard that would be applied going forward: it was up to the discretion of individual judges to decide whether to order disclosure of evidence. John had to hope he got lucky when cases were assigned.

In Cline's case, John did get a stroke of luck. The board remanded it for further consideration. While it had been on appeal, Fath had left the office; the case was assigned to Judge Clement Kichuk. Reevaluating John's discovery request, Kichuk wrote, "The importance of this issue in arriving at a just decision in this particular case cannot be overrated as it strikes at the very core of black lung benefit litigation." In creating the benefits system, Congress had stated that "all relevant evidence shall be considered," he noted, but withholding evidence might prevent that from happening.

The judge gave Jackson Kelly two weeks to turn over everything. John's hunch proved correct again. Wiot had read a series of Cline's X-rays as well as a CT scan thirteen years earlier, and he'd found that each was consistent with complicated pneumoconiosis. The case dragged on for years afterward, but Jackson Kelly eventually dropped its appeals. Cline won his claim twenty-eight years after he'd first filed, and he died two years later.

After the damaging forced disclosures in the Caldwell and Cline cases, Jackson Kelly seemed to modify its strategy, adding a wrinkle that could blunt efforts to uncover withheld documents. John encountered the approach while representing William Harris, who had retired after twenty-four years underground, filed a claim, and lost. When Harris tried again in 1997, John took his case and immediately filed a request for any medical evidence Jackson Kelly hadn't turned over in Harris's current or previous claims. Doug Smoot, a senior lawyer at the firm, responded with the usual claim of privilege.

Administrative law judge Lawrence Donnelly sided with John and ordered Jackson Kelly to turn over any medical reports it had withheld. This time, the firm didn't promptly comply as it had in the Caldwell and Cline cases. Maloy submitted a motion asking the judge to reconsider. Donnelly decided he'd settle the dispute at a hearing.

At the appointed hour on an April afternoon, John sat beside Harris in a room at the National Mine Academy in Beckley. Maloy was there to defend Harris's former employer, Westmoreland Coal Company. John had a difficult time containing his excitement; he had his own Perry Mason moment planned.

"We have found one piece of evidence that they've been withholding, and so, we're bringing it into the record today," he announced.

Maloy objected. "Your Honor, that is certainly very mysterious, but I don't know why it was withheld," she said. "It might have been withheld for a very sinister reason. It might have been withheld for a perfectly logical strategic reason."

John began ticking through the list of evidence he was introducing, finally arriving at the gem he'd unearthed.

The critical question in Harris's case, as in Caldwell's, was whether his X-rays and CT scan showed complicated pneumoconiosis. Jackson Kelly contended that he had the simple form of the disease and wasn't entitled to benefits because his breathing-test scores weren't bad enough for him to be considered totally disabled.

The case rested on a fine distinction. Defense experts had graded the X-rays as showing "coalescence," meaning the nodules had started to cluster together but had not formed a mass, which would be deemed complicated pneumoconiosis. Harris's experts had interpreted these same areas as evidence of complicated disease.

In Harris's previous claim, Wiot was among those who had seen coalescence, but he'd acknowledged during a deposition that it was a "close call" and that others might grade the films as showing complicated pneumoconiosis. "A CT scan can answer it in a minute," he'd testified. It would provide a sharper, more detailed image of the scarring.

Harris had undergone the scan and submitted reports by four doctors who had read it as complicated pneumoconiosis. Jackson Kelly, however, asked Wheeler to take a look, and he'd disagreed. The nodules were coalescent, not a mass, he'd said, before opining that they most likely weren't black lung at all. They were more consistent with healed TB or histoplasmosis.

That had been the deciding factor for Judge Jeffrey Tureck; he had determined that "Dr. Wheeler's opinion is entitled to the greatest weight. It is well reasoned and uncontradicted, and his expertise cannot be challenged. Accordingly, I find that the claimant does not have complicated pneumoconiosis, and his claim must be denied."

When John had first examined this record, he had wondered why Jackson Kelly hadn't gotten Wiot to read the CT scan. The firm had relied on his X-ray readings, and he'd testified that a CT scan could resolve the lingering uncertainty.

It seemed clear to John what must have happened, and he'd hoped the discovery request would confirm it. But he'd grown frustrated with Jackson Kelly's resistance to the judge's order, and with the hearing fast approaching, he'd tried a new tack. He sent Wiot's office a medical release of information form signed by Harris and a letter requesting a copy of the doctor's report on Harris's CT scan. John was guessing that such a report existed, and he was right. A few days later, he received the report that Wiot had sent Jackson Kelly four years earlier. Given the opportunity to evaluate the more detailed image, Wiot had changed his previous opinion and concluded that Harris had complicated pneumoconiosis. Once again, it appeared that Wiot had been the firm's first choice; his report was dated almost a month before Wheeler's.*

This was the report John now pulled out at the hearing before Donnelly. He would never forget the look of surprise on Maloy's face. She again objected, arguing that it was against the rules to spring new information on the opposing side at the last minute and that the firm deserved the opportunity to figure out how to respond.

"They've had four years to deal with responding to the substance of this report," John said. "They chose to withhold it."

As in Harris's previous case, John noted, the defense's consulting pulmonologist had relied on Wiot's X-ray readings without being informed that he'd changed his mind based on better-quality images. "This is a clear example of how this practice of withholding pertinent evidence skews the medical record in these cases and undermines the credibility of the medical reports and subverts the truth-seeking function of these proceedings," John argued.

John knew that the way he'd obtained Wiot's report wasn't a replicable

* Wiot died in 2010, so I was unable to speak with him.

gambit; this was probably the first and last time he could deploy it. In this case, though, it worked. Donnelly denied Jackson Kelly's request for reconsideration and again ordered the firm to turn over any other evidence it might have. That's when Maloy employed the strategy that would frustrate John's attempts to expose wrongdoing by the firm for years to come: She folded. A few weeks after the hearing, she sent a letter to Donnelly informing him that the company had agreed to pay the claim.

In Caldwell's case, Jackson Kelly had also folded, but that decision came *after* the firm had disclosed the withheld report. The firm's concession in Harris's case seemed to be a strategic decision of a different sort, John thought. By conceding, the firm could end the case and avoid complying with the judge's discovery order. Sure, Jackson Kelly and its client would lose *this* case, John reasoned, but the evidence that might reveal a larger pattern of wrongdoing would remain secret, leaving the firm free to deceive other miners and widows.

John didn't want to let Jackson Kelly off the hook, but he needed the permission of Harris and his wife to continue pushing. They gave it, and he filed a pair of requests. First, he asked Donnelly to force Jackson Kelly to disclose whatever else it had, which might include evidence that Harris was entitled to more back payments than Maloy had stipulated. Second, he asked the judge to impose sanctions that would deter the firm from withholding evidence in the future.

Donnelly refused, agreeing with Maloy that he no longer had any authority to do what John asked—by conceding the case, Jackson Kelly had taken it out of the judge's hands.

John still wasn't ready to quit. He filed a request for modification, which sent the case back to the initial stage before a Labor Department claims examiner. John again argued that Jackson Kelly should have to produce any withheld evidence. The claims examiner agreed and gave Maloy thirty days to comply.

Once again, Maloy folded rather than show the firm's hand. She said that Westmoreland Coal would pay benefits dating all the way back to 1991, when the CT scan that Wiot had interpreted as complicated pneumoconiosis was performed.

Now John was more suspicious than ever. Maloy was offering years of additional accumulated payments beyond what a judge could have awarded Harris under the system's regulations. John viewed it as an admission of wrongdoing, a payoff to make the whole thing go away. He wanted to continue pressing for sanctions.

To do that, though, he again needed Harris's permission. He talked through the options with Harris and his wife, and they decided they'd had enough. John was disappointed, but he could hardly blame them. They'd been fighting for twenty years, and Harris's health was deteriorating. Continuing to fight could jeopardize the additional payments Maloy had offered. The Harrises took the deal. John never found out what else Jackson Kelly might have been hiding.

John tried to convince lawyers it was worthwhile to wage discovery battles against Jackson Kelly, raising the issue periodically at conferences and meetings. But he didn't make much headway. Even among miners' lawyers, there was some uneasiness about the idea of having to reveal everything. A miner's medical records might include a shoddy report by a poorly qualified physician written before the miner had been able to find an attorney, and having to hand it over might muddy the key issues and hurt the case.

There were also practical obstacles. John was a salaried employee at the New River clinic; he didn't depend on fees collected after winning a case. Because of the benefits system's rules, lay representatives often received little or nothing in fees, and if John was paid anything for his time and expenses, it went to the clinic, not to him personally. Lawyers in private practice, however, relied on the fees, and lengthy discovery fights could bankrupt them.

Uncovering withheld evidence might not even be very helpful in some cases. Concealed reports could be vital in cases of miners with complicated pneumoconiosis, and John seemed to have a disproportionate number of these. Some lawyers, such as Tom Johnson in Chicago, had more cases in which the key issue was not what X-rays or CT scans showed but rather the severity or cause of the miner's disability. For them, withheld reports were less likely to alter the case's outcome. "Also," Johnson said of John, "he's just more tenacious than we are."

Cohen shared John's commitment to discovery fights, but he had a somewhat different perspective. "John was not as jaded as I was," Cohen said. "I expected this. John was always saying, 'How can they do this? It's not right.'" Cohen was focused on fighting the firm case by case. "There was no grand strategy," he said.

The community organizer in John thought in terms of causes and policy changes, and that's how he viewed the benefits system. The status quo was totally inadequate to protect miners, he thought. In most claims, there was little to check Jackson Kelly's behavior. Even if a miner found a lawyer who was willing to pursue discovery, and even if a judge ordered disclosure, the firm could still just concede the case. John thought the only solution was mandatory disclosure of all medical evidence by both sides, regardless of whether anyone asked for it.

He was convinced that what he'd seen in the cases he'd handled was part of a pattern of practice that was harming many miners, most of whom likely had no idea what had happened. He wanted to make it stop. He just didn't yet know how.

17

"All of the Evidence"

"Right-upper-lobe chest mass consistent with lung cancer."

The doctor's words reverberated in the minds of Mary and Terri as they watched the nurses wheel Gary away.

Over the past few months, Gary's coughing spells had gotten worse. Doctors had diagnosed him with pneumonia, then ordered an X-ray, and there it was: a lesion on his right lung. He needed part of his lung removed, they said. Now, on September 25, 1998, Mary and Terri sat in Raleigh General Hospital waiting, worrying, hoping the doctors' suspicions were wrong.

Gary lay on his left side in the operating suite, anesthetized, a tube down his throat. The surgeon made a curved incision and inserted metal instruments to separate the ribs and hold back the flaps of tissue. He located the mass and cut out a piece of Gary's lung where it had formed.

For eight days, Gary remained in a hospital bed, a tube in his chest. On the third day, the report from the pathologist, Dr. Gerald Koh, arrived. It was not cancer. Relief rushed over Mary and Terri. It was unclear, though, just what the mass *was*. Koh noted in his report that the tissue contained "anthracotic deposits," which just meant there was an accumulation of carbon. Someone who lived in an area with polluted air—a place like West Virginia, where more than 90 percent of power came from burning coal—might have such an accumulation just from going about daily life, and it wasn't necessarily cause for concern. There was no evidence that Koh knew Gary was a longtime coal miner.

Koh's final diagnosis was "inflammatory pseudotumor," a term for a benign lesion. *Tumor* was the word that stuck with the Fox family, and it was how they would describe the mass for years to come.

Soon, Gary returned to work at Elk Run, but he was thinking about getting out. His breathing had become more labored. Just going from the mouth of the mine to the bathhouse, he had to stop at least once to catch his breath. His scores on lung-function tests had worsened, and the state workers' compensation system considered him 15 percent disabled.

A month before his surgery, Gary had picked up form CM-911, a four-page grid of personal questions, checkboxes, and bureaucratese titled "Miner's Claim for Benefits Under the Black Lung Benefits Act," along with a two-page "Employment History" supplement. He'd filled them out but then set them aside.

As winter melted into the spring of 1999, he thought about filing them, and he discussed the possibility with Mary. She immediately supported the idea. She'd wanted him to get out of the mines for years. The fact that he was willing to consider it meant he must really be sick.

In May 1999, Gary sent the paperwork to the Labor Department office in Beckley, which then sent him a questionnaire. *Has your respiratory condition required you to put forth more effort to perform your job?* "Yes." *Do you miss work due to your respiratory condition?* "No." *Why do you continue to work?* "Because I have a family to support and I'm trying to get a daughter through college."

Terri had one year left at West Virginia University. By the time she graduated, the benefits claim ought to have come through, and Gary would be able to retire. For the moment, Gary and Mary hid from Terri how sick he was becoming. If Terri had known, she would have insisted on taking out student loans. "And I think that's why they never told me," she later said.

Gary selected Dr. Don Rasmussen to perform the exam paid for by the Labor Department. At his lab in Beckley, the revered coalfield physician, now seventy-one, ran Gary through the same tests he'd performed on generations of miners. Gary did have some breathing impairment, he determined, but not enough for him to be considered totally disabled under the benefits system. His X-ray, however, indicated complicated pneumoconiosis, category B.

If Rasmussen was correct, it was critical that Gary stop breathing the dusty mine air immediately. A miner with this advanced stage of disease might well become increasingly impaired even if he never breathed another speck of dust, but if the miner got out of the dust, he had at least a chance of avoiding the worst stages of the disease.

Based on Rasmussen's report and Gary's medical records, the Labor Department claims examiner issued an initial determination that he was entitled to benefits. Elk Run had a choice: It could begin paying Gary and Mary $704.30 a month plus medical expenses related to black lung, or it could contest the claim. It chose the latter and turned to its longtime defender.

The connection between Massey and Jackson Kelly ran deep. After all, if Don Blankenship's warning to a rival was true—"We spend a million dollars a month on attorneys"—the company had plenty of use for legal counsel, and Jackson Kelly was, as the firm's CEO later put it, "the coal industry's outside 'general counsel.'" In 2000, as Gary's case played out, a Jackson Kelly lawyer who specialized in defending energy companies against environmental lawsuits joined Massey as a top in-house lawyer, eventually becoming general counsel. Later, when residents of a small West Virginia community sued Massey, alleging it had dumped waste that fouled their water supply, Jackson Kelly defended the company and negotiated a settlement. And when Massey faced a class-action lawsuit alleging its coal silo perched above an elementary school had blanketed the building and the students' lungs with coal dust, Jackson Kelly lawyers convinced a jury to rule in the company's favor.

Naturally, Jackson Kelly was Massey's go-to firm for federal black lung–benefits cases too. In Gary's claim, senior lawyer Doug Smoot took the lead.

Gary was required to undergo an exam by a company-chosen doctor, and Smoot selected Dr. James Castle, a pulmonologist often enlisted by defense firms. Gary drove eighty-five miles southeast to Radford, just across the border in Virginia, for another round of poking and prodding. Based on the testing, Castle arrived at a conclusion similar to Rasmussen's regarding Gary's breathing impairment: the scores were lower than normal but not enough to qualify Gary for total disability. He differed, however, on what the X-ray showed. Castle acknowledged that Gary "possibly" had pneumoconiosis, but not the complicated form. Together, these findings, if credited by a judge, would amount to a loss for Gary.

Smoot, with help from colleagues Bill Mattingly and Mary Rich Maloy, began building the evidentiary pyramid. First up: Dr. Paul Wheeler and two of his colleagues in the Pneumoconiosis Section at Johns Hopkins. All three graded Gary's X-rays as negative for pneumoconiosis. They saw masses in his lungs but deemed other causes more likely—"inflammatory disease," "possible cancer," "old healed TB."

Smoot forwarded these reports to two more pulmonologists. One, Dr. Abdul Kader Dahhan, reached the same conclusion as Castle: simple pneumoconiosis, but no total disability. The other, Dr. Gregory Fino, went further: no pneumoconiosis at all and no breathing impairment. Then he went back to the trio at Hopkins with another X-ray for them to read. They said it didn't show pneumoconiosis; the lesions were more likely cancer or possibly TB or another granulomatous disease. Then all that went to a fourth pulmonologist, Dr. Kirk Hippensteel. His assessment: no complicated pneumoconiosis, no total disability.

Unusually for these cases, cigarettes weren't an available scapegoat. Gary had smoked for just a few years, while in his twenties.

Also unusually for these cases, the record included a pathology report. Among the different types of medical evidence used in a black lung–benefits case, pathology was considered the most useful in determining the cause and severity of a miner's disease. X-rays and CT scans were shadows on film; the pathology specimens were the actual lung tissue that cast the shadows. There was still room for significant disagreement among pathologists examining the same sample, but if there was a consensus, it would trump the radiographic evidence.

In Gary's case, it appeared there was such a consensus. Koh's finding of "inflammatory pseudotumor" was the sole pathology report, and the lawyers at Jackson Kelly built their defense around it.

On August 30, 2000, Smoot took Castle's deposition. Under questioning by the seasoned lawyer, Castle said he no longer believed Gary had even simple pneumoconiosis. "When I read the X-ray at the time, after I examined him, I felt that it did show changes consistent with that," he testified, "but I will have to tell you that based on the pathology from this case, there isn't any evidence of coal workers' pneumoconiosis there. These other nodular densities are most likely related to his pseudotumor and benign process that are present."

Smoot then used the pathology findings to discredit Rasmussen, who hadn't been given Koh's report when he examined Gary. "Do you think that Dr. Rasmussen would have been aided by having all of the biopsy medical evidence at his hand when he reviewed this case?" Smoot asked.

"I think that he would have," Castle replied, "and I would certainly hope so, because all of the evidence, as I've outlined, clearly indicates that this is not complicated disease."

One week later, Wheeler sat in the Johns Hopkins radiology department and answered questions from Mattingly as a court reporter took it all down. Reciting the criteria he had applied so many times before, the doctor explained why he didn't believe the masses in Gary's lungs represented pneumoconiosis. They weren't in the right location, and there weren't enough smaller spots in the background.

Mattingly then directed the doctor's attention to the pathology. Since reading the X-rays, Wheeler had been given Koh's report.

"Is it fair to say that the presentation on the chest X-rays as you've interpreted them is very consistent with the pathology reports inasmuch as they rule out the presence of a coal-mine-dust-induced lung disease?" Mattingly asked.

"Yes," Wheeler answered.

Moments later, Mattingly returned to the point, hammering it home: "Does knowledge of the pathology results also call into doubt any classification of this as a complicated pneumoconiosis or massive lesions consistent with exposure to coal-mine dust?"

"Yes," Wheeler again answered. Koh had ruled out cancer, and if pneumoconiosis was the cause, presumably he would have noted it, Wheeler reasoned. The logical explanation was that whatever inflammatory process—an infection, an autoimmune condition, or something else—had caused that mass had also caused the other lesions on the X-rays.

The radiology and the pathology agreed, Wheeler concluded: "In no way is this pneumoconiosis."

Everything, it seemed, depended on the story told by the small slices of Gary's lung tissue now frozen and tucked away in a lab and the one report that interpreted them—a report with shortcomings that would become clear only years later.

"This proceeding will come to order." Judge Edward Terhune Miller's voice echoed off the oak-paneled walls lining the courtroom within the federal building abutting Main Street in downtown Beckley. "This is a formal hearing of the claim for compensation under the Black Lung Benefits Act of Gary Fox."

Jackson Kelly lawyer Mary Rich Maloy sat at one table. Gary and Mary sat at the other.

"You've not been able to find an attorney, is that correct?" Miller asked.

"Yes, sir," Gary replied.

"You've made a significant and diligent effort to find an attorney?"

"Yes, I did."

"That being the case, are you willing and desirous of going forward with the hearing today without an attorney?"

"Yes, sir, I am."

Moments later, Miller continued: "I think we established you have a high-school education. You can read and write."

"Yes, sir."

"Have you ever been to a hearing like this before?"

"No, sir."

"You've never handled any kind of a legal hearing?"

"No."

For the past few months and right up to the date of this hearing, September 19, 2000, Gary and Mary had been calling around, trying to find someone to take a look at Gary's case. But no one would. Everyone they'd talked to had shunted them off upon hearing it was a federal black lung–benefits claim. Now they felt nervous and alone. They'd hoped Gary could just state his case and the judge would see the truth from the evidence. Lots of miners around here had filed claims, and many of them didn't have lawyers either. It wasn't like they were asking for much, just a few hundred dollars a month and coverage for the inevitable medical complications. Surely, they thought, the government had tailored the system accordingly.

That illusion quickly faded as Maloy began a recitation of contested issues and regulatory provisions.

Miller was sympathetic to the plight of miners like Gary. During hearings, he tried to make sure unrepresented miners got their evidence into the record, and he'd developed a list of questions to ask them to elicit key facts. He went through them with Gary—job, manual labor required, dust exposure ("Very constant"), work hours ("We run coal every day").

Any breathing problems? "I get tightness in my chest. If I try to walk or something like that, you know, I have to stop and rest." He used three different inhalers and slept with his head elevated by a thick pillow.

"Any hospitalizations recently?"

"Just a couple of years ago I had lung surgery."

"All right. And that was for the removal of a—"

"A tumor."

Maloy asked a few perfunctory questions about other medical conditions; Miller had a few more; and that was it. "This completes your portion of the case, Mr. Fox," the judge said.

After Maloy finished introducing fourteen exhibits into evidence, Miller turned to Fox: "Is it fair to assume that you're not going to undertake to have any responsive evidence? You don't have any desire to go to any other doctors to get an opinion?"

"No, sir."

Fifty minutes after it began, Gary's day in court was over. There were no dramatic closing arguments, just procedural tedium.

Maloy placed Koh's finding of an inflammatory pseudotumor at the heart of her written closing argument. "This lesion is not cancerous, and is *not* the result of coal dust exposure," she wrote. Unlike Rasmussen, the defense's expert pulmonologists had "considered all of the x-ray readings as well as the pathology... and unequivocally concluded that complicated pneumoconiosis is *absent*."

That, she wrote, was the inescapable conclusion based on "the most credible evidence."

When Judge Miller thought a case had merit, he worked with his clerks to assemble the evidence into a decision awarding benefits, then evaluated whether it stood up to scrutiny. The coal company's lawyers would seize on even the slightest oversights or mischaracterizations, and there was a good chance that the Benefits Review Board, which some judges considered overly conservative, would vacate the decision. To his frustration, Miller had felt legally compelled to write many denials that had started out as attempts to write awards.

"To come out with a fair result is a very tough problem," he later said. He was keenly aware of the resource disparities and lack of legal representation that tilted the playing field. "Justice depends on some recognition of the disparity," he said. "But if the miner, despite his disadvantages, simply hasn't got the evidence, then you've got nothing to work with."

Gary Fox didn't have the evidence, Miller concluded in a decision issued January 5, 2001. First, the X-rays: twenty-nine of the thirty-six readings were negative for pneumoconiosis. The judge noted the "superior credentials" of Wheeler and his two colleagues at Johns Hopkins and determined that, based on readings by "the majority of the best-qualified readers," the films did not show pneumoconiosis. There was just one report on the pathology, and it, too, did not show the disease. Because Rasmussen hadn't been able to consider all of the X-ray readings or the pathology report, his opinion carried less weight, Miller wrote. Jackson

Kelly's pulmonologists had considered most or all of this additional evidence, and their opinions were "well documented and based on comprehensive reviews of all the evidence of record."

The consensus that the X-rays were negative for pneumoconiosis was convincing, the judge wrote. "More persuasive, however, are the pathology results which proved the large mass in the miner's right lung to be a pseudotumor and neither cancer nor complicated pneumoconiosis."

Miller recognized that the case had hardly been a fair fight, but that unfairness was baked into the system as it currently existed; there were legal strictures he couldn't flout or alter even if he wanted to. He had no reason to think that what had unfolded was a reflection of anything more than the inherent asymmetry of the system. The lawyers at Jackson Kelly had appeared before him frequently. "My sense was that they were honorable and competent," he said. "There was no reason to smell a rat."

Gary and Mary were stunned. It had seemed so obvious that Gary had black lung. Rasmussen and the medical board for the state workers' compensation system thought he did.

It was as if the defense experts had been groping for anything that might explain the constellation of test results and symptoms, and when the attorneys at Jackson Kelly had given them the vague pathology report, they'd all latched onto "inflammatory pseudotumor." Gary and Mary wondered what they should have done differently. Should Gary have tried to get time off work to cross-examine Castle and Wheeler? Even trained lawyers often failed to wring concessions from these doctors during depositions.

Should they have hired more doctors of their own to write reports? How would they have known where to look and whom to ask? Could they have afforded it? Engaging in a legal arms race with the region's biggest coal company hardly seemed worthwhile.

"If we knew then what we know now," Mary later said, "yeah, we would have gotten details down to a pinpoint, probably. You just don't think of it at the time. We weren't lawyers. Didn't know anything about the law. We had never been in a courtroom before."

They hadn't told Terri the details of Gary's claim, nor did they tell her how it ended. She had finished college while the claim was ongoing, emerging with a nursing degree and zero debt, and she'd started working in the pediatric intensive care unit at the West Virginia University medical system's flagship hospital in Morgantown. It was everything that Gary had wanted for her and more. His seemingly insuperable pride in his daughter rose even higher.

But the family now faced a new challenge. Mary had been diagnosed with a chronic disease affecting part of the digestive tract, and she needed regular, expensive medical care. Going without health insurance was not an option.

The way Gary saw it, there was only one thing to do: Keep running coal for Don Blankenship.

Part Three

I went to the doctor, couldn't hardly get my breath
Went to the doctor, couldn't hardly get my breath
The doctor said you got something that could well mean
    your death.
With this pneumoconiosis, the black lung blues
Pneumoconiosis, the black lung blues
If you got one, you got the other, any way you lose.

I've always been a miner, breathed coal dust all my life
I've always been a miner, breathed coal dust all my life
Too old to learn a new trade, what can I tell my wife?
With this pneumoconiosis, the black lung blues
Pneumoconiosis, the black lung blues
If you got one, you got the other, any way you lose.

I'll tell nobody nothin', I'll just keep workin' on
I'll tell nobody nothin', I'll just keep workin' on
The kids need all their schoolin' before I'm dead and gone.
With this pneumoconiosis, the black lung blues
Pneumoconiosis, the black lung blues
If you got one, you got the other, any way you lose.

When I get to Heaven, St. Peter's going to cry
When I get to Heaven, St. Peter's going to cry
When I tell him the reason this poor boy had to die.
With this pneumoconiosis, the black lung blues

Pneumoconiosis, the black lung blues
I'll lay down my pick and shovel, lose these black lung blues.

—Mike Paxton, "Black Lung Blues"

By the spring of 2013, getting in and out of my cubicle required a minor feat of acrobatics. Overflowing cardboard boxes covered most surfaces, and coffee-stained papers and sticky notes competed for the remaining real estate. My fellow reporters at the Center for Public Integrity sometimes asked me what I was working on, and though they always listened courteously to my jumbled responses, I imagined they saw a colleague slowly descending into a rabbit hole.

Almost a year had passed since I'd first met John Cline. The complaints I'd heard from miners about the benefits system had piqued my interest, but the whole program seemed dauntingly complex. I was afraid it was a morass in which I would get stuck for months, then emerge with a story that shed little light on anything and was quickly forgotten.

Still, my nagging curiosity had convinced me to call John. When we'd talked, he had mentioned a grave injustice he believed he'd uncovered: the leading law firm retained by coal companies to defend benefits claims, he said, was undermining the system by withholding critical evidence. I was skeptical. Of course an advocate for a vulnerable and sympathetic client would see wrongdoing in the type of aggressive defense often mounted by corporate attorneys, even if the behavior was legal, albeit somewhat distasteful.

Nonetheless, I began learning more about how the benefits system worked so I could evaluate John's claims for myself. The more I dug, the more his allegations started to seem, at the least, plausible. The stakes, I realized, were high; if miners really were being wrongfully denied benefits, then many of them were being abandoned—relegated to years of financial hardship, forgone medical care, and immense suffering.

I decided I would try to find out whether there really was a pattern of withholding evidence by the lawyers at Jackson Kelly. I soon determined that this would be more difficult than I had anticipated. The final decisions by judges in the benefits system are public, but everything else in the case file is protected by privacy laws. There would be no heading to the courthouse to copy key documents; I couldn't even obtain them under the Freedom of Information Act. There was, however, another way to see them: I could review the full files if the miners or their surviving family members gave me explicit written permission.

Over the following months, I identified cases of potential interest and contacted miners, widows, and surviving children. Most of them gladly filled out and signed a form granting me access to the files. Some said they felt wronged by the system and by the firm and were happy someone was looking into it.

I submitted these signed forms to the Labor Department, and before long, I was carting stacks of cardboard boxes labeled with miners' names out of the agency's headquarters. Back at the office, I began going through the files and logging their contents. After a few months, I had created timelines for each of the fifteen cases I'd gotten. Among them were the claims of Charles Caldwell, Calvin Cline, and William Harris.

As I read through the files, certain trends became apparent. The most notable was the ubiquity of Dr. Paul Wheeler and the unit of radiologists he led at Johns Hopkins. I hadn't expected to see the letterhead of one of the world's most prestigious medical institutions at the top of reports submitted by coal-company lawyers in case after case. When I mentioned Wheeler and his colleagues to miners' advocates, the response was often a resigned sigh. Wheeler's predilections seemed clear, but despite some admirable efforts by miners' lawyers, attempts to discredit his opinions had been mostly futile. In many other cases, it appeared, his proclamations went largely unchallenged.

It certainly seemed as if he never saw black lung on a film, but what, I wondered, was his actual record? Surely there must be some sort of database somewhere with the answer. I asked various people and groups, including the Labor Department, and learned that no one was keeping a record of doctors' opinions in benefits cases—no such database existed. So I decided to build one myself.

Fortunately, the Labor Department posted on its website decisions by administrative law judges dating back to 2000 in searchable PDF format. In reaching their decisions, judges discussed the medical evidence in the record, usually mentioning doctors by name. Searches for variations of Wheeler's name ultimately yielded more than fifteen hundred cases over the previous thirteen years in which he'd interpreted at least one X-ray. Unfortunately, different judges presented the medical evidence in different formats, meaning there was no way to put together a database other than by going through each decision, finding the X-ray readings, and logging them by hand in a spreadsheet. Because Wheeler had interpreted multiple films in many cases, this meant creating a record for each of the thirty-four hundred readings he'd provided.

The method was crude and inefficient, but it seemed to me to be the only way. I began the months-long process of slowly building the spreadsheet—the digital analogue of the files engulfing my cubicle.

As I gradually pieced together whatever story the records would tell, I also traveled to West Virginia to meet miners and their families. In kitchens and living rooms, libraries and diners, I heard their stories. Back in Washington, DC, I cleared a spot on my desk for a lump of coal I'd scooped up from the side of the road during one trip. I began listening to old folk songs about the mines and reading books about the Mine Wars, the Farmington disaster, the 1969 wildcat strikes, and more. I had disappeared completely down the rabbit hole.

Amid all that I was learning, though, my attention kept returning to one thing. One of the timelines I'd built lacked a final resolution date. The ongoing case came to occupy my thoughts, as it clearly did John's, and called to mind something John had said when we'd first talked: "I've never encountered anything quite like the Gary Fox case."

18

"Creative Approaches"

For one of the few times in his life, John Cline felt out of place. It was the fall of 1999, and at age fifty-three, he was a first-year law student at West Virginia University. Many of his classmates were half his age and had their sights on plum jobs at prominent firms, maybe even on political posts. John never entertained the possibility of doing anything other than returning to the representation of miners in black lung–benefits claims.

In the past, John had been too restless to finish his bachelor's degree. He'd dropped out of Allegheny College in 1965 and earned an associate's degree from a two-year school in 1967, and since then he'd taken an occasional class or two but never stuck with it. When he was in a classroom, his attention often meandered to social observation—the professor's teaching style, the interaction of the other students—and away from the topic at hand. He'd wonder whether the material would be of any practical use. He'd sometimes wish he were back building houses.

But in the law—or at least the nether regions of it occupied by the black lung–benefits system—he'd found a subject that held his attention. "For the first time in my life," he'd written in his law-school application, "I really want to go to school."

The seed of the idea had been there since his youth. When he stood before a federal judge as a twenty-two-year-old and refused to serve in

Vietnam, it had been comforting to have a lawyer from the American Civil Liberties Union there standing beside him. He'd thought from time to time about his family's belief that knowledge could be used as a tool for good. He'd also thought of his great-grandfather who, after overcoming a hardscrabble childhood and wounds suffered while fighting for the Union during the Civil War, had become a respected lawyer and dean of the law school at Syracuse University.

As John had taken on more and more cases as a lay representative while at the New River clinic, he realized he'd stumbled into an area where there was a great need he could help meet. Keenly aware of both his abilities and his shortcomings, he didn't envision himself as a great legal mind or fierce litigator, but he felt that he nonetheless had something to offer.

His wife, Tammy, believed that too. She'd seen the all-nighters working on briefs, the indignation John unleashed when a miner lost, and the muted joy he couldn't hide when a miner won. She, too, believed he was uncovering something important, working toward something bigger. "There's such a camaraderie that miners have with each other," she said. "He developed that, and he was part of that, even though he hadn't worked in the coal mines. It was his passion to help, to win cases. I think it was finally where he found what he wanted to do, deep in his heart."

If John really was going to devote himself completely to handling black lung–benefits cases, though, he almost certainly couldn't do it without a law degree. The benefits system's rules allowed judges to order companies to pay fees to the miner's attorney in a successful claim, but the same was not usually true for lay representatives. As a result, lay representatives often collected little or no remuneration. While at the New River clinic, John had received a salary, and any fees he'd gotten went to the clinic.

John had always been content with frugality. But now he was in his fifties, and he figured that in fifteen years, when many of his contemporaries were retiring, he'd be looking at college tuition payments for his son Brooks. "I had never made more than, I would say, twenty-four thousand dollars a year in my life. I didn't have any retirement," he recalled, laughing. "So I thought maybe it would be a good idea to have something that I could do for longer."

In the past, John occasionally thought about going to law school, but he'd always dismissed it—he had to work; he had a baby to help raise; he hadn't even finished his undergraduate degree; the whole idea was crazy.

This time, Tammy had supplied much of the inspiration and encouragement that overcame his initial reluctance. She'd done something similar herself. Even while she was pregnant with Brooks, she started earning an advanced degree as a nurse practitioner. She'd worked, raised a newborn, and still finished her studies.

Buoyed by the discovery that a nearby college would accept many of the credits he'd amassed over the years, John picked up a few more while still working at the New River clinic. Thirty-five years after starting college, he earned a bachelor's degree.

Though his going to law school was sure to put a strain on their family life and finances, Tammy supported it. "You should do it," she'd told him. "Those years will fly by."

He applied to West Virginia University but assessed his chances for admission as iffy at best, given his underwhelming grades and score on the Law School Admission Test. What he lacked in academic credentials, however, he made up for with a decade of experience in the benefits system that had earned him admirers. In a letter of recommendation, attorney Bob Cohen wrote that John was "without doubt the best lay representative I have ever known." In a nod to John's unorthodox scrapping with Jackson Kelly, Cohen added, "Not only are his legal theories well grounded in the facts of the case, but he also takes creative approaches which more often than not are meritorious."

One of the administrative law judges who had seen John in action, Daniel Sutton, also urged the law school to admit him, praising his intellect and dedication and noting that "there is a disturbingly short supply of legal representation for the economically disadvantaged segment of the population for whom he is currently working."

John figured the glowing recommendations had nudged him from the margins to just inside the ranks of the admitted. Whatever it was, he was here now.

So, too, was his oldest son, Jesse, now twenty-seven years old. After dropping out of college in South Carolina and returning to West Virginia to become a raft guide, he had eventually earned a degree from West Virginia University. Unlike John, he'd scored high on the LSAT, and the intricacies of case law and legal reasoning came easily to him.

Father and son had different schedules and didn't study together much. John settled into a taxing routine: Every Monday morning, he left Piney View and drove three hours north to Morgantown, where he spent the next five days living in an elderly woman's basement. Every Friday evening, he drove three hours back home. His Subaru station wagon got a workout.

His classes exposed him to many areas of the law, some of which he found distasteful. He especially disliked a course on civil procedure; high-dollar lawsuits held no allure for him. Nonetheless, he was learning skills that would serve him well. Reminders of his reason for being here were all around him.

Jackson Kelly loomed large on the WVU campus. John sometimes walked past a conference room made possible by a donation from the firm. Students vied for the Jackson Kelly and John McClaugherty Law Scholarship, and they jockeyed for one of the coveted spots as a summer associate for the firm. There was a long-standing pipeline of top graduates to Jackson Kelly.

Bill Mattingly, one of the leaders of the firm's federal black lung unit, orchestrated coal companies' defenses from a nearby office tower, and he regularly taught a course at the law school. (John did not enroll in his class.) Another of the unit's lawyers, Kathy Snyder, was John's professor for appellate advocacy. John liked her and found her to be fair and accommodating in the classroom, though he bristled at some of her descriptions of legal principles.

John tried to tailor his education to his interests. For an independent study, he wrote a paper tracing the history of black lung law and critiquing its current inequities. John often played down his own academic abilities, but the professor who oversaw his independent study and taught him constitutional law, Bob Bastress, found him conscientious and sharp. "John has always been a genuinely modest individual," he said. "I wouldn't underestimate his intellectual capacity. He's sort of a mensch. He's unassuming and down-to-earth. He's like somebody who worked hard in the southern coalfields."

John sought the opinions of others on what he had uncovered about Jackson Kelly's withholding of evidence, curious how people steeped in the law would see it. Who better to ask, he figured, than his ethics professor Forest Bowman, a man who had taught at the law school for more than twenty years, given ethics seminars across the country, and held key positions on state disciplinary panels and the state bar association's ethics committee?

John scheduled an appointment with him and lugged a stack of papers—key documents from the Caldwell and Harris cases—to the professor's office. John laid out the facts he'd discovered, but Bowman seemed to shrug them off. John was miffed. Maybe he wasn't explaining them well. Maybe the professor didn't want to weigh in on something that might be ongoing and controversial, or perhaps he wanted to reserve judgment until he learned more. Maybe Bowman's ambivalence came from a principle of the law John didn't understand or a legal mind-set he didn't share. Still, as he left the professor's office, he couldn't help but wonder whether the muted reaction had something to do with the sign on the door: JACKSON & KELLY PROFESSOR OF LAW.*

This was not the last time Bowman would be asked to assess the firm's conduct. The next time, the setting was much more public and the stakes much higher.

* * *

---

* When I later asked Bowman about this meeting, he said he didn't recall it but added: "I remember John Cline. Very honest follow. I don't remember that, but he was an upstanding guy and if he says it happened, I have no reason to doubt that."

Though John now had far less time available, he remained involved in black lung advocacy during law school. Before starting, he'd helped secure government funding to hire and train a replacement at the New River clinic, and he'd helped assemble the training curriculum.

John also continued work on a few ongoing benefits cases during law school. One of them was that of William Harris. Jackson Kelly attorney Mary Rich Maloy had made her first attempt to concede the case shortly before John started law school. That was when he'd tried to keep the case alive to force her to turn over any withheld materials, filing a discovery request with the Labor Department district director. As John began his second semester, the official had granted his request.

By then, he had spent some time in the library reading case law. He'd scribbled in the margins of the Labor Department order the name *Great Coastal Express*. It was a reference to a 1982 decision by a federal appeals court, one of the few that addressed the relatively obscure legal term *fraud on the court*. This offense was more than mere fraud among the parties. It was, as the court put it, "a deliberate scheme to directly subvert the judicial process." If a party learned that an opponent in an already concluded case had obtained a favorable verdict through such a scheme, a finding of fraud on the court would invalidate the earlier decision and permit the cheated party to try again, which otherwise would not have been allowed.

John thought the doctrine held potential for miners. Jackson Kelly's withholding of evidence seemed to fit within the description of fraud on the court. The firm's lawyers hadn't deceived miners alone; they had deceived their own consulting physicians and, ultimately, judges. If he could persuade a judge of that, it might invalidate the denial in William Harris's previous claim and allow for years of additional payments for benefits that should have been awarded in the first place. A declaration of fraud on the court by a judge would also be the sort of stark finding that might create other legal and public relations problems for the firm, maybe even force it to change its practices.

But soon Maloy made her second and more lucrative offer, which provided Harris more back payments than John could have secured even if a fraud-on-the-court gambit worked. Harris's understandable decision to take the deal meant John didn't get a chance to try the strategy. He would have to wait for the right facts, the right judge, and, above all, the right miner.

19

"Go Get 'Em"

By 2002, John had a law degree, but his financial situation had hardly improved. He had more than fifty thousand dollars in student loans to pay off and virtually no money coming in. While in school, he hadn't been able to take new claims, so there was nothing in the pipeline. Even if he now accepted a prodigious caseload, it could be years before he saw any fees. Nor was he a licensed attorney yet. He continued as a lay representative until 2005, when he passed the bar exam on his third attempt.

Meanwhile, the impulse he sometimes called "the itch to build" returned. He and Tammy wanted a house that was a little larger and more private than what they had. They chose a 1.67-acre plot nearby in Piney View, and John drew up the plans. They hired a few subcontractors, but John and his two older sons, Jesse and Matthew, did most of the work. By March 2003, the 2,100-square-foot home was complete.

On the second floor—the law office of John Cline, soon-to-be Esquire (a term he never liked even after he'd earned it)—John began intensifying his efforts to force Jackson Kelly to change its tactics. He still wasn't sure how to do that, but he had a few ideas. If he exposed more examples of what he'd found in his earlier cases, maybe it would finally reach critical mass. Maybe he could convince a judge not to let the firm off the hook after it conceded a case to avoid disclosure; maybe he could even get a judge to issue sanctions. Continuing to press case by case might be a long shot, but it seemed like the only option. Given the current administration and Congress, it was laughable to expect any sort of regulatory or legislative reform.

One of John's first new cases presented an issue he hadn't seen before. LaVerne DeShazo had filed a claim as a surviving spouse after her husband, William, had died in 2001. The Labor Department district director had found she was entitled to benefits, but Jackson Kelly, representing Consolidation Coal Company, had appealed. John began representing her in 2003.

William had been receiving benefits for the decade leading up to his death. After thirty-nine years of work for Consol, he had filed a claim, citing frequent wheezing and shortness of breath and writing, "I feel like I keep a chest cold all the time." The medical exam paid for by the Labor Department had indicated complicated pneumoconiosis, and a Labor Department official had issued an award. Consol had decided not to continue fighting and had started paying benefits.

Jackson Kelly hadn't been involved in that case. This time it was, and it had submitted numerous reports from doctors contending that, if William had suffered from black lung, it hadn't been totally disabling nor had it caused his death.

"You think, *They can't do that.* But they did do that," said the DeShazos' son, also named William. "So they start throwing the paperwork at my mom, and I think they hope for one of two things: You run out of resources, or you can't find a lawyer to help you."

As John examined the files, he suspected that Jackson Kelly was hiding something. As part of its defense, the firm had submitted evidence originally obtained by a claims-management firm that Consol had hired to evaluate William's claim in 1989. Though doctors who had reviewed the X-ray taken during his Labor Department–paid exam had seen complicated pneumoconiosis, Jackson Kelly submitted readings that the claims-management firm had received from doctors who found only the simple form of the disease. Jackson Kelly gave these readings to two consulting pulmonologists, both of whom opined that, although William probably did have the simple variety of black lung, it had caused neither his disability nor his death.

John thought it likely that there were other medical reports in the claims-management firm's file that Jackson Kelly had decided not to submit; perhaps these potential missing pieces had convinced the company to stop fighting the earlier claim.

Case records hinted at what else might exist. The claims-management firm had sent William to be examined by Dr. George Zaldivar, and after a call from John, Jackson Kelly turned over this report, as it fell into the category of evidence that the firm acknowledged it was required to disclose if asked. John's suspicions again proved correct. Zaldivar had determined that William likely had complicated pneumoconiosis.

The records also showed that the claims-management firm had sent an X-ray to Dr. Jerome Wiot—the super-expert whose withheld readings in other cases had proved critical—but there was no report by him in evidence. John's guess: Wiot had seen complicated pneumoconiosis, and Jackson Kelly was withholding the report because it likely would trump the other readings of simple disease, dealing the defense a grievous blow.

John filed his usual request for any withheld documents, and Jackson Kelly attorney Doug Smoot responded with the firm's usual refusal. Administrative law judge Michael Lesniak agreed with John and ordered Jackson Kelly to turn over any withheld reports. Rather than comply, the firm conceded the case. John asked the judge to impose some form of sanctions on Jackson Kelly, arguing that something had to be done to keep the firm from manipulating evidence. Lesniak declined, and after John appealed, the Benefits Review Board upheld the judge's decision.

At the same time, Bob Cohen was also increasing the pressure on Jackson Kelly. He, too, suspected obfuscation in a widow's claim that had striking similarities to the DeShazo case. Longtime miner Mike Renick had won a benefits claim in 1994, but when he died in 2000, Jackson Kelly not only challenged the new claim by his widow, Edna, but also filed a request for modification—an attempt to reopen Mike's case and turn the now-deceased miner's earlier victory into a loss. After a judge granted Cohen's motion to compel discovery, Jackson Kelly conceded the case.

In two of Cohen's cases, the firm's lawyers took a particularly bold stance, refusing to comply with a judge's order granting Cohen's motions to compel discovery. Both Cline and Cohen began referencing other cases and arguing that the firm was engaged in a pattern of misconduct, but judges generally declined their requests to impose sanctions.

The miners eventually won, but Jackson Kelly mostly avoided damaging disclosures and formal rebukes. The firm seemed to have an answer for every strategy Cline and Cohen tried.

One judge, however, appeared to be growing increasingly exasperated with Jackson Kelly. Michael Lesniak had rejected Jackson Kelly's standard argument opposing disclosure during the Renick case in 2002, then again in the DeShazo case in January 2004. A few months later, another Jackson Kelly case and another discovery fight landed before him. By the case's October 2004 hearing in Beckley, he'd had enough.

Michael Lesniak's résumé was not that of the typical administrative law judge. A Brooklyn native, he'd attended the College of William and Mary on a football scholarship, then earned his law degree from the university. After working as a commissioned military police officer at an army installation, he'd joined the Federal Bureau of Investigation as a special agent, spending most of his time investigating organized crime in New York City. He accepted an assignment as a special attorney on a Justice Department strike force in Pittsburgh focused on organized crime and racketeering. He prosecuted bookies and extortionists.

He then spent almost eight years in private practice, but in 1983, wanting to return to government service, he applied for and won a post as a Labor Department administrative law judge in the Pittsburgh office, which handled a large number of black lung–benefits claims. Over the years, he'd developed a keen understanding of the obstacles miners faced—the difficulty finding a lawyer, the immense resource disparities. But the case that Bob Cohen presented to him in 2004 revealed something more, something that offended his sense of fairness and awakened his prosecutorial streak.

The case was that of Elmer Daugherty. Filed in 2000, the claim was not the longtime miner's first encounter with the lawyers at Jackson Kelly.

Daugherty had entered the mines in 1942 as a teenager. He had quit school after the eighth grade and gone underground, where he worked for more than thirty years. For much of that time, he ran a cutting machine at a Westmoreland Coal Company mine located near his home in West Virginia. In 1976, frequent bouts of breathlessness forced him to take a job aboveground at the company's preparation plant, but it provided little respite. "Well, they had just as much dust outside as they did in," he later said at a hearing. He'd picked slate, driven bulldozers, whatever Westmoreland needed. After a decade of outside work, he retired in 1986. "I just didn't have enough breath to do it," he said. "I was all washed up."

He filed a claim, and his lawyer, provided by the United Mine Workers, squared off with Westmoreland's counsel, Jackson Kelly. Doctors for both sides had agreed Daugherty had black lung, but they had disagreed over whether the testing indicated total disability. A judge had denied the claim in 1989.

In 2000, when Daugherty tried again, he was seventy-four years old and in rapidly deteriorating health. He stood five feet ten inches tall but had withered away to just 127 pounds. In his application for benefits, he wrote: "On level ground I can only walk about 300 ft then I must stop and set down to rest. I can only walk up 6 steps (out of 12) before I have to stop and rest. At night, I have to sleep in a recliner, propping my head up." Dr. Don Rasmussen examined him for the Labor Department and found "very severe pulmonary insufficiency, rendering this patient totally disabled." A radiologist saw complicated pneumoconiosis on the X-ray.

In January 2001, the Labor Department district director found that Daugherty qualified for benefits. Instead of paying him $500.50 a month, the company contested the claim, and Jackson Kelly attorney Doug Smoot started building a defense.

Daugherty began looking for a lawyer, and the judge to whom the case had been assigned put the proceedings on hold until Daugherty could find one. Almost three years later, Daugherty got in touch with Bob Cohen, who agreed to take a look at the case file, which was already a foot-high stack of paper. Flipping through the pages, Cohen saw that Jackson Kelly had sent Daugherty to be examined by Dr. George Zaldivar, the Charleston physician who was a frequent expert witness for coal companies. As Cohen read the exam report that Smoot had submitted for the record, he realized something was off. He reread it, scrutinizing each page, just to be sure. Puzzlement became realization, then excitement. *I've got them,* he thought.

There was no narrative report, the part where Zaldivar stated what he'd found—the same part that had been missing from miner Charles Caldwell's file when he'd brought it to John Cline ten years earlier. Someone at Jackson Kelly had removed the most damaging portion of Caldwell's report then, and it appeared that someone had removed it from Daugherty's report now.

There was one difference: The submitted portion did include the form showing Zaldivar's interpretation of the X-ray taken during his exam. Daugherty wouldn't have known what the *X* scratched across one box on the form meant, but Cohen did: complicated pneumoconiosis. Cohen also knew that, under Jackson Kelly's own legal theory, it would be obligated to turn over the rest of the report, which he assumed contained an explanation further supporting the finding of complicated disease, the sort of reasoned analysis that would be much more damaging to the defense than the X-ray reading alone.

On September 8, 2004, Cohen filed interrogatories, asking Jackson Kelly to turn over any medical evidence it had not submitted. Three weeks later, Jackson Kelly attorney Kathy Snyder submitted the withheld report that Zaldivar had written in 2001, and it contained much of what Cohen had suspected. Zaldivar had determined that Daugherty's "severe pulmonary impairment" was caused, at least in part, by complicated pneumoconiosis. A confounding factor was that another significant cause of his disability was his extensive emphysema. This could be caused by coal dust or cigarette smoking, but Zaldivar thought the latter more likely. Still, his report could have supported an award of benefits.

But Jackson Kelly hadn't left it at that. The missing piece of the original report was now a mere attachment to a "supplemental report" by Zaldivar dated September 20, 2004—twelve days after Cohen had filed the interrogatories. The firm had gone back to Zaldivar with reports it had obtained in the years since Zaldivar's exam, including negative X-ray and CT scan readings by Dr. Paul Wheeler, his colleagues at Johns Hopkins, and other doctors. There was also a report indicating that Daugherty had suffered some sort of serious respiratory condition, likely a fungal infection, in the 1960s. Because of this new information, Zaldivar wrote, he had changed his opinion: The lesions in Daugherty's lungs must be from this apparent fungal infection. He did *not* have pneumoconiosis, and the cause of his disability was smoking-related emphysema.

The fight now was about what else Jackson Kelly might have. The firm, as usual, had refused to turn over anything from non-testifying experts, and Cohen had responded with a motion to compel discovery. Soon, each side would get a chance to win over Lesniak at a hearing.

"I'd like to make an opening statement," Cohen said. "I know I don't normally do that, but I think it might be helpful in this case to do that."

From the start, it was clear that this hearing was going to be far from routine. These were normally uneventful affairs, easily handled by a single Jackson Kelly attorney. Today, both Dorothea Clark and Doug Smoot sat across from Elmer Daugherty as he watched his lawyer address Judge Lesniak.

Cohen ticked through the mass of evidence Jackson Kelly had gathered while Daugherty had struggled to find a lawyer: more than thirty X-ray and CT scan readings, eleven reports from pulmonologists, six depositions of expert witnesses. (Daugherty had filed his claim before the new regulations limiting the amount of evidence took effect.) "If this was all the record showed," Cohen continued, "I wouldn't be making an opening statement. However, it also shows a pattern of the Employer flaunting [*sic*] regulations and engaging in acts of gamesmanship by withholding evidence from the Department of Labor, from Mr. Daugherty, and from its own reviewers of evidence."

He detailed what the firm had done with Zaldivar's report originally and how it had tried to blunt the effect when he'd demanded disclosure. "Now, I strongly suspect that there still exists evidence which they're withholding," Cohen said.

Clark countered that the firm had done nothing wrong; it was free to choose which evidence it wanted to submit. The defense's decision not to provide some pieces of evidence to its medical experts "did not skew their interpretations in any way," she said.

Lesniak questioned Cohen about his allegations. Then he turned to Clark, and the former prosecutor started to bore in. Was it true, he asked, that Jackson Kelly had withheld part of Zaldivar's report?

Clark didn't answer directly, instead reiterating the firm's position that it was free to develop and submit evidence as it deemed best.

Lesniak cut her off. "Well, did you withhold Dr. Zaldivar's report of May 16, '01, from the Department of Labor? Yes or no?"

"Well, the Department of Labor did not ask for— "

"So you did withhold it?"

"Because we are free to develop medical evidence." Clark began to argue that the system's rules didn't require the firm to submit the report, but Lesniak again interrupted.

"Well, it may be," Lesniak said. "But the point is, is that you withheld the report, Ms. Clark."

"And we have— "

"And I could only assume that you didn't like what it had to say, so instead of paying the claim, you withheld the report."

"Your Honor, I— "

"These are the sorts of things we have to stop. And you can't win at any cost."

Later in the hearing, Smoot elaborated on the firm's thinking: "I think everything was submitted except the medical opinion of Dr. Zaldivar, because based on litigation strategy in this particular case, it was decided to wait and go back to Dr. Zaldivar after more information is available for the record to get a more complete opinion from him." This was not what the firm normally did, and nothing prevented it from submitting the full report initially, then soliciting another as new information came in. The firm did that routinely with pulmonologists. Smoot didn't explain what was different about this case.

Lesniak turned to Cohen's motion to compel discovery. The judge noted that he had dealt with this issue before. He cited his own ruling ordering disclosure in the recent Renick case and pointed to the precedent set by the Calvin Cline case that Cohen and John Cline had won. Any reports Jackson Kelly might have were not covered by the work-product doctrine, he found, and Cohen had met the requirements for mandating disclosure.

"So," Lesniak told Clark, "you're to turn them over."

A few days later, the judge memorialized his ruling in a written order, giving Jackson Kelly one week to submit any medical reports it had withheld. He also scheduled a hearing and instructed both sides to be prepared to argue Cohen's allegations of misconduct.

Two days before the disclosure deadline, the firm employed the tactic that had long frustrated Cohen and John Cline. Rather than turn over any withheld evidence, it conceded the case. Cohen called Daugherty and explained the situation. Daugherty had a choice, Cohen told him. He could accept the victory; the company would start paying him, and his long ordeal would be over. Or he could keep fighting. Cohen believed Jackson Kelly was concealing other evidence; he wanted to expose what the firm had done and pursue sanctions. But it was up to Daugherty—what did he want to do?

Daugherty didn't hesitate: "Go get 'em."

John Cline and Bob Cohen stood next to the copy machine in Cohen's Morgantown office as it hummed late into the evening of December 15, 2004. After repeated delays, the hearing before Lesniak was finally going to happen the next day in a Pittsburgh courtroom. John and Bob were putting together a sort of dossier they hoped to present to the judge.

It had been almost two months since the hearing in Beckley, and Cohen had fought to keep Daugherty's case alive so it could get this far. After Jackson Kelly conceded, Cohen had written Lesniak, urging him not to let the firm off the hook. Daugherty's case, he believed, was just "the tip of an iceberg," part of a "practice of illegally withholding evidence" that he suspected "has been applied on a widespread basis" for decades and "may be responsible for the loss of black lung benefits by hundreds, if not thousands, of other claimants."

Attorney Kathy Snyder had countered that, after the firm conceded the case, Lesniak no longer had jurisdiction, nor did he or any of his fellow administrative law judges have "the power to enforce their own orders." Lawyers from Jackson Kelly had made this argument in previous cases. In creating the Labor Department administrative bodies that heard black lung–benefits claims, Congress had not granted the system's judges the authority to issue sanctions or hold parties in contempt, the firm noted.

Responding to Cohen's arguments about misrepresentation, Snyder wrote that, even assuming for the sake of argument "that the Employer were 'skewing' the evidence to be favorable to its position, such action is neither improper nor unethical. Indeed, attorneys are under an ethical obligation to zealously represent their clients which includes presenting the evidence in the manner most favorable to their client's position."

Lesniak had rejected Snyder's arguments. "Adversarial or not, the attorneys, as officers of the court, may not mislead a court or this administrative body in the guise of an 'ethical obligation to zealously represent their clients,'" he wrote. The hearing was still on.

Cohen had seen an opportunity to make the case that Jackson Kelly was engaged in a pattern of misconduct. To do that, he'd enlisted John's help.

That was why the two men were now holed up in Cohen's office on the eve of the hearing, copying key documents from some of John's past cases. Cohen planned to have John testify about what he'd uncovered while representing Charles Caldwell, Calvin Cline, William Harris, and

LaVerne DeShazo, then introduce into evidence the massive stack of paper they were now compiling.

When they finished, they stepped out into the cold December night, lugging the files. They'd get a few hours of sleep, and come morning, they would drive to Pittsburgh, where finally, they thought, a judge might listen to the larger argument they'd been trying to make for years.

Lesniak quickly dispensed with the formalities and got to the point.

"Do you have documents that you're not turning over?" he asked Kathy Snyder.

She rose from her seat beside Bill Mattingly, who was on hand to help defend the black lung unit's actions. John Cline, watching from the front row just behind Bob Cohen, thought Snyder looked stiff as a board, and he sensed an uncharacteristic shakiness in her voice.

"There are other non-testifying expert reports in this case that Jackson Kelly has," she replied. But conceding liability had ended the case, she argued. Lesniak no longer had jurisdiction; he couldn't compel disclosure of the withheld reports.

"My December 6th order indicated that I was retaining jurisdiction. Is that correct?" the judge asked.

"Yes, sir."

"And you informed your client of that?"

"Yes, sir."

"And you still refuse to turn over those documents. Is that correct?"

"That is our position, Your Honor."

For the moment, that ended discussion of the discovery issue. Jackson Kelly had decided to flout Lesniak's order, and there was little he could do about it.

Lesniak turned his focus to the Zaldivar report. Under the judge's questioning, Mattingly put forward a creative argument: Zaldivar's "report" was not really one report; the narrative that the firm had withheld was somehow separate. Yes, it had all arrived at the firm's office on the same day in the same envelope, Mattingly said, but "I wouldn't say it was attached."

"Isn't this like taking a deposition, and because you don't like what the doctor has to say on pages 5 through 10, just tearing them out and submitting the rest?" Lesniak asked.

"No," Mattingly replied. The defense wasn't obligated to submit everything that arrived in the envelope, he said. What's more, he argued, it wouldn't have made a difference, since Jackson Kelly did originally include the X-ray reading that found complicated pneumoconiosis.

"It almost looks like you're churning evidence," Lesniak said, "that you're withholding evidence which is detrimental to your case, you are then hiring consultants, and then you're going back to, let's say, Dr. Zaldivar with those other experts, and you're trying to get him to change his mind."

Cohen saw an opportunity to squeeze in John's testimony on other cases. "I'm prepared to put on evidence that this is a practice of the firm, and they do this repeatedly in black lung cases," he said.

Lesniak cut him off. The defense had already conceded liability, the judge said, so what, exactly, did Cohen want him to do?

Four things, Cohen replied: Force Jackson Kelly to turn over any evidence it was still withholding. Issue a decision formally declaring that what Jackson Kelly had done was "illegal and unethical." Refer the case to a federal district court, which did have authority to impose sanctions. And order the firm to identify all cases in which it had done something similar.

"Judge," Mattingly interposed, "you don't have jurisdiction." The request to identify other cases was "just wishful thinking" by Cohen. "This whole theory of his that you have to show everything to every expert or you're purposefully skewing evidence is just flummery. It's utter and sheer nonsense. It's puffery by an attorney." A lawyer was allowed to make a strategic decision about what to give to an expert, knowing there was a risk that the judge would discredit a witness who hadn't reviewed the entire record, Mattingly argued.

But that was the point, Cohen countered. Because the firm was withholding evidence, the judge wouldn't know that the expert had reviewed an incomplete, skewed record. No one but the lawyers from Jackson Kelly would know. "I'm not saying that a lawyer in a black lung case has to show every single medical report to every doctor who offers an opinion," Cohen said. "What I'm saying you can't do is to show them only the stuff that favors you and deliberately hold back from them the stuff that doesn't favor you. That is skewing the record, and that is presenting what's, essentially, false evidence."

He tried again to turn to John. "And this is a pattern—we have an opportunity to put on John Cline as a witness today—which is repeated over and over and over again."

Lesniak pressed on, paying little attention to Cohen's attempted segue or the man with a stack of papers sitting just behind him.

Speaking to Snyder, the judge made his intent clear: "I'm no longer looking at the merits of this case. The reason I called for this hearing was to look into whether any ethics violations occurred before me. And that's where I think I have jurisdiction. Now with regard to that, are you saying that my order of December 6th, 2004, is unlawful?"

"Yes, Your Honor," Snyder answered. Mattingly agreed.

"So," Lesniak continued, "if any ethical misconduct took place, I can't look at it after you conceded liability. That's what you're saying?"

"Yes," Mattingly said.

As the proceeding was wrapping up, Cohen made a final attempt to get John's evidence into the record. He said he had forty-five exhibits that would show the pattern of misconduct he was alleging, referencing the fruits of their late-night copying. Mattingly objected. To Lesniak, Daugherty's case was complicated enough; he wanted to stick to the record before him. "Mr. Cohen," he said, "I'm not going to do that at this time."

The judge did have a few final words for Mattingly, however: "You do a lot of black lung cases. A first-class job, top-notch lawyers. But what concerns me is the sort of things that I talked about here today, that it just doesn't happen anymore.... This is a coal miner who worked for Westmoreland, it's been stipulated...for 26 years. And when Dr. Zaldivar's report, this is a doctor you've gone to many times, when he says complicated pneumoconiosis, that report was buried."

"No," Mattingly replied, "his x-ray interpretation was turned over."

On that defiant note, the hearing came to an end. It seemed to John that Mattingly and his colleagues were unable or unwilling to even entertain the idea that they might be wrong, that their actions might have serious consequences for people like Daugherty. He was still digesting what he'd just seen: Snyder and Mattingly had stood before a judge, openly defied his direct order, and argued that they weren't out of line—*he* was.

It was clear that the firm's actions had angered Lesniak, but it was less clear what he would do about it. Meanwhile, John hadn't gotten the chance to utter a peep. John and Bob's plan to present a larger case, to make the judge see the big picture, had fallen flat. They loaded the stack of papers back in the car and started the drive home to West Virginia.

The decision Lesniak issued on March 21, 2005, was indeed a strong condemnation of Jackson Kelly's practices. By refusing to turn over any withheld documents even after repeated orders to do so, the firm had "disobeyed a lawful order," the judge wrote. "My concern," he continued, "has been that all an attorney has to do to circumvent an investigation into allegations of misconduct is to concede liability."

Turning to what the firm had done with Zaldivar's report, Lesniak wrote that he found the separating of Dr. Zaldivar's original narrative "to be unconscionable and reprimand the attorney or attorneys responsible; this was a deliberate attempt to mislead to the Claimant, I expected more from this law firm. I find their defense of this practice (withholding Dr. Zaldivar's narrative, which was surely detrimental to Westmoreland's case) to be ludicrous. I admonish the attorneys involved not to tamper with exhibits, potential exhibits and/or any type of documents which may be entered into evidence in the future."

Nonetheless, Lesniak said he'd decided not to grant the remedies Cohen sought. Doing so, the judge reasoned, wouldn't change the outcome of the case and would only delay the start of benefits payments.

Cohen again called Daugherty to explain his options. Daugherty's position hadn't changed; he wanted to keep fighting. Cohen wrote Lesniak and urged him to reconsider his decision, relating what his client had told him: "He feels very strongly that Jackson Kelly, PLLC mistreated him in their attempts to prevent him from being awarded black lung benefits. Mr. Daugherty told me that he wanted me to 'go after them' to try to prevent this from happening to other black lung claimants."

Lesniak again refused. Though he was "troubled and chagrined" by the facts, he still believed it was "an isolated incident." He suggested the two sides try to resolve their differences informally. A month later, Cohen tried again. Jackson Kelly had not contacted him to work things out, and the firm "apparently believes it did nothing improper in this case," he wrote to Lesniak. Nor was Daugherty's case an "isolated incident," he added, referencing the testimony that John Cline would have delivered at the hearing.

This time, Lesniak reconsidered. Given the important issues at stake and the fact that Jackson Kelly continued to insist it had done nothing wrong, the judge wrote, he had decided to refer the case to the U.S. District Court for the Southern District of West Virginia.

Ten months later, that court's chief judge, David Faber, issued a decision dismissing the case without issuing sanctions. For primarily procedural reasons, he concluded, it wasn't possible to hold the firm in contempt.

"However," the judge continued, "in granting this motion, the court does not condone the conduct of Jackson Kelly lawyers before the administrative law judge." The allegations about the firm's conduct were "deeply disturbing," Faber wrote. He had decided, therefore, to refer the case to the West Virginia State Bar's Office of Disciplinary Counsel.

Now Cohen and Cline's hopes for sanctions rested with a handful of lawyers in Charleston. Attorneys in the Office of Disciplinary Counsel investigated complaints and presented cases to a panel appointed by the president of the state bar association's board of governors. This panel could then recommend that the West Virginia Supreme Court of Appeals, the state's highest court, impose penalties up to disbarment.

Cohen finally got to use the stack of files he and John had copied. On October 11, 2006, he mailed the chief lawyer in the disciplinary counsel's office the papers and a letter outlining the evidence he and John had amassed over the years.

The fight to reach this point, however, had consumed valuable time. Just over a month after Lesniak referred the case to the district court, Daugherty had died. Cohen had suffered a heart attack and undergone open-heart surgery. "Litigating is so hard, and I was never quite the same after that," he later said.

Still more waiting lay ahead; it would take time for the disciplinary counsel's office to investigate and decide whether to pursue charges.

Meanwhile, the Daugherty case had created a stir among the Labor Department administrative law judges, and John sensed momentum. Perhaps next time, his arguments would find a more receptive audience.

In June 2006, shortly before the district court's ruling, John had received a call from a potential client. He'd gone through his standard list of questions and jotted down the basic information that he needed to evaluate the case.

When the call ended, he'd tucked the paper in the manila folder where he kept track of contacts with miners and widows seeking his help. There was now one entry in the communications log labeled GARY FOX.

20

Lost Time

Gary Fox's days running a roof-bolt machine were over. He just didn't have the breath for it anymore. Not long after the denial of his benefits claim in 2001, he transferred to a job aboveground. His new duties included lugging supplies, running a front-end loader, and performing a variety of small tasks that kept the Elk Run complex, one of Massey's largest in the area, humming.

The work was less strenuous than roof-bolting, but the dust remained inescapable. It blew off the belt lines carrying the mountain's bounty to the surface and drifted from the piles of coal waiting to be loaded onto trains. The dust coated equipment, lifting into the air when Gary worked on the machinery. It accumulated on every flat surface in the bathhouse, where miners shed their dusty work clothes, and became airborne when Gary had to sweep.

"It's extremely dusty," Carter Stump, who worked at Elk Run at the time, said of the outside area where Gary transferred. "It was really no different than being underground, to be honest with you."

There was so much dust swirling around the Elk Run complex that people living in the nearby town of Sylvester worried about what it was doing to *their* lungs. In 1998, the company had expanded its coal-processing facilities to within about seven hundred and fifty feet of some homes. Residents complained that dust blew over and coated their houses and just about everything else in the small community.

"It looked like the sun was even covered," Pauline Canterberry, a longtime Sylvester resident who had lost her husband to black lung, told a reporter. "It would be in your hair, everything. It would come through your windows like they weren't even there." Canterberry and neighbor Mary Miller, both in their seventies, led a group that became known as the Dustbusters; its members publicly railed against Elk Run and its corporate parent, Massey Energy. They collected samples from homes and videotaped plumes of dust blowing through neighborhoods and the grounds of the local school.

After citations and an order from the state's department of environmental protection, the company agreed to erect a nylon dome to cover some of the coal piles and conveyors, a project it would complete in 2002. A trade publication dubbed the structure, which was longer and wider than a football field and 110 feet tall, the "Don Dome," a reference to Blankenship, Massey's combative chairman and CEO.

In 2001, about a hundred and fifty Sylvester residents—more than half the town's population—sued Elk Run, claiming the dust had created a nuisance and caused almost four million dollars in property damage. Naturally, Jackson Kelly came to the company's defense. Attorney Al Emch, who had been named the firm's CEO shortly before the case went to trial, told the Boone County jury that the company was working diligently to control the dust. Besides, was it really that bad? "One person's intolerable nuisance can be another person's welcome, or even embraced, inconvenience," Emch said. He reminded jurors of the area's economic reliance on the company. It provided five hundred jobs, he told them. "It's something we need to keep... something we need to be tolerating."

Boone County was Massey country. During the trial, a teacher at a nearby high school told a reporter that her students were expressing concern about what might happen if the Dustbusters won. "They're afraid that if Massey loses this, their fathers will lose their jobs," she said.

In February 2003, the jurors found that Elk Run had failed to comply with federal and state laws requiring dust control and had indeed created a nuisance for the people of Sylvester. But they declined to assess punitive damages, and they imposed economic damages of only $473,000—far less than what the residents had sought. Blankenship declared a moral victory in a press release after the verdict: "While we would have obviously preferred to have no damages awarded in this case, we were gratified that the jury did not feel the plaintiffs' estimate of $4 million in property damage was justified. The fact that the jury awarded no punitive damages also indicates to us that they recognized the significant and ongoing efforts we have made to reduce dust at our facility."

The company agreed to a dust-monitoring program for the community and promised to enclose some of the railroad loading area and rebuild the Don Dome, which had collapsed even as the jury had been deliberating and would be torn open by wind a couple of months later. Elk Run even agreed to buy the town a "state-of-the-art" street sweeper to help control the dust.

All of that did little to help workers, like Gary, who spent their days at the source. Around the time the Dustbusters filed their lawsuit, Gary went for another round of testing by the medical board in Charleston that assessed miners for the state workers' compensation program. His scores on the lung-function tests had dropped enough to qualify him for another lump-sum payment. Two years later, in October 2003, his scores decreased again. He was now, in the eyes of the state, 40 percent disabled. The state exam also included an X-ray, and on the 2001 film, a doctor had seen masses in both of Gary's lungs that looked like complicated pneumoconiosis. By 2003, they had grown slightly larger.

The pulmonologist treating Gary, Dr. Maria Boustani, was also becoming more concerned. After an appointment in February 2005, she wrote in his chart, "He definitely would not benefit from being around any coal dust of any significance."

Aside from his worsening breathing, Mary and Terri saw another change in Gary, this one more positive. It seemed easier for him to relax, to let things go, to enjoy life. One reason for this, they thought, was that he'd agreed to talk with someone about what he'd experienced in Vietnam. At

Mary's urging, he began attending group counseling sessions at a veterans' center. Sharing the memories that still gave him nightmares and seeing that he wasn't alone appeared to make a big difference.

"You could just see this weight taken off of his shoulders that maybe you didn't see before, but afterward, you could," Terri said.

On weekends, Gary looked for excuses to drive up to Morgantown to see Terri. They had always bonded over football, and he went up for games whenever he could fit it in with his work schedule. West Virginia University Mountaineers football inspires near-religious devotion throughout much of the state, with vibrant congregations in the southern coalfields. When Terri was growing up, she and Gary would sit together and watch games on television. For a few hours, the two introverts became screaming superfans. Mary sometimes teased them: They knew the people on TV couldn't hear their yelling, didn't they?

When Terri was a student at WVU, Gary occasionally came up for a game. Now, because his job as an outside utility worker had a somewhat less demanding schedule, he had a few more open Saturdays. Decked out in Mountaineer gear, Gary and Terri would proudly join the crowd of similarly clad revelers who arrived early to tailgate in the parking lot near the stadium. Somehow, despite being separated by a three-hour drive, Gary and Terri were closer than ever. Whether because of guilt, a perspective that came with age, or something else, Gary seemed intent on making up for lost time. He met Terri's friends and went to parties hosted by her coworkers. "Everybody loved him. A couple of my girlfriends would call him 'Dad,'" Terri said. "I think some people might think, *He's just a coal miner*. I never once thought twice about it.

"He could have a conversation with the smartest person I knew. He would sit down with any of the doctors I've ever worked with and have educated conversations with them. I took him everywhere with me."

By 2005, the game-day festivities included another Mountaineer devotee. The previous year, Terri had transferred from the pediatric intensive care unit to the cardiac catheterization lab, where she'd met a cardiology fellow named Rick Smith. The first time they talked outside of work, at a tailgate party thrown by their boss, they had hit it off.

Their first date was on New Year's Eve. Rick had called and asked Terri if she had plans for the evening. She was home in Beckley for the weekend, she replied. Rick asked if he could drive down to see her, and she said yes. They'd gone out to dinner, and Rick met Gary and Mary, who took a liking to him. "I thought that took courage," Terri said. "He was a keeper."

An avid Mountaineer fan himself, Rick would lay claim to a prime tailgating spot early Saturday mornings. Terri and Gary would join Rick and his friends later in the day, and they'd head over to the stadium. Gradually, Terri noticed that Gary seemed to be having trouble with this walk. He occasionally paused a moment to catch his breath. At first, the two of them tried to make light of it. "I'm an old man; you've got to slow down," Gary would say. "Come on, Pops, let's go," Terri would joke.

Sometimes, Gary would push himself to keep going, and Terri would try to convince him it was okay to stop for a minute. "No," he'd reply, "I'm not an invalid."

Terri became increasingly concerned. The pauses, it seemed, were becoming more frequent.

At the same time, Mary's chronic digestive-tract disease worsened. By 2006, it had gotten so bad that she sometimes had to get specialized treatment in Pittsburgh. Gary went with her. "If I got bad news and cried, he'd crawl into bed and cry with me," Mary said.

Gary tried to forestall discussion of his own health and focus on Mary, but he couldn't hide what was happening to him. He stopped mowing the lawn and walking down the driveway to pick up the newspaper. He had to pause partway up stairwells. He hacked and wheezed, and he kept a spit cup in his truck for the black mucus that he coughed up. When he blew his nose, the tissue would turn black. It was as if his body were fighting a losing battle to rid itself of coal dust.

Miners at Elk Run noticed his coughing fits. After one of these spells, a coworker approached him and said: "Gary, you need to get away. The dust is killing you."

"Yeah," Gary replied, "I know."

He often ended up at the VA hospital with respiratory infections he had trouble shaking. Doctors gave him antibiotics and steroids until he was well enough to return home. By early 2006, he was using a battery of inhalers to try to keep his airways open.

Boustani, his pulmonologist, tracked his declining weight and lung function with growing concern. From mid-2005 to mid-2006, he shed roughly twenty-five pounds. He was trying to eat, but he couldn't keep weight on. Tests of his ability to move air in and out of his lungs showed scores that were just above half of what they should have been. "Pneumoconiosis, advanced," Boustani recorded in his chart.

Finally, even Gary acknowledged that he couldn't go on. At the end of July 2006, after thirty-two years as a coal miner, he retired. He was fifty-six years old.

Elk Run had a small retirement party for him at the mine complex. There was a catered chicken dinner, and miners stopped by intermittently as they finished their shifts. Gary and Mary carried their plates to the area where the company had set up a few tables. Before they could sit, they had to wipe down the chairs—they were coated with coal dust.

A few months later, Gary filled out form CM-911, "Miner's Claim for Benefits Under the Black Lung Benefits Act," for the second time. Below a question about evidence of disability, he wrote, "Chronic cough," "wheezing," "short of breath," "infections of lungs, 4 to 5 times a year."

On November 7, 2006, he filed the form with the Labor Department's office in Beckley. The next day, he called John Cline.

21

Fraud on the Court

* DID J&K SUBMIT PATHOLOGY?
↳ IF NOT → SEND INTERROGATORIES.

That was the note John Cline scribbled on a notepad and placed in a manila folder after his first in-person conversation with Gary Fox, the meeting over John's kitchen table on January 10, 2007. Pathology reports—or rather, the lack of them. This was the crooked rafter, the missing piece that John saw as the two men had gone through the file from Gary's earlier, unsuccessful benefits claim.

John had flagged the potential importance of pathology evidence from the start. When Gary first called him in June 2006, outlining the basics of his case and telling him he planned to retire and file a claim in the coming weeks, he'd also mentioned the surgery he underwent in 1998 to have a chunk of his right upper lung removed. John had written down the diagnosis Gary had told him—"benign tumor"—then drawn an arrow to a question of his own: "CWP?" Was it possible that this piece of Gary's lung was actually evidence of black lung and the hospital pathologist had missed it?

John was suspicious, but the larger picture of Gary's case was still fuzzy. During their next three phone calls, which took place in November and December of 2006, John focused on gathering the basic information that would help him determine whether Gary had a viable claim.

As the two men reviewed records in John's home on that chilly January morning, John became convinced that the pathology was indeed a crucial piece of evidence. When they met again eight days later, John began to sketch a blueprint of the defense that Jackson Kelly had assembled in Gary's previous case. There were all the familiar names—Paul Wheeler, Gregory Fino, and others—with all the familiar contortions of the diagnosis "anything but black lung," including granulomatous disease and TB. But John drew an arrow branching off from this orderly grid of the usual suspects to the conspicuous gap in the design:

NO PATHOLOGY? → DISCOVERY REQUEST

In John's experience, Jackson Kelly virtually always had experts of its own choosing reevaluate key pieces of evidence, and, if their reports were favorable, the firm submitted them, even if they merely piled onto findings already in the record. Gary wouldn't have known it, but the lack of any additional pathology reports in his case was unusual.

There seemed to be a few possibilities. Maybe Jackson Kelly simply hadn't bothered to get one of the pathologists it normally used to review the slides of Gary's tissue. The finding by Raleigh General Hospital pathologist Gerald Koh—"inflammatory pseudotumor"—had served the firm's purpose. But Koh's report was somewhat puzzling. He hadn't explicitly ruled out black lung; he'd just offered another diagnosis without mentioning the disease. Presumably, he would have noted it if there was any indication that it could be present.

But why would Jackson Kelly leave it at that? An unequivocally negative report by one of the better-credentialed physicians the firm often enlisted would be more useful than Koh's. Sure, it would cost money, but that had never seemed to be an obstacle in other cases, including those in which the firm represented Massey Energy subsidiaries.

Assuming, then, that Jackson Kelly had sent Gary's slides to other pathologists for review, there seemed to be a few remaining possibilities. Perhaps these doctors had written unclear reports, and the firm had decided not to submit them, thinking they would only muddy the record. That was unlikely, particularly because the cover letters the firm sent to doctors asked them to specifically address whether black lung was present and, if so, how severe it was.

Perhaps the doctors had written reports finding that Gary didn't have black lung, but in that case, it seemed inconceivable that the firm wouldn't have submitted them for the record. There was another possibility, the one John considered the most likely: The doctors had written reports finding that Gary did indeed have the disease, and the firm had withheld them. He told Gary that he suspected this had happened.

Cases involving pathology were relatively rare, and any withheld reports on biopsies or autopsies would be particularly critical. In one recent case, John was convinced that Jackson Kelly had hidden a key pathology report, and the claim's outcome had filled him with resentment.

Edsil Keener, who had worked more than thirty-three years in the mines, filed a claim in 2001 but died the following year. The survivor's claim of his wife, Shirley, hinged on whether his autopsy showed simple or complicated pneumoconiosis. The doctor who performed the autopsy and a pathologist who reviewed the lung tissue had seen complicated disease, but two experts retained by Jackson Kelly had found only simple. John suspected the firm was sitting on other pathology reports, and he submitted a discovery request.

Administrative law judge Richard Morgan sided with Jackson Kelly and denied the request. In August 2005, Morgan denied both claims, finding that Keener had simple pneumoconiosis but that the disease hadn't been the cause of his total disability or death.

For years, John wondered what he hadn't been allowed to see, and the case haunted him. He saw in Gary's case an opportunity to expose what he hadn't been able to in Keener's.

But John also saw something different about Gary's case, something worse. Cheating retired miners or their widows was bad enough. But

Gary hadn't yet retired when he'd filed his first benefits case. He'd been just forty-eight years old, trying to put Terri through college. The denied claim had cost him much more than money; it had cost him a shot at getting out. The federal black lung–benefits program was a lifeline he'd tried to grasp, a chance to escape the dust that was ravaging his lungs and potentially halt the disease's progression before it reached the most advanced stage.

When the claim was denied, he'd continued to work. The masses in his lungs had grown, and his scores on lung-function tests dropped precipitously. Dr. Don Rasmussen had found moderate impairment when he'd examined Gary in 1999, but this time around, in November 2006, he found Gary's impairment to be "severe" and "irreversible." The purpose of the legal presumption of total disability for a miner who has evidence of complicated pneumoconiosis, as Gary's X-rays indicated he did in 1999, is to afford him an opportunity to leave the mine dust and prevent the sort of drastic decline that Gary experienced.

Given all of this, the implications of John's suspicions were almost unimaginable. Had the lawyers at Jackson Kelly really withheld evidence that might have allowed Gary to win, retire, and get the care he needed? If their own doctors told them that Gary had complicated pneumoconiosis, an indication that he needed to get out of the dust immediately, had they really just buried the reports?

When John first started representing miners fifteen years earlier, he would have doubted that apparently decent, professional people would be capable of doing that; he would have thought that even the zealous-advocacy justifications had their limits. He no longer harbored such illusions.

There was also something different about Gary himself, John thought. He reviewed Gary's scores on recent lung-function tests, and he knew from his years at the New River clinic what the numbers represented in human suffering. Gary was, as John later put it, "going to hell in a handbasket from a breathing standpoint." After their January 10 meeting, John noted that Gary was "very SOB [short of breath] w/exertion." Yet Gary didn't complain. John wondered how someone so sick could be so stoic.

After the two January meetings with Gary to review records, John had the information he needed to assess the case, and he agreed to represent Gary. Soon, they settled into a rhythm of periodic phone check-ins. John found Gary quiet and thoughtful. When asked a question, Gary would listen closely and think for a minute before answering, much like John himself. They often discussed more than just Gary's case. The two soft-spoken men developed an easy rapport and shared personal details. Gary, unlike some of John's clients, seemed interested in claims other than his own. John talked with him about the discovery process and what he'd been finding over the years in his fights with Jackson Kelly. "I felt he had a genuine interest in the issues beyond his own case," John said. "He cared about what happened to other miners."

They seemed to share a set of unspoken understandings. They sensed they were in a race with the disease that was consuming Gary's lungs. A victory in the benefits claim and a lifesaving lung transplant were goals in the distance, but they weren't sure they would reach them in time. Both also seemed to sense that they were working on something of broader significance, with ramifications beyond just Gary's case. John always kept notes of his interactions with clients, but in Gary's case, he did so meticulously and in greater detail. He had a feeling that one day he would want to have a comprehensive record of what had happened.

After agreeing to take Gary's case, John wasted no time setting in motion the discovery battle he knew was coming. Two days after his second meeting with Gary, he sent interrogatories to Jackson Kelly requesting any reports the firm had obtained but not submitted.

Almost three months later, on April 5, attorney Ann Rembrandt responded with the firm's usual refusal to turn over whatever it had, claiming privilege.

In June, a Labor Department claims examiner issued an initial determination that Gary was entitled to benefits. The decision cited Rasmussen's exam and Dr. Maria Boustani's treatment records; both physicians believed Gary had complicated pneumoconiosis and was totally disabled by it. Gary and John were pleased, but they knew this was only the beginning. They were not at all surprised when Rembrandt notified the department that Elk Run would not be paying Gary and Mary $876.50

a month plus medical coverage for his lung disease but would instead continue fighting the claim.

Gary's case file now moved to the chambers of administrative law judge Thomas Burke in Pittsburgh. On February 14, 2008, John filed a motion to compel discovery in which he tried to convey to Burke the larger scope of the fight.

"The undersigned has argued for the past fifteen years that federal black lung litigation should be an above-board process," John wrote. He summarized what he'd found in three cases that spanned much of his legal career, from Charles Caldwell in 1994 to William Harris in 1997 to LaVerne DeShazo in 2003, and cited the regulations governing discovery and the case law that he'd helped make over the years, applying it all to Gary's case.

In response, Rembrandt argued that there was no legal support for John's request for privileged documents.

On April 28, Burke issued the order John and Gary were hoping for. The judge found that, contrary to Rembrandt's assertion, the state of the law regarding discovery was clear: a judge could compel disclosure if he thought the claimant had met the legal requirements, and in this case, Burke believed John had. The judge gave Jackson Kelly seven days to turn over whatever reports it had withheld.

When those seven days were up, however, the firm did not comply. Instead, Rembrandt filed a motion asking Burke to reconsider. The judge declined and again ordered the firm to turn over any documents it had withheld. When the deadline Burke had set for Jackson Kelly to comply, August 4, arrived, Rembrandt called John with a question: If the defense accepted liability and agreed to pay the claim, would John and Gary still try to pursue discovery?

John had seen this coming. Three months earlier, as he was preparing his filings on the discovery issue, he had written in his notes that Jackson Kelly "could still accept liability to avoid disclosure." John had been here before, and what happened next had always left him frustrated. Conceding had proven to be an effective strategy for Jackson Kelly; it had brought nettlesome cases to abrupt conclusions. Either the miner was unwilling to risk the victory already in hand and endure still more protracted litigation or the judge was unwilling to go out on a limb, keep control of the case, and try to compel the firm to comply.

Even the stir created by the case of Elmer Daugherty hadn't broken this pattern. The ongoing investigation by disciplinary officials in West Virginia could lead to sanctions, but any withheld evidence that might exist seemed likely to remain hidden.

John had been waiting for years for the right confluence of facts, judge, and miner—the ingredients that might yield a different outcome. Gary's case, he thought, had all those things. But ultimately, the decision on how to respond to Rembrandt's question wasn't John's to make. He dialed Gary's number.

If ever two people could benefit from removing one item from their list of ongoing worries, they were Gary and Mary Fox. Both were seriously ill, and they now spent much of their time in hospitals. They stayed in Pittsburgh for weeks on end so Mary could get the care she needed. When they were back in Beckley, Gary was in and out of the VA hospital with lung infections or severe shortness of breath. Once, Mary's brother-in-law, Richard Roles, had picked up Gary to take him to the VA, and he'd had to drive his car across the lawn and pull up right in front of the door. "It was all he could do to get in the car," Roles recalled.

Boustani had referred Gary for a lung-transplant evaluation at the University of Virginia's medical center in Charlottesville. There he'd undergone testing to determine whether he was a good candidate. Among the procedures was a cardiac catheterization. For Terri, who had gone along, it was a strange experience to be on the outside, waiting nervously, rather than in the room helping with the procedure, which was her job back in Morgantown. Gary, of course, seized the opportunity to tell everyone in the hospital, "This is what my daughter does."

UVA turned him down. Performing a transplant on Gary, the doctors determined, was too risky, in part because of the effects of the 1998 surgical removal of much of his right upper lobe. When Gary told John about the decision, it was one of the few times John sensed disappointment in his voice.

As bad as the doctors' assessment itself was the lost time during the evaluation process. "Worsening in his symptomatology," Boustani noted after one appointment. "I continue to be concerned regarding his weight loss." He now had just 166 pounds on his six-foot-two-inch frame.

When John talked with Gary about his case, they also discussed how Gary was feeling. Next to notes about their conversations on discovery motions and hearing preparations were notes such as "still losing weight but OK" and "coughing up black stuff daily."

In July, the Fox family celebrated Gary's fifty-eighth birthday with lunch and cake on the banks of the New River. Though he was clearly having trouble keeping weight on, he still had some color to his complexion and could breathe without oxygen. They still had hope. With Boustani's help, he was trying to schedule a transplant evaluation at another hospital. The University of Pittsburgh Medical Center seemed to hold potential; maybe the doctors there would reach a different conclusion than those at UVA.

When John called to relay Rembrandt's question—if the defense agreed to pay the claim, would Gary drop the case?—Gary was in Pittsburgh, staying with Mary while she received treatment. John explained the situation: Jackson Kelly was folding. Gary could accept the win, end the legal fight, focus on his and Mary's pressing health needs. No one would blame him for making that decision.

The other option was to press on. They could try to convince Burke to keep the case alive and force the firm to comply with his discovery order. This path, however, was likely to be unpleasant, and there were risks. Though John strongly suspected Jackson Kelly had withheld critical pathology reports, he couldn't be sure. The firm could be standing on principle, preserving its overall litigation strategy. There might be nothing damning in its files. Continuing to fight could turn a win into a loss. The choice, John explained, was Gary's to make. Gary listened quietly, thought for a moment, then spoke the words John had been waiting for years to hear: Keep fighting.

Mary supported the decision completely. She later said, "Gary's opinion, which I agreed with, was it may not help him, but he wanted to stop the lawyers from doing that to other miners. It wasn't helping him, but to help the next person."

John immediately fired off a letter to Burke requesting that the judge force Jackson Kelly to comply with his order to turn over any withheld reports.

He also brought up a mystery he'd been trying to solve for months: Where were the pieces of Gary's lung that had been removed in 1998? John wanted to have a pathologist examine them to see whether they actually showed black lung and not an inflammatory pseudotumor. The pieces should have been on slides kept in storage at Raleigh General Hospital. But when John had called to ask, no one there could find them or a record of what had happened to them. These crucial slivers of evidence were, apparently, just gone. In mid-May, John had sent a letter to Jackson Kelly asking if perhaps the firm had sent the slides to a pathologist for examination, which the firm was entitled to do. No response. John's subsequent written requests for information also received no answer. Now, in August, he asked Burke to order the firm to cooperate.

A definitive answer on the slides' whereabouts would come later; for now, Rembrandt's response focused on the judge's order to turn over withheld documents. She argued, as the firm's lawyers had many times before, that conceding the case took matters out of Burke's hands; he had no authority to force disclosure. Gary's case, as far as Jackson Kelly was concerned, was over.

On August 28, Burke issued his decision. John's arguments, the judge wrote, had merit. If Jackson Kelly had withheld key documents, they might support an earlier entitlement date or even a reopening of the previous claim. His order was the one John had been hoping to get for years: Burke was retaining jurisdiction. Jackson Kelly had until September 19 to turn over whatever it had.

The small printer that doubled as a fax machine in John's home office hummed to life at 3:48 p.m. on September 19, 2008. The cover sheet, with Jackson Kelly's letterhead, inched into view.

```
For immediate delivery to: The Honorable
Thomas M. Burke
John Cline, Esquire
From: Ann B. Rembrandt, Esquire
Total number of pages, including this cover
sheet: 24
```

John scanned the first page: "In accordance with the discovery orders, the Employer hereby produces to the Claimant and the Court..."

This was it; Jackson Kelly was complying. The missing pieces from Gary's case started to pile up in the tray of John's printer.

First came the X-ray readings—sixteen of them. Two were negative readings by a Virginia pulmonologist from Gary's earlier case that the firm hadn't used. There was no need to—the negative readings by Dr. Paul Wheeler and his colleagues at Johns Hopkins had been much more useful. The rest of the readings were from doctors whom Jackson Kelly had consulted in Gary's current claim. Thirteen of them were by West Virginia physician Joseph Renn. In 2007, he'd read a series of X-rays that the firm had sent him and found rapid disease progression that mirrored the symptoms Gary had experienced: negative for pneumoconiosis in 1974, complicated category A in 1997, category B in 2001, and category C in 2006. The final X-ray reading was from the revered radiologist Jerome Wiot. He'd written in a 2008 report that Gary's June 2006 film was "very abnormal" but that "the findings are not those of coal worker's pneumoconiosis." He saw a large mass but suspected sarcoidosis. He noted, however, that part of Gary's lung had been removed and concluded that "the pathology would give an answer as to whether there is any evidence of coal worker's pneumoconiosis."

It was all interesting but did little to change the overall case—dueling X-ray interpretations, with everything ultimately resting on Dr. Gerald Koh's pathology report.

Then the final four pages of the fax lurched into the tray. John saw the name on the letterhead: Richard L. Naeye, MD, the pathologist who had cowritten the 1979 standards still used to identify pneumoconiosis. He had a conservative view of the disease, and Jackson Kelly solicited reports from him regularly, holding him up to judges as a sort of super-expert. John saw the date of the report: April 20, 2000—one month after Gary's first claim had landed before an administrative law judge.

"The mass removed from Gary Fox's lung in 1998 is almost entirely comprised of fibrous tissue with small amounts of admixed black pigment and birefringent crystals of all sizes," Naeye had written. The findings were enough "to suggest at least a partial silicotic origin. At its edge the lesion appeared to have been actively expanding at the time it was removed. This man's many years of working at the coalface and as a roof bolter increase the risk of such a lesion. There is no way of knowing if this man has smaller silicotic lesions in other lobes of his lungs.... If his pre-surgical findings met the legal definition of complicated CWP the surgeons have removed it. In short it has had a surgical cure. Will findings that might meet the criteria of complicated CWP return? There is no way of knowing."

It was a tortuously written report that Rembrandt described in the cover letter as "inconclusive," but stripped of jargon and addled syntax, Naeye's conclusion was much more straightforward than Rembrandt claimed: The piece of Gary's lung removed in 1998 was not an inflammatory pseudotumor. The scarred tissue contained remnants of the dust Gary had breathed for years—coal and, especially, silica. Though Naeye framed the statement as a conditional, he was essentially saying the mass removed from Gary's lung was complicated pneumoconiosis.*

Naeye's report wasn't the last. John's printer spit out one more before it grew quiet. The author was P. Raphael Caffrey, another prominent pathologist on whom Jackson Kelly sometimes relied. His description of the tissue from Gary's lung was similar to Naeye's; he saw scarring containing "anthracotic pigment" and "birefringent spicules"—likely indicators of coal and silica dust. But he stated his conclusions more clearly:

* Naeye died in 2013, so I was not able to talk with him.

I. Large, 5 cm mass consistent with complicated pneumoconiosis.
II. Areas of simple coal worker's pneumoconiosis identified.*

The first thing John felt was relief. His suspicions had been correct. But relief soon gave way to anger. Reviewing the records from Gary's first claim, John now filled in the missing pieces in the timeline, the parts of the picture that Jackson Kelly hadn't allowed Gary, the judge, or its own experts to see.

After the company had appealed the initial decision by the Labor Department to award benefits, the firm had started building its defense. In the following months, it had obtained the slides from Gary's 1998 surgery and sent them to Naeye, then Caffrey, but it had withheld both doctors' reports.

The slides were now in John's possession; Rembrandt had sent them to him with virtually no explanation of where they'd been or why she hadn't responded to his previous requests for them. A few months later, another Jackson Kelly attorney finally provided an answer: The firm hadn't concealed them; it didn't know where they were, and after John had asked, it had begun searching for them. Eventually, they turned up in the office of Dr. Grover Hutchins, a pathologist at Johns Hopkins. The firm had sent him the slides one week after receiving Caffrey's report. Hutchins had never written a report, the firm said.

John had a guess about what had happened; he suspected that someone from Jackson Kelly had spoken by phone with Hutchins, learned that the pathologist's opinion echoed those of Naeye and Caffrey, and told him not to bother writing a report.†

Even physicians who tended not to find complicated pneumoconiosis

---

* Caffrey declined to talk with me.

† In a brief, Jackson Kelly attorney Kathy Snyder later called John's theory "idle speculation" and wrote, "To counsels' knowledge, Dr. Hutchins never reviewed those slides." Years later, however, the cover letter sent from the firm to Hutchins came to light. In it, a legal assistant asked Hutchins to review the slides and write a report answering specific questions, including whether the slides contained evidence of coal workers' pneumoconiosis and, if they did, whether it would be considered totally disabling. Nonetheless, there was no indication that Hutchins actually wrote a report. Hutchins died in 2010, so I was not able to discuss the case with him.

had seen it on Gary's slides, so how had Koh missed it? One possible answer was that they were simply better qualified. Aside from his contribution to the 1979 standards, Naeye had more than 260 published articles to his credit. Both Naeye and Caffrey had practiced for decades in a coal-mining state—Pennsylvania and Kentucky, respectively—and had often evaluated slides sent to them by defense attorneys. By contrast, Koh had spent much of his career at children's hospitals in non-mining regions. He'd taken a job in West Virginia just five years before he interpreted Gary's lung tissue.

Another explanation was that Naeye and Caffrey had more information. They knew that Gary had worked underground for about twenty-five years; there was no evidence that Koh even knew Gary was a coal miner.

Nor did it appear that Koh had appreciated the significance of silica. Naeye and Caffrey had both noted that scarred pieces of tissue contained telltale crystals, and this had been an important part of the final findings by both. Koh hadn't even mentioned silica. The mineral is not always obvious under a normal microscope, but seen under polarized light, the crystals light up. Caffrey had stated explicitly in his report that he'd used polarized light, and the medical terms Naeye had used, along with his emphasis on silica, strongly indicated he'd done the same. Koh's report contained none of this.*

It looked to John like Jackson Kelly had carried out a cruel deception: The firm's lawyers had known the pathology was the most critical element of the case, and yet when two, or maybe three, of the experts they had relied on repeatedly for years determined that Gary's tissue showed complicated pneumoconiosis, they buried those findings and built the case around Koh's inflammatory-pseudotumor report.

The three pulmonologists who reviewed the records that Jackson Kelly chose to send them made clear in their reports that the pathology was the cornerstone of their determinations. The pulmonologist who had examined Gary for the defense originally believed Gary had at least simple

* Koh died in 2009, so I was not able to talk with him.

pneumoconiosis, but after reviewing Koh's report, he changed his mind and concluded that Gary didn't have black lung at all.

Now, reviewing the Jackson Kelly attorneys' questioning of witnesses and arguments to administrative law judge Edward Terhune Miller in Gary's first claim, John saw the firm's actions in a much darker light.

Doug Smoot prompting one pulmonologist to discredit Rasmussen's report because he had not reviewed "all of the biopsy medical evidence."

Bill Mattingly eliciting testimony from Wheeler that the negative pathology evidence bolstered his negative X-ray readings.

Mary Rich Maloy, in her written closing argument, calling Koh's report "the most credible evidence."

And, ultimately, Miller's reliance on these representations in denying Gary's claim, finding that the defense experts' reports were "based on comprehensive reviews of all of the evidence of record."

Jackson Kelly's defense was that "all of the evidence" meant "all of the evidence *in the official record*"—in other words, whatever the firm chose to withhold didn't count as evidence. To John, this seemed, at best, highly misleading.

Both the firm's own consulting physicians and Miller certainly seemed to be under the impression that they were reviewing everything. Letters from Jackson Kelly to three pulmonologists it had asked to review records and render an opinion began, "Enclosed please find copies of all of the medical records that we have been able to develop concerning the above-referenced black lung claimant."

Years later, I spoke with Miller, who by then was retired, about Gary's case and told him about the reports by Naeye and Caffrey that Jackson Kelly had withheld.

"I'm utterly dumbfounded," he said. "I just cannot conceive of attorneys doing that. We know that there are legal hacks, but with a reputable firm and reputable attorneys, I just cannot conceive of that.... That's really misleading the court. It's misleading the witnesses. It's tainting the witness testimony."

He continued: "I frankly think that, when you get to that point and you are offering evidence of a certain kind and you know material is there which clearly makes that evidence false or incomplete—you just don't do that; that's wicked."

I asked whether he thought the consulting physicians in the case had assumed that there were no other pathology reports besides Koh's. "Of course," he replied. "What else would they have thought?" (None of the doctors themselves would discuss the case with me.)

Would knowing what Naeye and Caffrey had written have changed the judge's decision from a denial to an award? It was very possible, he said, not wanting to offer a definitive answer without the benefit of reviewing the full file. "What, it seems to me, should have happened in Fox is that the proper thing for the attorney to have done, if he had that kind of evidence and it came from doctors they usually relied on, they should have paid benefits," Miller said.

He had been following the ongoing proceedings related to Jackson Kelly's conduct in the case of Elmer Daugherty. I asked if he wondered whether something similar had happened in other cases before him. "Sure," he replied. "I don't sleep better because of it." But, especially in cases like Gary's, ones in which the miner didn't have a lawyer, "there really isn't any way in that context that I or any other judge would have been likely to ferret that out."

What, then, would be a solution? I asked. "There's no question there could be a regulatory change," he said. "The employers would fight you tooth and nail, but I would think that, if both sides were required to disclose all of the medical evidence that they may have generated in connection with a case and make it available to the other side, I think it might be very beneficial."

John had been arguing for just such a regulatory change for years, but as Gary's case unfolded, the idea still had little traction among Labor Department officials. The documents that had just come across his fax line, he thought, might change that.

John wrote and rewrote a brief outlining an argument he'd been waiting years to make. Jackson Kelly, he asserted, had committed fraud on the court.

This was the little-used legal tool that John had happened upon in law school, the one that he'd wanted to try in William Harris's case eight years earlier but hadn't been able to when Harris and his wife opted to take the deal Jackson Kelly offered. If Burke found that the firm's conduct was egregious enough to meet this heightened legal standard—not merely fraud, but "a deliberate scheme to directly subvert the judicial process"—he could invalidate Gary's loss in 2001 and order Elk Run to pay years of additional accumulated benefits. Otherwise, if Gary won his current claim, he would likely be entitled to benefits dating only to 2006.

John also thought such a stark repudiation of Jackson Kelly's tactics could reverberate beyond Gary's claim, perhaps compelling the firm to change its practices or sparking increased scrutiny from other judges.

The case law addressing fraud on the court was relatively scant. John cited the seminal 1944 case in which a glassmaking company had ghostwritten an article extolling one of its machines and gotten a prominent union official to sign it for publication in a trade journal. The company had then used the printed article to win a patent application and an infringement suit against a competitor. The victimized competitor hadn't uncovered these facts until after its appeals had been exhausted. It had filed a petition to reopen the case, but a federal appeals court ruled that it was too late—the decision was final. The U.S. Supreme Court, however, ruled that the law's preference for finality was not absolute and that, in this case, the company behind the ghostwritten article had committed fraud on the court. The Supreme Court ordered the lower courts to vacate the earlier award, writing, "The public welfare demands that the agencies of public justice be not so impotent that they must always be mute and helpless victims of deception and fraud."

John contended that Jackson Kelly's handling of the pathology reports in Gary's claim was a similar scheme. "This kind of intentional misrepresentation severely undermines the credibility of the adjudication process, so it is not just a wrong against Mr. Fox but also against the administrative agency that Congress established in order to safeguard miners," John wrote. "It is virtually impossible for the average claimant to contend with the kind of deception and intentional misrepresentation practiced in this case and others."

He asked Burke to find that Jackson Kelly had committed fraud on the court and award benefits dating back to September 1998, when Gary had undergone the surgery that yielded the pathology slides.

Jackson Kelly attorney Ann Rembrandt countered with a brief citing more recent cases in which judges had found that even actions such as offering perjured testimony and fabricated evidence did not amount to fraud on the court. "As the litigation of federal black lung claims is adversarial, the parties do not have a duty to generate 'neutral' evidence or disclose all relevant evidence to opposing counsel," she wrote. "Rather, counsel for the parties have an ethical obligation to zealously represent their clients by developing evidence and presenting evidence that supports their clients' positions." She asked that Burke deny John's requests and award benefits dating back only to June 2006.

Now, in November 2008, the decision was Burke's. All John could do was wait nervously.

As Gary and Mary also waited to see what Burke would rule, they received a piece of good news: the University of Pittsburgh Medical Center would evaluate Gary as a candidate for a lung transplant. The decline in his health, however, seemed to be accelerating. The complications and infections that required hospitalization were now constant, and they threatened to forestall or eliminate the possibility of a transplant.

"The patient is feeling miserable," Boustani noted after examining Gary in October. In December, she recorded, "He has an appointment to go to Pittsburgh on 01/05/09 so he is eager to get better." The treatment at the VA no longer seemed to be helping much. Boustani now assessed the severity of his disease as "advanced, end stage."

By Christmas, Gary was on oxygen. He and Mary drove up to Morgantown and spent the holiday with Terri, who was on call at the hospital. She was amazed at how much worse her father looked—pale, gaunt, lugging a tank that was his lifeline.

Still, when the new year came, he was well enough to begin the battery of tests that physicians at UPMC would use to determine whether he was a viable transplant candidate. He and Mary once again took up temporary residence in Pittsburgh.

On February 3, John's oldest son, Jesse, called. He'd aggravated an old ankle injury and thought he ought to get it X-rayed, but he didn't think he could drive himself to a clinic. John went over to the house in Beckley that Jesse, now thirty-six years old, shared with his girlfriend and her young daughter from a previous marriage.

After graduating from law school, Jesse had clerked for a judge in nearby Wyoming County and spent some time in private practice. He now worked for the consumer advocacy division of the West Virginia Offices of the Insurance Commissioner. He'd kept up his love affair with the waterways of the state, kayaking whenever he could. He often dropped by to see John and Tammy, and they talked regularly. Jesse took a keen interest in John's work, and they sometimes discussed Gary's case. They were of one mind in their view of Jackson Kelly's behavior.

When John arrived at Jesse's house, it was clear to him that there was something wrong beyond his ankle. He'd been having flu-like symptoms and had barely slept, but that didn't seem like anything particularly severe or unusual. John took him to a clinic for an X-ray of his ankle.

Back at Jesse's house that afternoon, John left Jesse in the care of his girlfriend while John went to an appointment. When he returned about forty-five minutes later, Jesse's girlfriend was panicked. There was something really wrong with Jesse, she told John. She'd gotten him up into bed.

In the bedroom, John found his son shivering. He wrapped him in more blankets and called 911. He held him in his arms to try to warm him, slapped him in the face when he seemed to be losing consciousness, but he kept fading. John lifted him onto the floor and started performing CPR. "But I believe he had already died in the bed," John said later. "In my arms, actually. I think I felt the life go out of him."

The ambulance arrived and took Jesse to the ER, where doctors pronounced him dead. Tammy and Brooks arrived moments later and found John in shock.

They later learned what had taken Jesse: an infection by a bacteria known as *Haemophilus influenzae.* It was relatively common and typically caused mild symptoms, somewhat like the flu. In Jesse's case, however, the infection had gotten into his bloodstream, and he'd died of septic shock. Why or how that had happened to a young, healthy man was a mystery.

The memorial service took place near the New River Gorge at a pavilion that was owned by the rafting company where Jesse had worked. John had some of Jesse's ashes buried beneath a marker in a cemetery in East Aurora. The rest he had buried in a cemetery in Beckley; John chose a spot behind the other gravestones in a wooded area backing up to the cliff overlooking Piney Creek.

John told Gary about Jesse's death during a phone check-in. Mary later told John how much the news had bothered Gary. John didn't normally share such personal information with clients, but Gary was different. They had known each other little more than two years, but they had an intense bond of the sort sometimes forged by shared struggle.

John found work a welcome diversion, and in it, he knew he was doing something Jesse had believed was important.

Six days after Jesse's death, Burke issued his decision. John thought the judge's previous orders had signaled a willingness to at least consider his novel and far-reaching arguments. He was right.

Burke didn't buy Rembrandt's argument that the withheld pathology reports of Drs. Naeye and Caffrey were "inconclusive"; they were clearly findings of complicated pneumoconiosis, the judge wrote, and were "more definitive" than the report by Koh.

He recounted the bitterly ironic scenes from Gary's first case—the Jackson Kelly lawyers' questioning of doctors about "all of the evidence"—and wrote that, by building its case around Koh's report while withholding contradictory reports from more qualified doctors, the defense had "skewed the medical opinions of its own reviewing physicians" and "deliberately misled each physician from whom it requested an expert opinion."

What's more, he continued, when John requested any withheld documents, the defense had "engaged in a course of conduct designed to conceal its actions; first denying the presence of the reports, then conceding liability to prevent their disclosure. While perhaps initially not concocted as such, Employer's knowledge and behavior is tantamount to a scheme intended to defraud its experts, the *pro se* Claimant, and the court."*

Burke then rejected the justifications Jackson Kelly had put forward—that the benefits system was intended to be adversarial, that there was no duty to disclose every piece of medical evidence, that lawyers had an ethical obligation to zealously represent their clients. "An expert's report cannot be considered to be solely a reflection of the evidence selected and provided by a party," the judge wrote. "If such were the case, an expert medical opinion could never be accepted as a reliable diagnosis."

Congress created a remedial system that, while adversarial, was intended "to be liberally construed" for the benefit of miners, Burke wrote. "Employer's 'zealous' representation strategy instills uncertainty and cynicism into a program intended to compensate miners disabled from black lung disease."

The firm's actions constituted fraud on the court, he concluded. The denial in Gary's first claim was invalid, Burke found, and Gary was entitled to benefits dating back to January 1997, when doctors had seen complicated pneumoconiosis on an X-ray.

It was a resounding victory, but John knew the case was not over. Rembrandt promptly appealed. The decision would have to make it through the Benefits Review Board and possibly a federal appeals court.

Still, John was glad to be able to call Gary with a bit of good news.

---

* In a later court filing, Jackson Kelly attorney Kathy Snyder argued that conceding the case was not an attempt to avoid disclosing withheld evidence. Rather, she wrote, the decision to fold "was due in part to economic reasons." Elk Run's insurer had informed the firm that it would grant Gary's state compensation claim, which could offset some of the federal payments, Snyder wrote.

* * *

March brought a glimmer of hope for the Fox family. UPMC notified Gary that he was a viable candidate; he was now on the list for a lung transplant.

Maybe the plans the Fox family members had discussed would come to fruition after all. They'd build a garage behind the house where Gary could do woodworking and fix up an old Volkswagen Beetle like he and his brother-in-law, Richard, had always talked about doing. He and Mary could travel. He'd finally get to spend more time with Terri.

That same month, Terri and Rick got engaged, putting another one of Gary's dreams tantalizingly close. He desperately wanted a grandchild, and he'd joked with Terri on occasion that "you know, these days you don't have to be married to have kids."

Once again, however, Gary and Mary were trapped in a sort of medical purgatory. If a set of lungs became available, they would have to be nearby, prepared for surgery on short notice. But that set of lungs might come now, in two weeks, or in six months. Time, again, was the enemy. Gary was so thin that he was "a little bit of nothing," as Mary later put it. Every time he had to be hospitalized, he was taken off the transplant list until he recovered.

When Gary wasn't in the hospital, he and Mary stayed in a building of apartment-style suites with a communal kitchen set up by the non-profit Family House, which provided an alternative to hotels for patients needing prolonged care and their loved ones. They would sit around playing cards to pass the time; there wasn't much else Gary was able to do. Once, the building's fire alarm went off and the elevators went out of service, and Gary had to slide down four flights of stairs on his butt; he didn't have the breath to walk.

Terri visited as often as she could. If Rick was able, he joined her. Gary sometimes scolded her about missing work; it hadn't been an option for him. "Dad, I got it," Terri would answer. "I've got days off. This is more important to me." When she was back in Morgantown, she'd call every day. Often, Gary couldn't carry on a conversation for more than a few minutes without becoming breathless.

On March 20, John recorded in his notes the key points of a voice mail from Gary: "Out of hospital, feeling better & *on transplant list*." Then, on March 28: "Still waiting."

The infections and breathing troubles seemed to be worsening. Hospitalizations now included stints in the intensive care unit. The nurses told Mary that Gary wouldn't let himself fall asleep; he feared he wouldn't wake up.

On Easter Sunday, April 12, 2009, the whole family gathered in Gary's hospital room. Terri brought the dinner she'd made, but Gary could barely eat. Mary gave Terri an Easter basket—part of a plot she and Gary had developed. Terri and Rick were in the midst of wedding planning, and Terri had told her parents that they didn't need to pay for anything. Gary, however, was bent on buying her wedding dress, and he and Mary had figured out a way to make her accept the money: They'd stuffed it inside the plastic eggs that Terri now opened.

The next day, Mary called Terri to tell her that Gary had taken a turn for the worse and the doctors were going to intubate him. Terri and Rick left work and drove up. When they arrived, Gary had a plastic tube in his throat to aid his breathing, and he could communicate only by writing notes on a piece of paper. On Tuesday, Mary had an early-afternoon appointment, and Terri told her to go—she and Rick were there; it'd be fine.

Terri stepped out for a moment, and when she returned, she saw Rick standing at the bedside talking to Gary. She overheard what her fiancé said before he realized she was there. "I heard Rick tell him he would always take care of me and my mom no matter what," she said.

Around two p.m., Gary's blood pressure dropped. Doctors administered the maximum dose of drugs to bring it back up, and he stabilized for a moment. But it didn't last, and Gary's heart rate began to slow. Monitors started alarming, triggering a frenzy of activity among the hospital staff.

By this point, Mary had returned. She, Terri, and Rick watched the team of doctors and nurses at work. Done right, CPR can appear violent. Terri had performed the procedure herself plenty of times, but it was different watching a stranger beating hard on her father's chest as he lay there, not responding.

Minutes passed; the doctors exhausted every option they had. They asked Mary if she wanted them to continue trying to bring Gary back. Mary and Terri talked. At 3:34 p.m. on April 14, 2009, Mary told the doctors they could stop.

Part Four

You'd better listen to the voices from the mountains
Tryin' to tell you what you just might need to know,
'Cause the empire's days are numbered if you're countin'
And the people just get stronger, blow by blow.

You'd better listen when they talk about strip minin'
It's gonna turn the rollin' hills to acid clay.
If you're preachin' all about that silver linin',
You'll be talkin' till the hills are stripped away.

You'd better listen to the cries of the dyin' miners,
Better feel the pain of the children and the wives.
We gotta stand and fight together for survival,
And that's bound to mean a change in all our lives.

In explosions or from black lung they'll be dyin',
And the operator's guilty of this crime.
But the killin' won't be stopped by all your cryin'.
We gotta fight for what we need, let's seize the time.

You'd better listen to the voices from the mountains
Tryin' to tell you what you just might need to know,
'Cause the empire's days are numbered, if you're countin'
And the people just get stronger, blow by blow.

—Ruthie Gorton, "Voices from the Mountains"

As John Cline's Subaru station wagon wound through the southern West Virginia coalfields, I watched the rolling landscape unfold from the back seat. Riding shotgun was eminent pathologist Dr. Francis H. Y. Green. We were on a pilgrimage to the small town of Sophia.

We had spent the past few days at the Pipestem Resort State Park for the annual conference put on by the state association of black lung clinics, hearing presentations on everything from how to navigate the benefits system to what was driving the recent resurgence of disease. Green, a lung specialist, had come all the way from the University of Calgary to attend, and two days earlier, I watched him deliver a presentation on a study he and eight fellow physicians had just finished, the first one to evaluate what this seemingly new form of black lung looked like under the microscope.

"CWP is changing," Green had told the crowd of clinic workers, lawyers, lay representatives, and miners as he clicked through PowerPoint slides, projecting images of diseased lungs on the screen behind him. The team of researchers had looked for subjects whose X-rays showed rapid progression of the disease and for whom tissue samples, either from a biopsy or an autopsy, were available. Ultimately, thirteen cases met these requirements, and almost all of them were miners from West Virginia. Most were continuous miner operators or roof bolters. Their median age was fifty-six. The findings were so consistent that the authors believed the results were representative of the larger population of miners.

The study's results reinforced the burgeoning scientific literature that researchers at the National Institute for Occupational Safety and Health were producing. Epidemiologist Scott Laney and his colleagues had proposed a few hypotheses to explain the rise in disease rates. One theory was that miners were working longer hours, meaning they spent more time breathing dust, and their bodies had less recovery time to clear that dust. Government data showed that the average number of hours a miner worked during a year had increased from about 1,800 in the early 1980s to about 2,400 in 2008.

Another possible contributor was the behavior of some coal

companies—cheating and exploiting legal loopholes when taking dust samples. Over the years, scientists had developed a reliable way to predict disease prevalence based on dust-sampling data. But when Laney and his colleagues looked at the prevalence of black lung in central Appalachia from 2005 to 2009, they'd found that there was more than twice as much disease as the dust samples suggested there should be. An obvious inference was that companies were submitting manipulated samples that didn't reflect how much dust miners were actually breathing.

Perhaps the leading theory: Miners were breathing dust that contained more silica, which can cause a nastier, more aggressive type of disease. Coal companies had mined out many of the thickest coal seams years earlier, but thanks to the economics of the industry and the advent of more powerful machines, it had become feasible and profitable to go after thinner seams. That often meant that miners had to cut through more rock above and below the seams, releasing finely ground silica crystals into the air.

Evidence of the increase in silica dust appeared to be showing up on miners' X-rays. A specific type of shadow on a film is generally regarded as a marker for a nodule commonly caused by silica. Laney and his colleagues compared data on surveillance X-rays taken during the 1980s with those taken since 1999, and they found that the proportion of films with this particular marker had almost quadrupled.

This last theory had found particular support in the study Green presented. A panel of pathologists had evaluated tissue samples from the thirteen miners in that study, and what they'd seen in most cases was different from the textbook description of coal workers' pneumoconiosis. In general, the deposits of coal dust were less obvious, and there was a heavier load of silica dust. The dust was more finely ground, meaning it was more harmful to the lungs and more difficult to see in the tissue samples. Only four samples showed classic CWP. Most had multiple types of scars—some caused by coal dust, some caused by silica dust, and some caused by both. The combination of more silica, less coal, and finer particles was a recipe for both rapidly progressing disease and misdiagnosis, Green told the audience.

Green then clicked through slides of tissue samples. One image showed a pink slab with swirls of white scarring. Another zoomed in on a particular area with flecks of black dust; when it was viewed under polarized light, tiny, jagged crystals glowed purple-white.

"This," Green told the audience, "is the new face of lung disease."

Slide by slide, Green told the story of an evolving epidemic as revealed by the lungs of thirteen miners—one of whom, unbeknownst to the audience, was Gary Fox.

Now, as John steered his car onto Main Street in Sophia, I talked more with Green about the disturbing findings he had recounted. Soon we arrived at our destination: the home of Dr. Don Rasmussen. The revered physician, eighty-seven years old and still working, was a regular presenter at the conference in Pipestem, so his absence this year had been palpable. Everyone wanted to know: Was Don okay?

He had fallen, but after a stint in the hospital, he had recovered and was back home. We found him in a wheelchair on his patio, eating lunch next to a flower garden.

"Francis got lost on the way back to Calgary and ended up in Sophia," John joked as we pulled chairs into a semicircle around Rasmussen. The bruising around Rasmussen's right eye and down his cheek was turning from purple to red-orange. He pointed to his face and named the bones he'd fractured in his fall. His injuries, he said, were "not a big deal."

The conversation quickly turned to the resurgence of severe black lung. Green, in jeans and a black T-shirt, repeated for Rasmussen the key findings of the study he'd presented at the conference. A pained look spread across Rasmussen's face. "It's a lot like acute silicosis," he said. Green nodded.

They searched for some explanation—maybe it was the increasingly powerful machinery grinding up finer particles or thinner seams leading to more silica exposure. As Rasmussen spoke, one of the cats that had been roaming the garden came over and leaned against his wheelchair. After a moment, it swiped at the half-eaten turkey sandwich on the tray in Rasmussen's lap, and without a pause in his speech or so much as a glance down, Rasmussen parried the cat's strike and continued his thought about silica.

The conversation seemed to enliven him, and I thought perhaps I was getting a small glimpse of the fire that had helped force lawmakers and coal-industry officials to recognize and compensate miners for black lung more than forty-five years earlier. He railed against those doctors—usually the ones working for coal companies—who performed blood-gas tests ("the Rasmussen test") in ways that masked disability. "That's a great way to screw a guy," he said. "I was saying that back in 1969."

He was eager to return to his work seeing miners. "I've got a big pile of cases on my desk at the office," he said. "When I go for physical therapy, I'm going to sneak in there."

Throughout the conversation, Green leaned forward as if he didn't dare miss a word. A physician of significant repute himself, Green seemed to be soaking up the teachings of a sage.

When we rose to leave more than an hour later, Rasmussen said, "It's good of you to come by."

"I couldn't come to West Virginia without taking advantage of this opportunity," Green replied.

Four days later, back in Washington, I received an e-mail from John: "Don had a mild stroke last Saturday. He was flown to Morgantown, had some speech impediment but apparently no other effects.... They expect him to be released soon and are hopeful that he will fully recover."

Six weeks later, John sent another update: "When I saw Don Rasmussen last Friday, he seemed to be improving although still weak, but on Saturday morning, they couldn't arouse him. When I saw him at the hospital tonight, he seemed to know I was there but couldn't speak, and his breathing was labored."

Two days later, John sent a one-sentence e-mail: "Don died about 30 minutes ago."

In the *New York Times* the following week, a picture of Rasmussen placing an X-ray on a light box appeared below the headline "Dr. Donald L. Rasmussen, Crusader for Coal Miners' Health, Dies at 87." The article detailed his role as a reluctant hero for miners and quoted James Green, labor historian and professor emeritus at the University of Massachusetts, Boston: "In the annals of American labor history, there is no one, no union official or a physician, who exceeds the accomplishments of Dr. Rasmussen in substantially reducing the causes of such a widespread and deadly disease as black lung or in enhancing the treatment of a group of afflicted workers."

Craig Robinson, who had known Rasmussen since the heady days of the 1969 wildcat strikes and worked with him to deliver the science on black lung to the miners themselves in rallies across the coalfields, gave a speech at his memorial service. "The black lung movement would not have happened without him," Robinson said. He described Rasmussen's compassion for his patients, his generosity with his time, and his commitment to scientific rigor. Rasmussen had endured withering criticism, but time had proven him correct. "The only remaining criticism of Don, one can imagine, is heard in the executive suites of coal companies and among their attorneys," Robinson said.

To me, it seemed fitting that just days before his death, Rasmussen had still been looking for answers—searching for a reason for the needless suffering of miners like Gary Fox and a way to put an end to it. Indeed, that day in his backyard, he, John, and Green had spoken of Gary. John had described the rapid disease progression he was seeing more often among his clients, and in no case, he'd said, had the deterioration been faster or more alarming than in Gary's. Rasmussen had documented that deterioration in exams just seven years apart. Green had observed the seeds of it in Gary's lung tissue. And John had seen the end result. Each now had a connection to Gary, and as they discussed his case, they seemed to share a collective sense of frustration and regret that went beyond any one man. Gary was a flesh-and-blood instantiation of an unfulfilled promise.

I had come to a similar realization about Gary. Everything I'd learned during the past few years, all the ways that the 1969 law had been systematically undermined, seemed to converge in the life of this one man. The political maneuvering, the chicanery and outright fraud in dust-sampling, the sway of well-credentialed and well-paid physicians, and the intentional concealment of potentially life-altering evidence in the benefits system—Gary had endured it all and chosen to fight. His death, it turned out, was far from the end of his story.

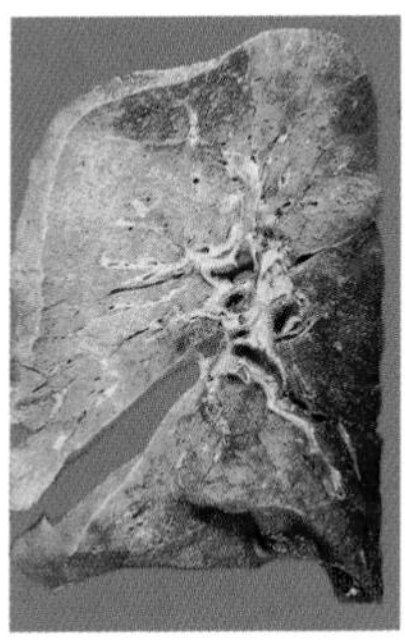

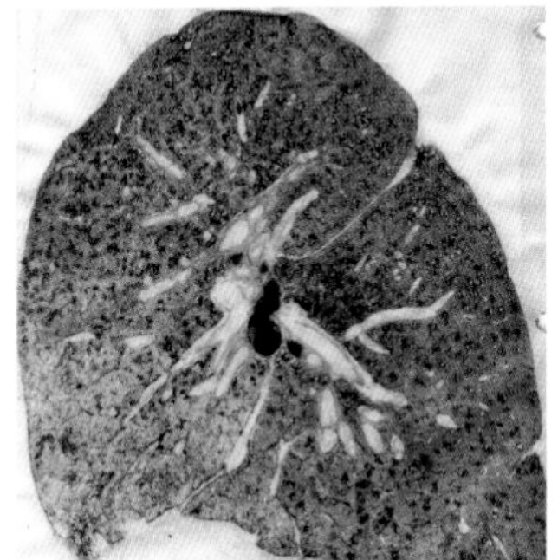

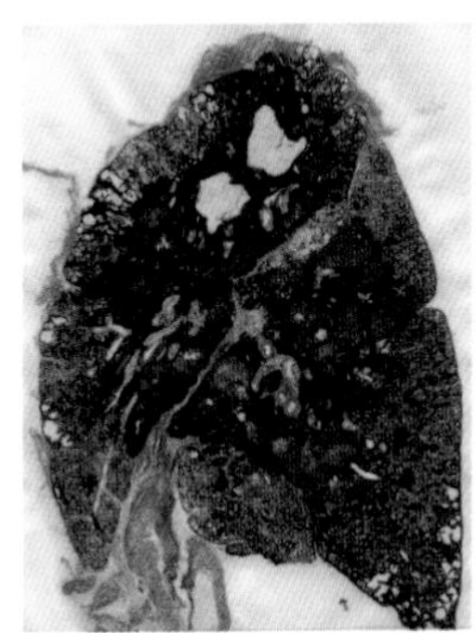

A normal lung next to those of miners with the simple and complicated stages of black lung. As coal dust accumulates in a miner's lung, it steadily destroys healthy tissue; the damage is irreversible. *(Courtesy of the National Institute for Occupational Safety and Health)*

John Cline at his desk at the New River Breathing Center in Scarbro, WV, which received federal funding to diagnose and treat miners with black lung. John worked as a benefits counselor, advising miners on how they could file state and federal black lung–benefits claims. *(Courtesy of John Cline)*

John Cline on a home construction site with carpenters Richard Gilbert and John Matthews and John's oldest son, Jesse. From roughly 1976 to 1987, John and his father, Crawford, ran a construction business in southern West Virginia that primarily built homes for people of modest means who qualified for low-interest loans through a government program. *(Douglas Yarrow, Courtesy of John Cline)*

John Cline's law school graduation photo, taken in 2002. He was 56 years old. *(Courtesy of John Cline)*

An aerial photo of part of Massey Energy Company's Elk Run mine complex near Sylvester, WV, where Gary Fox worked from 1993 to 2006. *(Lyntha Scott Eiler, courtesy of the American Folklife Center, Library of Congress, AFC 1999/008: CRF-LE-C037-07)*

An aerial photo of a sludge dam at Massey's Elk Run site. Similar impoundments dot the coalfields, holding back a toxic black liquid known as slurry that is the byproduct of washing coal in a bath of chemicals. *(Lyntha Scott Eiler, courtesy of the American Folklife Center, Library of Congress, AFC 1999/008: CRF-LE-C038-14)*

Massey Energy CEO Donald Blankenship during a Senate committee hearing on May 20, 2010 — about one month after an explosion at the company's Upper Big Branch mine in southern West Virginia killed 29 men. *(Andrew Harrer/Bloomberg via Getty Images)*

West Virginia miner Charles Caldwell. Missing evidence in his case sparked John Cline's suspicions and led to decades of fighting with the law firm Jackson Kelly PLLC over withheld medical reports. *(Courtesy of Patsy Caldwell)*

THE TRAGEDY OF BLACK LUNG

HON. ROBERT E. WISE, JR.

OF WEST VIRGINIA

IN THE HOUSE OF REPRESENTATIVES

Thursday, March 5, 1992

Mr. WISE. Mr. Speaker, I would like to intro-
ce for the RECORD a statement written by
ke South, a constituent from Beards Fork
V. Those who have been exposed to the
vastating effects of black lung realize how it
fects the lives of victims and their families.
For those who have not been exposed to in-
viduals with black lung, please read Mr.
outh's account of what it is like to live with
is disabling condition. I encourage my col-
agues to keep Mr. South's comments in
ind as Congress considers important legisla-
on that will affect the lives of thousands of
iners across the Nation.

STATEMENT OF MIKE SOUTH

To those who are members of this commit-
, we, the living and dead victims of Black
ng, appeal to your sense of humanity.

Those who do not suffer from lung disease
an in no way know the agony that it puts
amilies through. When you mention a per-
on suffering from lung disease it involves
he whole family. The spouse who tries to
ake over the tasks once done by her hus-
and who once did all the heavy physical
hores.

The children who sit and watch their fa-
er pant and gasp for breath from such sim-
e tasks as eating or speaking; and the man
mself who suffers even more than his fam-
y realizes.

He suffers in ways that others may con-
der foolish, especially his wife and chil-
ren. He feels that he is no longer an asset to
is family. He can no longer provide mone-
arily for the support of his family. He is
othing! He goes to doctors, but with little
no results, for his lungs worsen with time.
e takes his breathing treatments four
imes a day and stays on oxygen as rec-
mmended by his physician but yet he still
s his condition worsen as time goes by.

here are times during the long breathless
ights that he lies awake thinking how
uch longer he has to endure the suffering
e is going through. Times when he gasps for
reath and is asked if he is all right and he
esponds "yes". When in truth he often won-
ers if this might be his last gasp of life.

I would not be afraid to wager that not a
erson in this room knows what it is like to
et up from your bed and walk 10 ft. to your
athroom and be breathless before you get to
he toilet. To take a shower and have to rest
everal times during the procedure. To step
ut of the shower and into a thick terry
loth robe because you haven't the breath to
owel yourself dry. And, when you dress, it
eems like it takes forever to pull on your
rousers and especially to try to tie your
hoes.

The longing to be able to do at least a modicum of the things that you used to do in the past before death took hold of your life. A slow and agonizing death that takes away so many of life's simple pleasures. Not being able to play with your children or pet. The fire and passion that was so much a part of your life has been replaced by sedentary depression.

Many breathless hours are spent trying to do tasks that used to take minutes to accomplish. No more cutting the lawn, because you cannot push, or even less, walk behind the mower. Maintenance on the cars and home is out of the question.

Your life now consists of oxygen tubing and its 50 ft. life line. A line that you curse day after day. Your world consists of a 50 ft. radius in which you drag your life line like an extension cord. A cord that you sometimes wish were attached to the coal company executives and members of the Department of Labor.

If only they could spend 24 hours in your shoes. To get a taste of how worthless and lifeless your existence is. I wonder if then they would change their attitude towards those who suffer from lung disease.

I think not. They sit back and take their apathetic stance hoping the victims will die before any Black Lung claim is settled. And when the victim dies the claim goes with him, for the widow stands no chance to prove the existence of Black Lung in her dead spouse.

The parties involved know the hardships and years that are spent trying to prove that they are the "walking dead".

Some men spend anywhere from eight to sixteen years being shuffled from doctor to doctor trying to obtain evidence that company doctors say does not exist. I often wonder how these physicians can sleep at night, but I guess they just "blanket" themselves with the money given them by the coal companies.

It has to be the love of money and greed that fuels these physicians and companies, for compassion has no part to play. Human suffering (physical) is supposed to be alleviated by the healing compassionate hands of a physician; instead, these hands are stained "green" from the dyes of money and greed.

This stain has put many a miner in an early grave. A stain that has spread and engulfed a whole nation that has turned its back on the suffering that exists in the "death" of a miner. A "death" that means nothing to any one except the miner's family and friends. A nation that has put a man on the moon and won countless wars, yet the suffering still continues for the coal miner. A miner who has helped in all the endeavors this county has put forth. Yet when he is down with failed health, he is spurned by the nation that he helped lead to such greatness. A nation that is complacent in its attitude, that it does not affect me, it does not exist.

To the powers that be; listen I beseech you. Take a walk in my weary shoes and pass HR 1637; for without it countless numbers of deserving men and their families will suffer in the quagmire of red tape involved in the present system. A system established for the behest of big business and not the men, without whom they would not exist.

And, as they reap their enormous profits, they hire lawyers to protect their greed. A greed that does not encompass compassion for the men who die for their dollars. So, in reality, they trade "dollars for death" and think none the less for it. Has the nation become so callous that greed overrides everything that is supposed to be the make-up of human existence? Has common decency and compassion gone by the way of the grave? I would hope not, but from my point of view, it has; for it seems that the plight of the coal miner is forever to exist in poverty and suffering.

Could this distinguished body exist on $600 a month? I think not. Yet, that is all the monthly benefits that a miner receives from Black Lung. Some gentlemen pay $600 or more for their suits, yet a miner is asked to survive a month on that amount.

Members of Congress say they cannot sustain their lifestyle on less than $115 thousand dollars a year.

Slip on my size 8 shoes and live on my yearly income, and then ask for a raise. It comes to mind the words, "I once complained of no new shoes, till I saw the man who had no feet".

So remember, without good lungs *you* cannot perform your daily tasks, for without them the gift of speech means nothing. Pass HR 1637 and let those who deserve their right to breath, breathe a little easier. Thank you.

Statement by former coal miner Mike South that was introduced into the Congressional Record in 1992. South, also a gifted painter and illustrator, became president of the National Black Lung Association. *(Congressional Record)*

West Virginia miner Elmer Daugherty at age 80 in 2005. Withheld evidence in his benefits claim led to an ethics case before the West Virginia Office of Disciplinary Counsel. *(Courtesy of the Daugherty family)*

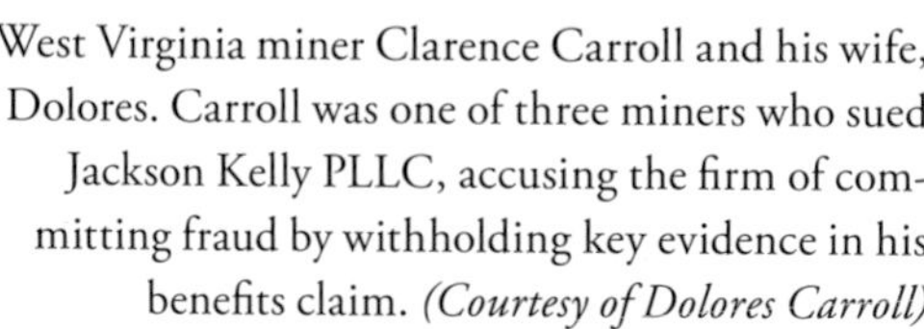

West Virginia miner Clarence Carroll and his wife, Dolores. Carroll was one of three miners who sued Jackson Kelly PLLC, accusing the firm of committing fraud by withholding key evidence in his benefits claim. *(Courtesy of Dolores Carroll)*

West Virginia miner Norman Eller underground in 1994. Eller also sued Jackson Kelly, alleging fraud. *(Courtesy of the Eller family and Shifoto Studios, Coal City, WV)*

Coal miner Steve Day served in Vietnam for the air force, married Nyoka Fortner soon after returning home to West Virginia, and spent more than thirty years underground, doing some of the dustiest jobs at the coalface. He eventually contracted advanced black lung. *(Courtesy of the Day family)*

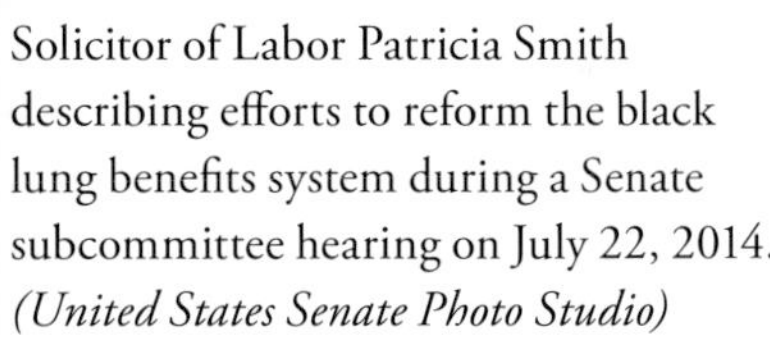

Solicitor of Labor Patricia Smith describing efforts to reform the black lung benefits system during a Senate subcommittee hearing on July 22, 2014. *(United States Senate Photo Studio)*

Former miner Robert Bailey Jr. speaking to senators during a subcommittee hearing on July 22, 2014. Bailey, who suffered from black lung and eventually received a lung transplant, became an advocate for reforms to the benefits system. *(United States Senate Photo Studio)*

John Cline speaking to senators during a subcommittee hearing on July 22, 2014. *(United States Senate Photo Studio)*

John Cline with Appalachian miners on Capitol Hill, July 2019. John frequently traveled with miners on lobbying trips to Washington, DC, walking the halls of Congress. *(earldotter.com)*

22

"An Affront to Justice"

John had seen clients die before. It always bothered him, but Gary's death felt different. "Something about Gary really got me," he said. "For a while, I couldn't talk about it without breaking down."

His client was now Mary Fox, and she was just as dedicated as Gary had been to pressing forward. In the months after Gary's death in 2009, Mary applied for, and was awarded, benefits as a surviving spouse, but Jackson Kelly continued to fight Gary's claim. The question was no longer what his lungs showed; an autopsy revealed a clear and extensive case of complicated pneumoconiosis. The question now was what the legal system would do about the firm's conduct.

As gratifying as administrative law judge Thomas Burke's decision finding fraud on the court had been for John, he worried it might not survive Jackson Kelly's appeal to the Benefits Review Board. In September 2012, the board issued the decision John had feared it would. By a two-to-one vote, the three-judge panel vacated the heart of Burke's ruling.

The majority concluded that the firm's conduct "did not rise to the level of fraud on the court," which was a grave finding reserved for "the most egregious of cases, such as the bribery of a judge or juror, or improper influence exerted on the court by an attorney, in which the integrity of the court and its ability to function impartially is directly impinged." Jackson Kelly's actions didn't meet the required standard of threatening broad harm beyond Gary's individual case, the majority found. Because the firm had conceded liability, the two judges concluded, the award stood, but benefits would date back only to 2006, not to January 1997, the point set by Burke.

Both judges in the majority had first been appointed to the board during the Reagan administration. Dissenting judge Betty Jean Hall, first appointed during the Clinton administration, wrote that she did believe Jackson Kelly's actions cleared the high bar for fraud on the court. The firm's fraud was "directed at the judicial process," and "its effects go beyond the parties involved in the case," she wrote. It was just the sort of conduct that the U.S. Supreme Court, in a seminal case on the issue, had deemed a "wrong against the institutions set up to protect and safeguard the public."

The question now was whether to appeal to the Fourth Circuit Court of Appeals. John was hesitant. "The last thing I wanted to do was argue before the Fourth Circuit," he said. "I didn't feel like I had the training or the confidence."

There was, however, reason to try. Whatever final decision came out of Gary's benefits case might have an impact on another legal action currently unfolding.

For years, Bob Cohen, John's compatriot in combat against Jackson Kelly, had been talking with fellow Morgantown lawyer Al Karlin about the problems with the black lung–benefits system, especially the firm's tactics. Karlin had a small private practice that mainly represented workers in employment cases, and he was troubled by what Cohen was describing. Cohen put Karlin in touch with John as Gary's case was playing out. The facts as John described them were compelling, Karlin thought. With Gary's permission, John shared updates with Karlin.

John had called Gary in Pittsburgh shortly before his death to see how he felt about a possible civil suit. The conversation was similar to the one they'd had when John asked Gary if he wanted to continue fighting after Jackson Kelly conceded liability in the benefits claim. As he had then, John described the cons to Gary first—it would be a long, unpleasant process with no guarantees that anything positive would come of it. Then he gave him the pros, the main one being that it might prevent similar behavior by the firm in the future. Once again, John asked Gary what he wanted to do—did he want to sue? He'd recorded Gary's response in his notes: "Yes. No one should have to go through what I have gone through."

Mary shared that commitment. Five weeks after Gary's death, John and Karlin had filed suit in state court accusing the firm of engaging in fraud to defeat Gary's first benefits claim. The goal of *Mary L. Fox v. Jackson Kelly PLLC* went beyond winning money damages. The complaint asked for court orders "declaring that the conduct of Jackson Kelly is unlawful" and "prohibiting Jackson Kelly from engaging in similar conduct in the future."

Karlin knew better than anyone else involved how difficult and time-consuming it would be to take on the state's oldest and largest law firm. But, as he later said, "We hoped that it would contribute to a change in what was seen as acceptable conduct in black lung litigation."

The priorities of Karlin's practice reflected his upbringing as a self-described "child of the '60s." During his adolescence in a Chicago suburb, his parents had instilled in him many of the same values that Crawf and Kay Cline had in John. "My parents taught me that life is supposed to be about more than making money, that we had a responsibility to others," Karlin said. After graduating from Yale University, he volunteered as a VISTA and was sent to Texarkana, Arkansas, where he'd worked on organizing efforts around various issues, including welfare rights. He had then earned a law degree from the University of California, Berkeley, and moved to West Virginia. "If you were a lawyer and you had a sense of philosophy or politics in your work, it was a place that was attractive where you thought you might be able to go do something and be useful," he said.

He had joined a nonprofit legal-aid organization in Morgantown and helped locals with everything from bankruptcies to divorces to applications for welfare or food stamps. In 1981, he started his own practice and made a name for himself representing workers who said they had been wrongfully fired or had endured sexual harassment or discrimination. He'd also represented injured workers, including victims of mine accidents.

Through his conversations with Cohen, Karlin had gained an appreciation of how complicated and adversarial the black lung–benefits system had become, and when Karlin spoke with John, he was impressed. "I think John is saintly in the way he devotes himself to his clients," he later said.

John and Al were now a team, both in Gary's benefits case and in the civil suit. The latter also included fraud allegations by additional miners. For many of the other benefits cases in which John had uncovered withheld evidence over the years, the statute of limitations had expired. But there were two that had unfolded at roughly the same time as Gary's. Both of those miners had also given the lawyers the go-ahead to sue Jackson Kelly.

Clarence Carroll was eighty-five years old when John began representing him in 2006. He'd retired after spending more than forty-five years in the mines, and he was trying to enjoy the time with his wife, Dolores. Clarence was a "snazzy dresser," Dolores said. "He didn't smoke. He didn't drink. He was just a gentleman." They enjoyed line-dancing, and they always held hands when they went for walks around Rainelle, their southern West Virginia home.

He'd filed for benefits previously and lost. The key question in his case, as in many of the others John handled, was whether the masses seen on X-rays were complicated pneumoconiosis or something else. Carroll's doctors saw complicated pneumoconiosis, but Jackson Kelly had submitted many more readings that were negative for the disease, including a few from the radiologists at Johns Hopkins. A judge had credited the defense experts and denied the claim.

When John took up Carroll's new claim, he filed his usual interrogatories. Even under Jackson Kelly's interpretation of the law, the firm had to turn over withheld reports if it had submitted others by the same doctors, and there were a number of these in Carroll's case. Among them was a report by Dr. Jerome Wiot—the same highly regarded radiologist whose withheld reports had played key roles in other cases—that classified one X-ray as consistent with complicated pneumoconiosis. Wiot had interpreted other films as negative for the disease or positive for the simple variety of it only, and the firm had submitted these readings but withheld the report indicating complicated disease.

Jackson Kelly had exhibited particular dexterity in its handling of the interpretations by radiologist Harold Spitz. The doctor reviewed a series of Carroll's X-rays and determined he had complicated pneumoconiosis, but Jackson Kelly withheld the report. The firm then returned to Spitz two years later and showed him only the last film in the series. This time, viewing the X-ray without the benefit of the context he'd been provided before, he saw simple disease with coalescence, meaning the nodules had started to cluster together but had not reached the complicated stage. The firm submitted that report.

Determining whether a pattern on film represents coalescent nodules of simple pneumoconiosis or masses of complicated pneumoconiosis can be difficult, even for experienced radiologists. One way to judge the reliability of an interpretation is to consider what other information the doctor was given. A reading of a series of films taken over the course of years is viewed as more accurate than a reading of a single film because it allows a radiologist to see the progression of the disease. The firm's lawyers had argued exactly that when they submitted serial readings by other experts in Carroll's case. But they did not submit Spitz's interpretation of the serial readings; they withheld that and submitted only his report on the single film.

In other words, Jackson Kelly concealed the report that, by the firm's own argument, was more accurate and asked the administrative law judge to rely on the less accurate one.

John filed a motion to compel discovery, asking for whatever else Jackson Kelly might have withheld, and the firm conceded the case. Not long after Carroll finally won benefits, his health began deteriorating rapidly. He couldn't eat, and his body weakened. The coughing spells became longer, the hospitalizations for lung infections more frequent. There were no more walks, let alone line-dancing. He died in 2011, leaving Dolores to continue the civil suit on his behalf.

The other case was that of Norman Eller. He, too, had lost a previous claim before John began representing him in 2007. After almost forty years in the mines, the last nine of them at Massey's Elk Run complex, where Gary had worked, Eller retired. He filed a benefits claim in 2001, and a CT scan performed that year became a crucial piece of evidence.

Jackson Kelly first sent the scan to Wiot, who noted in his report that some of the detailed lung images were missing from the scan. "Based on this single study," he wrote, "coal worker's pneumoconiosis cannot be excluded. In summary, this study is incomplete. Therefore, evaluation for the presence or absence of pneumoconiosis cannot be made."

Rather than submit that report or try to obtain a more complete scan, Jackson Kelly sent the same scan for interpretation by Dr. Paul Wheeler and another member of the Johns Hopkins black lung unit, Dr. William Scott. Wheeler was less explicit than Wiot had been, writing, "No pneumoconiosis but lung settings are incomplete." Scott's report, in its entirety, said, "Emphysema. No evidence of silicosis/CWP. Limited images at lung windows."

Jackson Kelly attorney Mary Rich Maloy presented Scott's report to pulmonologist George Zaldivar on the morning of his deposition and asked him about it. "There was no mention of any nodules that could represent retention of dust," Zaldivar said.

Maloy quoted the exchange in her closing argument and wrote, "Clearly, Dr. Zaldivar has obtained a complete picture of the miner's radiographic, CT scan, and clinical status." She concluded, "The CT scan evidence is negative."

Thus, Maloy cast the CT scan as affirmative evidence that Eller didn't have black lung when, in fact, Wiot had stated that it was basically useless for rendering a diagnosis one way or the other.

The judge credited the opinions of Zaldivar and another of Jackson Kelly's consulting pulmonologists, who had reviewed Wheeler's report but not Wiot's, and denied benefits.

When Eller filed another claim with John as his lawyer, John sent interrogatories. Because Jackson Kelly had previously submitted X-ray readings by Wiot, the firm turned over Wiot's report on the CT scan. This time, a judge awarded benefits.

By this point, however, Eller was growing increasingly breathless. His retirement dream—buying a mobile home, hitching a boat to the back, and working his way across the country one fishing spot at a time—was no longer realistic. He had to stop singing in the gospel choir at the Baptist church where he was a deacon.

He, too, died in 2011 as the civil suit trudged on. His wife, Linda, took up the cause until she also died. It eventually fell to Eller's son Norman to continue the case.

"My dad told me he wanted me and my sister to continue it," Norman said, referring to the lawsuit. "And it's not about money or glorification. It's about the precedent. He wanted to set the precedent to show what they do to coal miners.... His main thing was to bring it out, bring it to attention because maybe it wouldn't happen to the next person."

Eller had known that he didn't have long, and the case gave him a sense of purpose, Norman said. "This was a big deal for him, this case.... He'd already made his funeral arrangements and stuff, and then he decided—he donated his body to science. He told me, 'Look, my body's going to be in the ground. I'm going to heaven. Why not give them my lungs and anything they can do research on and find something that might help the next man?' It was after he got with the case—with John Cline and all them—that he decided to donate his body to science."

Each of the three cases—Fox's, Carroll's, and Eller's—cited other claims John had handled over the years and accused Jackson Kelly of engaging in fraudulent misrepresentation as a pattern of practice. A judge consolidated them and treated them as essentially one case going forward.

When he and Al first filed the case, John had dreamed of getting to look behind the curtain, of scouring Jackson Kelly's files on an array of cases. By the time I met John, in mid-2012, he was just hoping the case wouldn't get thrown out. As the reality of the undertaking became clear, John recalled why he'd disliked his class on civil procedure in law school.

Jackson Kelly tried a variety of tacks to hamstring or kill the lawsuit before it got started. The lead attorney for the firm's defense team was

Jeffrey M. Wakefield, a top lawyer at another of the state's largest firms, Flaherty Sensabaugh Bonasso. A seasoned litigator, Wakefield commonly defended businesses in what his biography referred to as "high exposure cases." He filed multiple motions outlining an array of reasons why the court ought to dismiss the case.*

John and Al knew Wakefield's arguments would be enticing for many judges. Allowing the case to go forward likely would mean signing up for years of tedious litigation in a case with complex and unfamiliar questions of law, made all the more fraught by the elite status of the defendant in state legal and political circles. Tossing the whole thing on jurisdictional grounds would be an appealing way out.

To overcome this natural tendency, they would need to present a compelling human story and land before the right judge. Some had the reputation of being plaintiff-friendly, but a sympathetic ear was not the trait John and Al thought most important. They hoped the lawsuit would be assigned to a judge with a sharp and independent mind, someone who could understand the complexities of the case and who wouldn't be beholden to or intimidated by Jackson Kelly.

They were encouraged, then, when the case was assigned to Raleigh County circuit judge H. L. Kirkpatrick III. He had grown up in the coalfields and attended the same high school that Gary Fox had. After graduating from West Virginia University's law school in 1976, he had spent more than twenty years in private practice before becoming a judge.

Relatively early on, John and Al got the sense that Kirkpatrick understood the case and was taking it seriously. His response to Wakefield's contentions seemed to confirm as much. "At first blush," Kirkpatrick wrote in a 2012 ruling, "this court finds Jackson Kelly's argument to be a persuasive and attractive resolution to these cases, at least on a temporary basis.... Frankly, this court does not relish the prospect of deciphering complex and highly technical black lung regulations that have

* Wakefield did not respond to my requests for an interview.

been charitably characterized as 'Byzantine' in nature.... Nonetheless, this court is no stranger to administrative appeals of all types and complex civil actions that are steeped in nearly incomprehensible rules and procedure."

Indeed, Kirkpatrick himself had handled a variety of administrative proceedings in his days of private practice. That included some state and federal black lung–benefits claims, and he'd represented both miners and companies. He rejected each of Wakefield's arguments and noted that there was another case that further undercut Jackson Kelly's defense. The state disciplinary investigation of the firm's Doug Smoot over his behavior in the benefits claim of miner Elmer Daugherty that had begun in 2006 was finally over, and the conclusion had a clear relevance for the lawsuit that Kirkpatrick could not ignore.

"So there's going to be some tension in this room over the next two days," Jackson Kelly's general counsel Stephen Crislip told the three-person panel seated before him. "It's just a fact. Because here you have a guy with a spotless record and his professional life on the line. There will be tension."

It was June 18, 2009—five years after Bob Cohen discovered the missing part of a key medical report in Daugherty's case, five years after attorneys from Jackson Kelly had stood before administrative law judge Michael Lesniak and refused to comply with his order to turn over any withheld documents that might exist, three years after a federal district court judge had referred the case to the West Virginia Office of Disciplinary Counsel. That office had finished its investigation, and an investigative panel had charged senior Jackson Kelly attorney Doug Smoot with multiple ethics violations. Sanctions up to disbarment were possible.

Now, in a hearing room in the state judicial building in Charleston, some of Jackson Kelly's top lawyers mounted a vigorous defense of one of their own, a mainstay of the firm for almost three decades and mentor to many of the lawyers who had come through the black lung unit over the years.

This was not the usual disciplinary case, Crislip told the panel, which was composed of two lawyers and one layperson, a professor of education. Smoot was not some hack who had shown up drunk in court or bilked his clients. He was a lawyer of national repute and a pillar of the community. He was active in his church, gave his time and money to organizations serving the homeless, sang in a traveling gospel chorus that had given almost eighty performances the previous year. He was a regular on the list of Best Lawyers in America, was admitted to practice in multiple federal circuits, and had sat at the counsel's table before the U.S. Supreme Court. He'd lectured at multiple universities and given training seminars. His disciplinary record was unblemished.

"I think what you're going to be dealing with here in the next two days is also very different for you in the world—the legal world in which Mr. Smoot lives and operates," Crislip said, "because he is a member, which means he's one of the owners, of a firm that opened its doors here in 1822 in this city and has practiced for 187 continuous years in this city. Now, that's 46 years after the country started up. It's 41 years before the state started up. I tell you that because its lawyers, almost 600 now in that period of time, were involved in the formation of the state, the first justices on the Supreme Court, the formation of a voluntary Bar in the 1880s, the formation of the mandatory Bar that we all belong to in the 1940s, and they pride themselves—they have prided themselves all these years that ethics and professionalism are their biggest assets. That is what they're the most proud of."

The three-person panel was to determine, after this two-day hearing, whether Smoot had violated any ethical rules and then issue a report with its findings and recommendations to the West Virginia Supreme Court of Appeals, which would ultimately decide whether to impose sanctions. The basic facts were not in dispute. Both sides agreed that Smoot had sent Daugherty to be examined by Dr. George Zaldivar, that he had received all of the doctor's findings in one envelope, and that he had submitted only the test results and not the doctor's report stating his conclusions. The question for the panel was whether this conduct ran afoul of the state's ethical code.

The core of Jackson Kelly's defense: Under the rules of the federal black lung–benefits system, Smoot was not required to turn over everything he'd received from Zaldivar unless the other side requested it; therefore, he could not possibly have violated ethical rules.

For the Office of Disciplinary Counsel, the key question was not whether Smoot had violated the rules of the federal black lung–benefits system. Regardless of any disclosure requirements, the office argued, Smoot had misrepresented Zaldivar's opinion by withholding a critical portion of his report and putting forth the partial report as if it were the whole thing. This conduct, the office alleged, clearly violated the provisions of the state's ethical standards for "fairness to opposing party and counsel" and "misconduct."

The office's first witness, testifying under subpoena, was Bob Cohen, who no longer represented miners in benefits claims. In 2008, the U.S. Senate had confirmed his nomination for a post on the Federal Mine Safety and Health Review Commission, the body with final authority to rule on contested citations for safety and health violations. Yet he had not lost his burning outrage (or his mop of curly hair), and his command of the benefits system's intricacies remained sharp.

Jackson Kelly argued that Smoot had, in fact, provided the best evidence to Daugherty: the test results and the X-ray interpretation form. The narrative report, the firm contended, was equivocal and inconsistent; there was no attempt to mislead Daugherty or hide something from him.

Cohen explained that, to the contrary, the narrative summary was "enormously significant" and "much more" useful than a mere *X* marked on a box of an X-ray form. It was the sort of detailed medical opinion that was often critical in deciding claims. Withholding the narrative, Cohen said, fit within a pattern he'd seen in cases against Jackson Kelly. He outlined the pyramiding—providing consulting pulmonologists with only favorable reports and withholding any others—that he said he and John had uncovered over the years. This practice, Cohen said, "occurred in case after case after case after case."

The next witness was Smoot himself, and he had a few defenses prepared. First, he hadn't separated Zaldivar's report. They might have come in the same envelope, but "I considered them two different types of reports"—an objective set of test results and a subjective narrative. Because he found the narrative was "not only equivocal, but contradictory from one page to the next," he decided not to use it, so "there was nothing in the federal Black Lung Act or regulations that required me to submit Dr. Zaldivar's narrative report unless requested by one of the parties."

Did it seem likely, Andrea Hinerman of the disciplinary counsel's office asked, that a seventy-five-year-old retired coal miner with an eighth-grade education and no lawyer at the time would know to file a request?

"But in the federal black lung practice," Smoot said, "I do not make a distinction and there is no legal requirement to make a distinction when you are dealing with a pro se claimant versus a claimant with a lawyer."

When Crislip's turn to question Smoot came, he asked how the disciplinary proceedings had affected him personally.

"Well, it seems like it's dragged on for years, this whole process," Smoot said. "It's taken a toll on my health. And yet a case where I have done nothing wrong legally, morally, or ethically in my opinion in following the black lung regulations and in litigating a black lung claim.... And to be convicted of an ethics charge for something that you're doing within the meaning of the rules, I mean, I would rather personally give up my law license than get an ethics conviction."

Many of the arguments he put forward, however, strained credulity. In the cover letter to his first submission—Zaldivar's test results only—he had called the document the "exam report." But when submitting a different doctor's report in Daugherty's case, Smoot had used the same term to refer to both the test results and the narrative. Even that doctor, who had also examined Daugherty and reviewed other records, wondered whether something was missing from the copy of Zaldivar's report he'd been given, noting, "The usual summary letter from Dr. Zaldivar is not in the packet." Zaldivar's narrative itself hardly seemed as equivocal as Smoot contended. The doctor stated: "In my opinion... Mr. Daugherty has a combination of simple and complicated pneumoconiosis, emphysema, and old tuberculosis," and his total disability "is the result of a combination of emphysema and coal workers' pneumoconiosis."

These inconsistencies and their implications, though, might well have been lost on a panel of three people who had never been exposed to the vagaries of black lung–benefits law. The disciplinary counsel had to hope that the panelists had followed Cohen's testimony.

When it was Jackson Kelly's turn to present its defense of Smoot, the firm focused on the murky legalities of the benefits system. To evaluate Smoot's conduct, Crislip had told the three panelists earlier, they would need to interpret this system's rules, "so we need to educate you a little bit." The firm now turned to the list of witnesses it had lined up to deliver a crash course, calling on some of its well-placed connections to undercut the disciplinary counsel's case.

At Jackson Kelly's request, retired Labor Department administrative law judge Rudolf Jansen had reviewed records and written a report with the ultimate conclusion, "I see no misconduct or wrongdoing of any type." He now described for the panel the black lung regulations, interpreting them somewhat differently than Cohen, and repeated his conclusion.

Under cross-examination by Hinerman, Jansen acknowledged that his first contact with Jackson Kelly about the possibility of testifying in this case came during his retirement party three years earlier when the firm's Bill Mattingly, his former law clerk, mentioned it. He also acknowledged that, for preparing his written opinion, Jackson Kelly had paid him "something in the area of $19,000," and that he would be billing the firm three hundred dollars an hour for his testimony. For his travel time, he was charging a "reduced rate" of two hundred dollars an hour.

Privately, some current administrative law judges were disgusted by their former colleague's actions. They circulated his written opinion by e-mail, and a few offered their assessments.

One: "Seems like Ethics 101 that when a lawyer decides to submit evidence, he does not alter it; that's called tampering. Shame on Ruddy [*sic*]."

Another: "I strongly suspect that given the length and complexity of Jansen's letter, he did not write it himself and it was probably written by Bill Mattingly, his former law clerk, or some other attorney for Jackson and Kelly. I wonder how much Jackson and Kelly paid Jansen for this tendentious letter purportedly representing his impartial view of this case."

And another: "I have read his paper twice now. Very sad and disappointing. If a lawyer had made those arguments to Rudy as a sitting judge he would have gone ballistic. I believe we will be discussing this procedure and JK for a long time."*

On the second day of the disciplinary proceedings, Jackson Kelly called on Forest J. Bowman, whom Crislip had earlier described as "easily the leading expert in the east coast on ethics and professionalism." He had served for eleven years on the state bar association's legal ethics committee, chaired the disciplinary board's investigative panel for a few years, given ethics seminars in more than thirty states, and taught ethics at the West Virginia University College of Law for twenty-three years. Many of the people sitting in the hearing room had learned ethics from him.

He had also taught John Cline. It was Bowman who had seemed to shrug off the documents John showed him, leading John to wonder if it might be because he held the title Jackson Kelly Professor of Law.

Addressing Smoot's conduct, Bowman told the panel: "If it's perfectly proper to do this within the black lung law, then it's perfectly proper to do it. It's not a violation of the ethics code." Pressed by Hinerman, he acknowledged that he had never been involved with a federal black lung–benefits case before and that his knowledge of the system was based on conversations with Jackson Kelly attorneys. His role as an ethics expert, he said, was to reach a conclusion based on a given set of hypotheticals. The hypothetical in this case: Assuming that the rules of the benefits system didn't require disclosing the entire report, would Smoot's conduct constitute a violation of the state ethics code? His answer was, emphatically, no.

Bowman, too, had connections with Jackson Kelly and was being amply compensated for his testimony. Aside from the professorship, he had worked for about six years as a consultant for Acacia Business Solutions,

* Jansen did not respond to my messages asking to discuss his work on the Smoot disciplinary case.

a company owned by Jackson Kelly. He gave speeches and seminars on business ethics and had an office in Jackson Kelly's Morgantown location, he said, but he was separated from the firm. "I had a wall built around me," he said.

He'd already received $8,000 for his time preparing a written report and meeting with Crislip, he said, and he would be charging three hundred and fifty dollars an hour for his time traveling, sitting in the hearing room the previous day, and testifying. "I bill portal to portal," he said.*

When Bowman was finished, Mattingly testified in Smoot's defense, reiterating the firm's main arguments. Jackson Kelly concluded with three witnesses who had frequently been Smoot's opponents in benefits cases, two lawyers and one lay representative. All three praised Smoot's character and expertise, and the lawyers said they believed Smoot had done nothing wrong in the Daugherty case.

Jackson Kelly's unapologetic defense of Smoot's actions over the course of the two-day hearing raised a question: How many other times had he or another of the firm's lawyers done something similar? During his testimony, Smoot estimated that he was involved in at least a hundred and fifty cases in an average year, and Hinerman asked him, "Now, was this an isolated incident on your decision to withhold part of Dr. Zaldivar's report in your practice, since that's your entire practice?"

"No," he replied, "it's not an isolated incident."

The panel issued its report on March 30, 2010. Jackson Kelly's defense strategy, it seemed, had worked. The panel credited the defense witnesses' testimony that it was not unusual to withhold documents or submit only parts of them in black lung cases. "This Panel is bothered by this practice but is constrained by the evidence in this case," the report said.

---

* When I spoke with Bowman almost a decade later, he said that he had since retired and didn't remember specifics about the Smoot disciplinary case. I asked whether his connections to Jackson Kelly or the money he'd received for his testimony had influenced his evaluation. "No," he said. "I had too much at stake for my reputation to do that. I wouldn't cave in for him [Smoot] or anyone else."

Because Smoot's conduct didn't violate the rules of the black lung–benefits system, the panel concluded, it also didn't violate state ethical standards. The panel recommended dismissing all charges against Smoot.

The Office of Disciplinary Counsel filed a brief urging the state supreme court of appeals not to accept the panel's recommendation and instead sanction Smoot. Meanwhile, the small community of miners' advocates, fearing a precedent that would further embolden Jackson Kelly, decided they could no longer watch from the sidelines.

They filed two *amicus curiae* briefs, one on behalf of the National Black Lung Association and the Appalachian Citizens' Law Center and the other on behalf of the United Mine Workers of America. Both argued that Smoot's conduct actually was far from a commonly accepted practice and that he had misrepresented critical evidence, a clear violation of ethical rules. If the state's high court allowed the panel's recommendation to stand, both briefs contended, gamesmanship and misrepresentation would become the new norm. The court's decision, the UMW brief concluded, would determine "whether any person whose life's work has caused debilitating mortal disease can appear before a system of justice in West Virginia that prohibits officers of the court from misleading him. Beyond the Black Lung arena, this case will set the standard: whether justice is a right available to even the most vulnerable citizens, or a game tilted in favor of unscrupulous players."

The court issued its decision on November 17, 2010. It didn't buy Jackson Kelly's argument that Zaldivar's report was actually two reports, nor did it believe the firm's claim that taking apart reports was commonplace. Nothing in the black lung rules made it acceptable to submit only part of a report and present it as if it were the entire thing.

The court also rejected Jackson Kelly's argument that Smoot had not tried to mislead anyone because he had submitted the test results and the X-ray interpretation form. Those things, the court wrote, "would not immediately inform either the ALJ [administrative law judge] or Mr. Daugherty, an unrepresented claimant with no expertise in the area of black lung evaluations, of the conclusion that was plainly stated in the withheld narrative portion of Dr. Zaldivar's report."

The court continued: "We have little difficulty concluding that Mr. Smoot's conduct was deceitful, dishonest, a misrepresentation, and prejudicial to the administration of justice." It found that Smoot had violated the ethics rules for "fairness to opposing party and counsel" and "misconduct."

Turning to its consideration of the penalty to be imposed, the court wrote:

> It is apparent that he lacks remorse and has refused to acknowledge the wrongful nature of his conduct. The claimant who was deceived by Mr. Smoot's conduct was quite vulnerable. Mr. Daugherty was a seventy-four-year-old man with a limited education who was acting *pro se* at the time of Mr. Smoot's misconduct. We also find the seriousness of the conduct to be an aggravating factor. Submitting an altered report to a tribunal is an affront to justice that simply cannot be tolerated. Finally, notwithstanding the excuses that have been provided by Mr. Smoot to explain his conduct, we find the evidence is sufficient to establish that he acted with a dishonest and selfish motive by advancing the interests of his client above the integrity and fairness of the litigation process.

The court concluded, "Mr. Smoot's license to practice law is suspended for a period of one year."

On the day the court released its decision, the associate chief of the Labor Department administrative law judges e-mailed a copy of it to his colleagues and wrote, "The Court provides a very welcome decision to clearly refute Mr. Smoot's argument that this was an acceptable practice in Black Lung litigation."

Other judges weighed in to applaud the court's decision. One called it a "powerful deterrent to future attorney misconduct." Smoot's conduct, another wrote, was "unethical and unconscionable."

Though the court's decision was a bit of vindication for Bob Cohen, he derived no joy from the outcome. "The legal ethics system is built so that, if somebody does something really bad, you can make him a fall guy, and the system goes on as before," Cohen said. "Smoot was the fall guy. It's too bad. I like Doug.... I never wanted to hurt Doug. I was hoping to get a judge to say that this system of not disclosing evidence to the other side and suppressing evidence was illegal and to enjoin it. It was an attack on Jackson Kelly, not on Doug Smoot, the strategy that Jackson Kelly used."*

It remained unclear what deterrent effect Smoot's suspension might have. Three other Jackson Kelly attorneys—Bill Mattingly, Kathy Snyder, and Dorothea Clark—had also been under investigation by the Office of Disciplinary Counsel because of their involvement in the Daugherty case. That investigation focused on the lawyers' refusal to comply with Lesniak's orders to disclose withheld reports. Jackson Kelly's defense was that the only way to challenge the judge's order before an appeals court was to disobey it and face sanctions. The investigative panel declined to file charges but warned the three attorneys not to disobey discovery orders in future cases.

Attention now turned to the two Fox proceedings. The firm had filed another motion to dismiss the civil suit, and Judge Kirkpatrick had put his decision on hold, waiting to see what the Fourth Circuit Court of Appeals would decide in the benefits case. In a brief to the appeals court, the firm's Kathy Snyder asked the court to hold oral arguments because of "the significance of this issue, not only to this claim, but to many pending and future administrative claims."

John and Al poured their energies into an eleven-thousand-word brief that they hoped would convince the appeals court to overrule the Benefits Review Board and reinstate administrative law judge Thomas Burke's finding of fraud on the court. John was happy to have Al handle the oral argument, which was scheduled to take place on October 29, 2013. Al began honing his words and practicing.

* * *

* As of November 2019, the West Virginia State Bar's website indicated that Smoot had an active law license, but Jackson Kelly's website indicated that he was no longer with the firm. Smoot did not respond to my interview requests.

"I represent the widow of Gary Fox, who, when he was fifty years old, had a hearing to try and get black lung benefits so that he could get out of the coal mines and exposure to dust," Al began.

He stood behind a podium facing a three-judge panel of the U.S. Court of Appeals for the Fourth Circuit, located in Richmond, Virginia. A few minutes earlier, he and John had ascended the steps of the granite-and-limestone edifice that had endured since its construction in 1858, even though much of the rest of the city burned when the Confederate Army fled the capital in the final days of the Civil War.

With John watching, Al started to tell Gary's story for the judges. He was barely three minutes in, however, when Judge J. Harvie Wilkinson III interrupted. Wilkinson, sixty-nine, had occupied a seat on the Fourth Circuit since 1984, when President Ronald Reagan appointed him. He'd been one of President George W. Bush's finalists for the Supreme Court seat that ultimately went to John Roberts in 2005.

"I don't think what the company did here was admirable by any stretch of the imagination," Wilkinson began in his soft Virginia accent, "but the question I have is, should the courts really get into how a party handles its expert witnesses?"

"Most of the time," Al replied, "the answer would be absolutely not. But there are occasions that go well beyond that." He tried to elaborate, but Wilkinson cut him off.

Already, Wilkinson was steering the argument in a direction not wholly unexpected. In general, lawyers had a possessive attitude toward expert-witness evidence, and judges were usually reluctant to get in the middle. Each side would attack the other's experts, and the truth would emerge—that was the core of the American system of justice. Jackson Kelly had hammered that point in its brief, writing, "The Claimant is essentially seeking to remove the adversarial aspect of the federal black lung system."

Al understood that the position he and John had staked out would make some lawyers and judges uncomfortable. But they weren't arguing for an end to the adversary system; they were arguing that it had its limits, that there was a difference between legal combat and fraud. Now,

Al tried to elucidate that distinction. Jackson Kelly's own pathologists had essentially told the firm it had no case, he argued, but the lawyers had nonetheless pressed on and presented an argument they knew to be misleading. They had done this to an unrepresented miner, "sending him back into the mines where six years later he came out needing a lung transplant and died just a few years later," Al said.

Wilkinson was not swayed. Wasn't there some way to address what Jackson Kelly did other than the extreme finding of fraud on the court?

Attempting to assuage the judges' fears that ruling in his favor would unleash a torrent of other cases alleging fraud on the court, Al tried multiple times to articulate the "very narrow and tight opinion" he hoped they would write, but the judges kept cutting him off with increasingly skeptical statements phrased as questions. He could feel the argument veering dangerously off track but couldn't find a way to pull it back.

The light on the podium indicating he was almost out of time was on. He tried to return to his central point, but Wilkinson again cut him off: "Why don't you bring this up on rebuttal."

The microphone on the podium picked up Al's deep sigh. He took his seat. Now the podium belonged to Al Emch, the former Jackson Kelly CEO who had helped argue in Smoot's defense during the West Virginia disciplinary proceedings.

"Judge Wilkinson, you hit the nail directly on the head immediately," Emch said. Al was asking the judges, Emch said, to impose a sweeping new duty on lawyers to disclose every report from every expert to all other witnesses, to the other side, and to the judge—an obligation that most judges would find distasteful. That was not what Al was arguing, but he would have to wait for his few minutes of rebuttal to try to set the record straight.

When Emch's time elapsed, Al had a few more precious minutes at the podium. By this point, the judges had clearly signaled how they were likely to rule. The fraud-on-the-court argument seemed lost. Al's mission now was damage control. He feared that the court's opinion might state or imply that what Jackson Kelly had done was really not that bad. Such a ruling could have serious consequences for both future black lung claims and the ongoing civil suit. "I want to emphasize, because I'm concerned not just about the outcome but how the opinion is written, that there are serious problems within the black lung system with how some attorneys litigate cases," Al said. Moments later, he added, "Your Honor, my concern is that you not write an opinion that encourages the continuation of this."

He tried again to describe the facts and clarify that the real problem wasn't nondisclosure but rather misrepresentation. But Wilkinson continued to cut him off, his voice rising higher: "I mean, what are we supposed to do? Sit up here and tell a party how it's to litigate its case?"

"No, Your Honor, if I can explain," Al replied. Amid the questions, he tried to press on. "But our point—"

"Thank you, Mr. Karlin, you're out of time."

After a few final arguments from Emch, the gavel brought the proceedings to a close.

When Al was on, he was as good as they came. He had a knack for weaving humanity into legal reasoning, a trait honed during years of telling workers' stories. Today, however, he'd felt unable to right a ship that was clearly foundering. It was, he said, "not one of my better days in court." John recalled, "He was in a funk for a while after."

A couple of months later, the Fourth Circuit issued the decision that the judges had telegraphed during oral arguments. The unanimous opinion, written by Wilkinson, found that the firm's actions, "while hardly admirable," hadn't cleared the high bar of fraud on the court. In the court's view, Jackson Kelly's conduct didn't undermine the integrity of the legal system but rather affected a single litigant, Fox. It was the sort of conduct that was supposed to be detected by the "self-policing" adversary system.

What Wilkinson wrote next infuriated John: "Fox had the right to cross-examine Dr. Koh regarding his qualifications and conclusions.... He had the right to cross-examine Elk Run's other experts to test their understanding of and reliance on Dr. Koh's report. He had the right to question the apparent lack of additional pathology reports. He had the right to present a contradictory medical opinion from a pathologist of his own choosing. That he did none of those things is not so much an indictment of the adversary system as it is a statement that he did not fully avail himself of it." In other words, it was Gary's fault. If the benefits system was unfair to miners like Gary, Wilkinson contended, it was up to Congress, not the courts, to do something about it.

In a minor victory, Al's damage-control efforts seemed to have affected at least the phrasing of the court's conclusion: "We bestow no blessing and place no imprimatur on the company's conduct, other than to hold that it did not, under a clear chain of precedent, amount to a fraud upon the court."

John, Al, and Mary could appeal to the U.S. Supreme Court, but the odds of that court taking the case were beyond slim. More likely, the attempt would eat up precious time and money and end in disappointment. They decided to accept that it was over. They had technically won the benefits claim, but when it came to the larger goal of forcing a change in Jackson Kelly's practices, they had lost. Worse, the Fourth Circuit decision gave the firm ammunition that it put to good use in the civil case, placing it at the heart of another brief seeking dismissal of the lawsuit.

Almost five years after Gary, Mary, John, and Al had committed themselves to ensuring that what had happened in Gary's case would never be repeated, the civil suit seemed to be their last remaining hope. They had to find some way to keep it alive.

23

"Lack of Scientific Independence"

From Beckley, I drove west, following the path Gary Fox had taken on his way to the Birchfield mine—past the dense forest, the denuded landscapes left by strip-mining, the cycle of post offices, gas stations, churches, and board-and-batten homes. Eventually, I steered the car to a roadside patch of gravel near the base of Bolt Mountain. A few minutes later, a dark sedan appeared in my rearview mirror. It was obvious from the portable oxygen tank sitting in the passenger seat that this was the man I'd come to meet.

His name was Steve Day. He was sixty-six years old, with close-cropped gray hair and a reserved demeanor. Nine years had passed since he'd reluctantly taken his doctor's advice and retired after working more than thirty years underground, much of that time spent running a continuous mining machine for a subsidiary of one of the nation's top coal producers, Peabody Energy. A year later, as his breathing worsened, he filed a benefits claim.

The outcome of his case appeared to be a stark example of a pattern I'd been documenting. When I met Steve in April 2013, I was partway through the process of populating a database with the 3,400 X-rays that Dr. Paul Wheeler of Johns Hopkins had read in cases decided since 2000. As I logged these readings, I also compiled a list of cases in which miners seemed to have particularly strong evidence in their favor but had lost primarily because of Wheeler's readings. I located as many of them as I could, called, and discussed their cases and their health with them.

Among the numerous conversations I'd had with miners, my call with Steve stood out. After filing his benefits claim, he had chosen Dr. George Zaldivar to perform the exam paid for by the Labor Department, not knowing that the doctor was often the choice of defense firms. Yet even Zaldivar had found complicated pneumoconiosis. An X-ray showed masses on Steve's lungs, and pulmonary function tests revealed total disability.

The company's lawyer, Paul Frampton of the firm Bowles Rice McDavid Graff and Love, had sent Steve to be examined by Dr. Robert Crisalli. Based on the X-rays and CT scan, Crisalli, too, had determined that Steve had pneumoconiosis.

Frampton, however, submitted negative readings of the X-rays and CT scan by Wheeler and his colleagues Drs. William Scott and John Scatarige. In a deposition, Wheeler rattled off his usual criteria, used the same metaphors, told the same anecdotes, and explained that Steve most likely had a healed case of tuberculosis, or maybe histoplasmosis or mycobacterium avium complex (bird TB). Not black lung.

Zaldivar stuck to his guns during a deposition. Frampton suggested other explanations for the masses on Steve's X-rays and CT scan. Could it be old TB or histoplasmosis? It was possible, Zaldivar answered, but given the size of the masses, Steve likely would have been extremely sick at some point and required hospitalization; there was no record of that happening.

Crisalli, though, had changed his opinion after being given the Johns Hopkins reports. At his deposition, he said he believed Steve did not have black lung. "The basis for the conclusions," Crisalli said, "primarily centers around the imaging."

In his written closing argument, Frampton emphasized the readings by the "exceptionally credentialed" Wheeler and his colleagues "at the world renowned Johns Hopkins Medical Institutions." Steve had managed to find an attorney, but that lawyer's written closing argument contained typos and concluded by urging the judge to award benefits to "Cecil Reed."

In August 2009, administrative law judge Richard Stansell-Gamm credited the readings of Wheeler and his Hopkins colleagues and found that Steve did not have black lung. He did find, however, that Steve was totally disabled. It wasn't clear what the cause of this disability *was,* but it apparently *wasn't* the disease for which Steve's doctor was treating him.

Steve's lawyer appealed, and the Benefits Review Board remanded the case for technical reasons. In 2011, Stansell-Gamm issued essentially the same decision for the same reasons. Three days later, the Labor Department sent Steve a letter demanding $46,433.50. Because a district director had initially awarded his claim and the company had appealed, the government trust fund had made monthly benefits payments for years as the case had dragged on. Now, the department expected Steve to pay that money back.

Steve asked the department to waive the so-called overpayment, writing:

> *Overall, we fully or in part support nine people, and what money we do have is spent in the best possible way to benefit this family. Clothing, detergents, fuel for vehicles, diapers, food, and many other necessities come from this money. Not everyone in the family has a vehicle so we share. Each person of age tries to help but overall it isn't enough to survive on without borrowing. It has been very humiliating to have to do so, when everyone knows that I worked my life away from my children and my wife, in order to end up on full time oxygen for a company who isn't descent [sic] enough to acknowledge the damage "their" job done to my body, my life, and my family.*

Eventually, after a review of Steve's financial records showed that basic living expenses consumed almost all of the family's monthly income, the department granted the waiver, but when I first called Steve, the unfathomable sum hung over him, an additional weight stacked atop the denied claim. His circumstances—both financial and medical—looked dire, but as we spoke, I sensed in his voice not self-pity but quiet anger. I asked if I could come meet him in person, and he invited me to his home in Glen Fork. His address, like those of other miners I'd visited, stumped Garmin and Google, so he offered to meet me along a more clearly marked thoroughfare in the coalfields of Wyoming County.

That's how I came to be parked on the roadside in the shadow of Bolt Mountain in April 2013. I followed Steve around hairpin turns in pine forests that periodically opened on small valley towns, the road's path dictated by the mountain's contours, eventually crossing a creek and arriving at the four-bedroom, one-bathroom house that he'd lived in as a child and that he now shared with his wife, Nyoka; his daughters, Stepheny and Patience; Patience's husband; and the couple's two children.

Inside, Steve disconnected from his portable oxygen tank and hooked into the machine set up in the living room. It pumped and hissed rhythmically as Steve and Nyoka, sitting side by side on the sofa, spoke of their life together.

Both had grown up nearby, the children of coal miners. They started dating shortly after Steve returned from a tour in Vietnam, where he'd worked base security for the air force. After just a few weeks, Steve asked her to marry him. They eloped on July 25, 1969.

Their personalities proved complementary. Steve was quiet, though he did get fired up if anyone spoke ill of Chevy trucks, and he could, on occasion, transform into a storyteller with luminous facial expressions. Nyoka was the cutup, gently needling Steve and keeping things lively. "Watching my parents," Patience later said, "I learned how to appreciate another person. I learned how to love another person completely."

Nyoka had begged Steve not to go into the mines, but the pay proved too enticing. In September 1969, he went underground, taking a job with Eastern Associated Coal, a Peabody subsidiary. He spent the next three decades shuttling among the company's mines throughout the southern West Virginia coalfields, performing the dustiest jobs. He ran the machinery at the coalface—continuous miners, roof bolters, shuttle cars. Over the years, rockfalls busted his back and tore open the flesh on his hands and arms. He sometimes worked in low coal, crawling for entire shifts. He'd come home with his knees swollen up like melons, covered in coal dust, and stop to toss the ball around in the yard with his children.

By 2004, his breathing problems had gotten bad enough that his doctor told him he had to retire, which Steve did, grudgingly. As the claim he filed in 2005 lurched toward its deflating conclusion in 2011, his condition began to deteriorate rapidly.

Steve now slept in a recliner; if he lay flat, he felt like he was being smothered. Nyoka slept lightly in the next room over, waking often to listen to his labored breathing. Many nights, his breathing slowed so much that it seemed about to stop altogether, at which point Nyoka would jump out of bed, run into the living room, bend Steve over, and beat his back until he coughed up a chunk of black phlegm. After that, his breathing would improve, and he could drift back to sleep.

Nyoka struggled with health problems of her own. She had rheumatoid arthritis and a disorder that caused her body to retain too much iron; the conditions left her in constant pain and sapped her energy.

Steve and Nyoka weren't ready to give up on the black lung–benefits system, but they didn't know what else to do. Steve knew well why he'd lost his claim, and he scowled when I mentioned Wheeler's name. Even if he did file a new claim, it was difficult to envision a different outcome; Wheeler and his colleagues would still be there. And Steve's options for proving Wheeler wrong were limited. Given his age and poor health, undergoing a lung biopsy would likely be too risky. That left one other way. He'd repeatedly told Nyoka to have an autopsy performed on him if he died.

As they spoke of future prospects, Steve sat stoically, and Nyoka wept. Their hands met in the center of the sofa. Like other miners and their family members, Steve and Nyoka had tracked the progression of Steve's disease with a crude metric: pillows. As miners' breathing worsens, they have to be more and more upright to sleep.

"You start out with one pillow," Nyoka said. "Then you go to two pillows. Then three pillows, and that's supposed to be your top. Well, he went through that, and he got to where he couldn't breathe. So he got in the recliner, and he's just lived in that recliner for..."

"Years," Steve interjected, staring at the verdant landscape outside the window.

"And now the recliner..." Nyoka paused to gather herself. "It's not enough."

A couple of months after meeting Steve, I finished logging Wheeler's X-ray readings in a database and ran a few queries. Now the doctor's record came into focus. In the roughly 1,500 cases in which he'd been involved, Wheeler had never once found complicated pneumoconiosis, even though other doctors looking at the same films had seen this severe disease in 390 of those same cases. In about 4 percent of cases, Wheeler had seen early stages of pneumoconiosis, but in about half of those, he had ultimately concluded that the miner more likely suffered from another disease.

The analysis also showed his tendency to attribute lesions he saw on X-rays to causes other than black lung. He saw a different disease, most often tuberculosis or histoplasmosis, on 34 percent of the X-rays that others had graded as positive for some form of pneumoconiosis. This tendency was especially pronounced in the cases of miners who had signs of the most severe form of the disease. On two-thirds of the X-rays that other doctors graded as showing complicated pneumoconiosis, Wheeler saw TB, histoplasmosis, or a similar disease.

The consequences: When Wheeler weighed in, miners lost about 70 percent of the time. The number of denied claims over a thirteen-year period was more than 800, and that included 160 cases in which other doctors had found complicated pneumoconiosis.

Pathology reports—the "gold standard" for diagnosis, as Wheeler repeatedly preached—existed in about 19 percent of the cases, and in about half of those, the findings were in dispute or unclear. But in more than one hundred cases, there was no dispute: autopsies or biopsies had definitively proven Wheeler wrong.

The doctors who evaluated Steve had seen complicated pneumoconiosis on his X-rays and CT scan, but I wondered what someone who specialized in identifying black lung—and who was not part of Wheeler's unit at Johns Hopkins—would see on the films.

I called Dr. Jack Parker, who had been chief of pulmonary and critical care medicine at the West Virginia University School of Medicine for more than a decade. He previously ran the department at the National Institute for Occupational Safety and Health that was in charge of the X-ray surveillance program and the certification tests that doctors, including Wheeler, had to pass to read films as part of the benefits system. For years, he had taught physicians across the United States and around the world to read X-rays for pneumoconiosis on behalf of NIOSH and the American College of Radiology.

He graciously agreed to look at a few X-rays and CT scans and grade what he saw, at no charge. With Steve's permission, I sent the films to Parker and told him only Steve's name, age, mining history, and the fact that the interpretation of the films was contested.

In August 2013, he sent me a report detailing his assessment of the films, which included some that Wheeler had graded as negative. Steve's films showed a classic case of complicated coal workers' pneumoconiosis, he wrote. "Mr. Day's history and findings are so characteristic that I am confident that Mr. Day has lung disease related to his coal mining," he wrote. "I am so confident, that I am indeed certain a biopsy or an autopsy, if performed, would confirm my diagnosis."

He added: "Based on my findings in reviewing this case, and the classic nature of the medical imaging and history, I am deeply saddened and concerned to hear that any serious dispute is occurring regarding the interpretation of his classically abnormal medical imaging. If other physicians are reaching different conclusions about this case...it gives me serious pause and concern about bias and the lack of scientific independence or credibility of these observers."

During the reporting process, my bosses at the Center for Public Integrity talked with the investigative team at ABC News, and they joined as a partner for the story about Wheeler and Johns Hopkins. I shared with them the data analysis I'd done and the research I'd compiled about the myriad ways Wheeler's views were at odds with the medical literature and the opinions of other prominent radiologists. They conducted on-camera interviews of people central to the story, including Steve, and we coordinated to figure out what remained to be done.

The most important thing left was to try to interview Wheeler. Johns Hopkins agreed to let us come to the hospital in Baltimore and talk to the doctor. We knew we would get only one opportunity, and as much as I wanted to question Wheeler myself, I understood why ABC wanted on-air investigative correspondent Brian Ross to conduct the interview. I sat down with Ross a couple of times to prepare. We reviewed documents he could use to confront Wheeler. We discussed key questions, and I more or less played the role of Wheeler, responding as I anticipated the doctor would—no great feat, given the consistency and predictability of his statements over the years.

On July 24, Ross, producer Matthew Mosk, and I (along with an ABC camera crew and Johns Hopkins media officials) walked down the hallway of the Johns Hopkins Pneumoconiosis Section with Wheeler. The septuagenarian doctor had a full head of short gray hair and wore a short-sleeved button-down shirt with a gray-blue-green plaid pattern accompanied by a red tie. The clothes hung off his slim frame.

We passed a storage closet containing stacks of file folders with the names of law firms and coal companies written in Sharpie on their sides. At one of the unit's workstations, there were papers bearing the letterheads of prominent firms. There, Wheeler flicked on the X-ray light box machine and put up a few films. What followed was a live recitation of the descriptions and metaphors I'd read so many times in his reports and depositions. The lesions on the film didn't have the "birdshot pattern" he wanted to see or their location was "out of the strike zone."

"I'd classify it as compatible with coal workers' pneumoconiosis," he said of one X-ray. "But I'd also say it certainly could be or more likely is histoplasmosis." Ditto for another film.

Though his comments were almost identical to those he'd made so often over the years, I hadn't anticipated that he would make them about these particular films. The X-rays he was describing were the International Labour Organization standard films—the archetypes in the system Wheeler was supposed to follow. When determining how to grade

X-rays, doctors compared them to these ILO films, which were meant to depict the classic appearance of varying stages of disease. To pass the NIOSH certification exam, which Wheeler had done every few years for decades, he would have to grade these X-rays as positive, but he was now saying he didn't believe they really represented pneumoconiosis.

In a few depositions I had read, Wheeler mentioned his skepticism about the ILO system. Now he was being more explicit, saying that the standard films were "not proven" and that the classification system had "some quality issues." He seemed to be saying that he graded films one way to pass the NIOSH test but another when coal companies were paying for his opinion. NIOSH arguably had little authority to prevent this; it was charged with overseeing the certification exam and the readings performed as part of its surveillance program, not readings performed within the benefits system.

We walked to a nearby room for a sit-down interview. Ross described what the data analysis of Wheeler's record showed and asked if he stood by that record. "Absolutely," Wheeler replied. He added, "I have a perfect right to my opinion.... I found cases that have masses and nodules.... In my opinion, those masses and nodules were due to something more common."

When Ross showed him some of the records of cases in which he'd been proven wrong by biopsies or autopsies, he questioned the qualifications of the pathologists. "I know my credentials," he said. "I'd like to make sure that the people proving me wrong... have... credentials as good as mine."

He insisted he was following accepted medical practice by offering differential diagnoses and suggesting biopsies, which were "very safe, very safe." He questioned why a miner who truly thought he had black lung wouldn't undergo a biopsy. "I think if they have coal workers' pneumoconiosis, it should be up to them to prove it."

When Ross pointed out that many doctors considered it dangerous to perform the procedure on a miner with damaged lungs and that the federal law governing the benefits system didn't require it, Wheeler snapped, "I don't care about the law."

* * *

On October 29, 2013, the Center for Public Integrity released the first installment of the series "Breathless and Burdened." In an image beside the opening paragraphs, Gary Fox smiled at readers, captured in a photo underground wearing a hard hat, work boots, and knee pads, kneeling beside a roof-bolting machine. In the story, I detailed the evidence I had gathered about Jackson Kelly's practices during the sixteen months since I'd first met John, Terri, and Mary.

Before the article's publication, I had tried repeatedly to contact individual Jackson Kelly attorneys to arrange interviews, but none had agreed. The firm's general counsel, Stephen Crislip, eventually replied on behalf of all the firm's attorneys and said he could not discuss the Fox case because it was ongoing. "As you have learned in your job," he wrote in an e-mail, "we lawyers have overarching duties to clients and courts that prevent us from discussing cases even when we want to do so in order to defend our good name and reputation of 191 years in duration."

I suggested that we could still discuss other cases and the general practice of withholding evidence, but Crislip replied: "We certainly would like to talk to you, and look forward to doing so, when the Fox case is over and we have had time to talk with all involved parties. Reporting on it would be incomplete if not misleading, in my opinion, until then."

The day after publishing the piece on Jackson Kelly, the center released part two, which detailed the record of Wheeler and the Johns Hopkins unit and told the stories of Steve Day and other miners whose cases looked eerily similar to his. That evening, ABC aired its piece on Wheeler and Johns Hopkins on *World News Tonight* and, using a longer version, *Nightline*.

After the interview with Wheeler, members of the Johns Hopkins communications team had questioned our findings and asked for documentation. If we supplied evidence, they said, they would answer our questions about it. After we'd mailed them a stack of documents and a list of questions, however, they responded with a statement that said Wheeler was "an established radiologist in good standing in his field." It also said, "To our knowledge, no medical or regulatory authority has ever challenged or called into question any of our diagnoses, conclusions or reports resulting from the . . . program." When we pointed out to the communications officials that the statement didn't contain answers to most of our questions, they sent the same statement again.

The final installment of the series went out on November 1. I described what was emerging as the next scientific battleground in the black lung–benefits system. Physicians and researchers were growing increasingly confident that breathing coal dust could cause a previously unrecognized form of lung disease. The response by the industry and its allies in the medical and legal professions was yet another iteration of the deny-and-contain strategy they had used for more than a century. They dusted off the old script once more and began reciting it in response to growing evidence that an abnormal pattern of linear scarring in the lower lobes of the lungs—as opposed to the rounded scarring concentrated in the upper lobes more typical of classic coal workers' pneumoconiosis—could be caused by coal dust.

The story highlighted the benefits claim of retired miner Ted Latusek, who had this pattern of disease and who had been fighting his former employer, Consol Energy, in an ongoing case that already had consumed two decades. (In 2018, a federal appeals court would uphold a judge's finding that Latusek's disease was indeed caused by coal dust, giving him a final victory twenty-four years after filing a claim.)

I knew, based on conversations with colleagues and personal experience, that a news story, even a large investigative project, often has the staying power of a slightly underinflated balloon. Some people grasp hold of it for a moment, but before long, it either flitters off or falls in a shriveled clump to the floor to be swept away. But every now and then, a story breaks through, readers fume, leaders act, and the reporter's tank is refilled with enough of the peculiarly cynical optimism to fuel another dozen duds. Though I did not expect it, I hoped the series we'd just released would find such a reception.

24

Making Good

On the evening of November 1, 2013, I called Steve Day with news: Johns Hopkins had just announced that it was suspending the work of Dr. Paul Wheeler's Pneumoconiosis Section pending an internal review. Steve let out a joyous howl unlike anything I'd heard from him before. He gathered his family around the phone, and as I told him the details, he excitedly relayed them. Before long, though, his voice returned to its customary reserved tone. He asked, "I guess maybe I helped some people?"

The stories of Steve, Gary Fox, and miners like them seemed to strike a chord both throughout the coalfields and beyond. Over the past few days, I had received messages from an array of readers, from miners and advocates in Appalachia to doctors and lawyers in cities far from coal country.

The editorial boards of the *Charleston Gazette,* the *Lexington Herald-Leader,* and even the traditionally conservative *Charleston Daily Mail* weighed in, expressing alarm over the alleged conduct of lawyers at Jackson Kelly and doctors at Johns Hopkins. The *Herald-Leader* called their actions "shocking and repugnant." The articles also noted the resurgence of black lung and urged the Labor Department to finalize a rule that it had proposed in 2010 designed to prevent future disease. "We hope this new uproar brings further industry reforms to reduce the curse," the *Gazette* wrote. The *New York Times* editorial board later penned a similar article under the headline "Miners Battle Black Lung, and Bureaucracy."

Officials in the federal government were listening. In the wake of the Upper Big Branch mine disaster in 2010, an informal group of both House and Senate staffers—primarily those whose bosses were coal-state Democrats or held top positions on the committees overseeing workplace safety and health—had met periodically to discuss possible improvements to mine-safety laws. They now turned their attention to the black lung–benefits system and started crafting legislation. Meanwhile, Labor Department officials began evaluating what potential reforms they might already have the authority to implement.

Leading the bureaucratic effort was Solicitor of Labor Patricia Smith, the department's top lawyer and essentially its third in command overall. She was in charge of a team of hundreds of attorneys who represented the department's various agencies in legal proceedings.

Plainspoken and hard-nosed, Smith had been raised in southwestern Pennsylvania coal country. After earning a law degree from New York University, she'd worked for about a decade handling employment law for a legal-aid group, twenty years in the Labor Bureau of the New York Attorney General's Office, and three years as the state's labor commissioner. In 2010, the U.S. Senate confirmed her as solicitor of labor after a bruising, yearlong fight.

Republicans had held up her nomination because of their concerns that she was too close to labor advocates, and they accused her of making inaccurate or misleading statements during a hearing about a community-outreach program she'd started as labor commissioner. Two of them asked President Barack Obama to withdraw her nomination.

"I think I've been called a liar on the Senate floor more than anyone else in the twenty-first century," Smith said.

Because Democrats held a majority in the Senate, she'd cleared the chamber on a party-line vote. The Upper Big Branch mine disaster occurred just two months after her confirmation, and the subsequent investigations occupied much of her time during her first two years on the job.

Now, in late 2013, she, too, shifted much of her attention to the black lung–benefits system. She convened the team of lawyers assigned to the system, she recalled, and told them: "Here are the problems that are laid out. How do we fix them? I want you to come back to me with proposals."

With the approval of the Labor Department's top two officials, Smith and her staff began working on the proposals with the agency that oversaw the benefits system, the Office of Workers' Compensation Programs. That agency's top officials reported to Smith's office weekly to provide updates on their progress, she said.

During this period, Smith recalled, "I probably devoted twenty percent of my time to black lung reform." It was an extraordinary commitment of resources for her office, which represented an array of agencies charged with enforcing more than one hundred and eighty federal labor laws.

Meanwhile, the coalition of miners' advocates sensed a rare opportunity—a confluence of public pressure, keen congressional interest, and a receptive Labor Department. The last time such a window had opened was twenty years earlier, and the collapse of many of the reform efforts during the final days of the Clinton administration still stung. Now John Cline and other miners' advocates began having conference calls and compiling a list of proposed reforms.

One long-sought reform had been accomplished already, although not quite in the way they had anticipated. They had been lobbying for years to restore the fifteen-year presumption—the rule stating that if a miner had a totally disabling breathing impairment and had worked at least fifteen years in the mines, it was up to the coal company to prove that his disability was *not* caused by black lung. Congress had added the provision in the first amendments to the law in 1972. A decade later, the amendments pushed by the Reagan administration removed the presumption, and this reversal had been one of the chief reasons that claims approval rates had plummeted.

In 2010, West Virginia's longtime senator Robert Byrd proposed to restore the presumption by adding amendments to the capacious health-care reform legislation that came to be known as Obamacare. The amendments also specified that widows of miners who were receiving benefits at the time of their deaths were automatically entitled to benefits themselves.

The president of the West Virginia Chamber of Commerce had warned that Byrd's proposal would be a "job killer," and the National Mining Association opposed the provisions, claiming they would cost the industry between $332 million and $697 million.

Byrd countered, "Despite the rhetoric to the contrary, this language will not cost one additional dime, unless employers took insufficient precautions to protect their workers from black lung disease."

The provisions remained in the bill that Congress passed and that Obama signed into law in March 2010.

When the miners' advocates began assembling a list of priorities in late 2013, though, there was much that they thought still needed to be done. They soon arrived at a consensus list of key proposals. These included increases in staffing and training for administrative law judges as well as more incremental changes that they believed would lead to faster and fairer decisions.

The last proposal—the biggest ask of all—was what John had sought for years: require both sides to turn over all medical reports.

In conference calls with Labor Department officials and a meeting in DC with congressional staffers, the advocates made their pitch.

Mary Fox added her voice, describing Gary's case in a March 2014 letter to Secretary of Labor Tom Perez. What happened in her husband's claim, she wrote, "is not unique from the standpoint of another coal miner claiming to have suffered health-related issues as a result of working years underground in the mines. It is quite unique however, from the standpoint that it has helped to uncover a pattern of legal practice that blatantly withholds and manipulates credible, life-altering evidence, as a matter of financial gain for the coal company."

She continued: "I respectfully ask that you and the Department of Labor join forces with motivated Senators and Congressmen . . . to enact necessary laws preventing this unjustly aggressive 'too big to fail' arrogance from entering our courtroom ever again.

"Gary always said that justice would prevail, as long as all of the facts and evidence could be heard. Sounds like a reasonable and fair place to start."

The head of the Mine Safety and Health Administration, Joe Main, greeted longtime friends and colleagues as he made his way to the stage set up at one end of the atrium in the National Institute for Occupational Safety and Health's Morgantown office on April 23, 2014. He'd hoped for this day, worked toward it, for more than thirty-five years.

Main had grown up on a farm in southwestern Pennsylvania coal country and entered the mines at age eighteen. In 1976, Main had gone to work for the United Mine Workers' department of health and safety. By then, it was already clear that there were serious problems with the dust-sampling program intended to prevent black lung. At a conference that year, another safety official had postulated that the solution was a sort of "black box"—a sampling device that would measure dust levels accurately and in real time and would confound companies' attempts at cheating. The idea had struck Main. Developing such a device, along with closing the loopholes that allowed companies to submit unrepresentative samples, would force the coal industry to control the concentrations of dust in the mines, not just on paper.

These reforms became his mission for much of the next three decades. He became head of the union's safety and health department, a post he held for twenty-two years, and he was a leading voice on the Labor Department advisory committee that had recommended sweeping changes in a 1996 report—recommendations that had failed to become policy when proposed rules collapsed in the final days of the Clinton administration.

President Obama had appointed Main to run MSHA, and the Senate had confirmed him in October 2009. Soon after taking office, he made it clear that enacting changes that would eliminate black lung was his top priority. In 2010, his agency published a proposed rule containing many of the reforms that Main and others had sought for years. Most notably, the proposal would lower the standard for the concentration of dust allowed in mine air and close loopholes in the sampling regulations.

Unsurprisingly, the reaction from the coal industry was fierce. During the three years following the publication of the proposal, Main and the officials at his agency had responded to reams of comments warning that the rule could kill the coal industry. They held seven hearings, rewrote parts of the rule, and pushed to get it through the White House's Office of Management and Budget, which had a reputation as a graveyard for controversial rules. They withstood efforts by members of Congress to kill the regulation with language tucked in a budget bill or stall it with multiple mandated studies by the Government Accountability Office.

Now, on this April morning in 2014, Main got to announce that the long-sought rule was final. He took his spot on the stage alongside Secretary of Labor Tom Perez and longtime U.S. senator Jay Rockefeller, a West Virginia Democrat. The atrium was filled with rows of folding chairs, almost all of which were now occupied; spectators spilled into the standing area in the back of the room. For some in the crowd, the rule was validation of years of work. Among them were the NIOSH staffers who, almost two decades earlier, had marshaled the scientific evidence into a tome that now supported key provisions of the new rule. Also present were the agency epidemiologists who had been documenting the resurgence of severe black lung, which had brought a sense of urgency to finalize the rule.

The rows of chairs closest to the stage, however, remained empty; they were marked RESERVED. Shortly before the ceremony began, the audience members' attention turned to the atrium's side door as the people for whom the seats had been saved, the guests of honor, slowly filed in—a procession of miners and their family members, a few of them limping or dragging an oxygen tank. Beside the doorway, pointing them to their seats, was the man who had helped organize the bus trip that brought them up from the southern West Virginia coalfields: John Cline.

When all had taken their seats, Perez told the audience, "The nation made a promise to American miners back in 1969, and, at long last, we are making good on that promise."

Rockefeller also noted the historical significance but added a cautionary note to the optimism permeating the gathering. He had seen numerous well-intentioned policies prove disappointing during his decades of political life in West Virginia. A member of the prominent Rockefeller family and great-grandson of the Standard Oil magnate, he had arrived in West Virginia as a VISTA in 1964 and decided to stay—much the same path John Cline had followed. He'd served in state politics, including two terms as governor, before being elected to the Senate in 1984. Recently, he had announced that he would retire when his current term ended the following year. His office was now among those working on reforms to the benefits program, and he considered his efforts to accomplish prevention of and compensation for black lung part of the legacy he hoped to leave.

"You've got to watch us and watch my colleagues in the federal government," Rockefeller said, his eyes scanning the rows of miners seated closest to the stage. "It's one thing to make an announcement; it's another thing to actually make it happen."

Indeed, the original black lung insurgents had made history in 1969, but forty-five years later, their ultimate goals remained unrealized. If there was an overarching lesson of those intervening years, it was this: There was no grand solution, just lurches forward and backward; progress was possible, but only with vigilance and scrapping.

Alongside the government officials on the stage was retired miner Dewey Keiper, there to give a few brief remarks; he offered a stark reminder of the stakes. He was a client of John Cline, and the two had become particularly close. Keiper, just fifty-four years old, had complicated pneumoconiosis and had gone for a lung-transplant evaluation the previous week.

He leaned forward, and the microphone picked up the audible gasps that came with each breath, hoarse ellipses that punctuated the one sentence he was able to utter: "I worked for twenty-eight years... never really had any serious injuries... and now... I can't work anymore." He dropped his head, and the yellow bill of his West Virginia University hat hid his face.

"We love you, Dewey," Main said. Keiper raised his head, his bottom jaw quivering. The audience erupted in applause.

When it was Main's turn to address the audience, his body language registered the internal struggle between his stoic demeanor and the gravity of the moment. He looked down, stroked his chin, tapped his fingers on the podium. After a few seconds, he cleared his throat and turned toward Keiper. "That's why I took this job," he said.

Main described the reforms his agency was implementing—changes intended to prevent future miners from ending up as ill as Keiper. The new rules would take effect in three phases over the next two years, he said. Phase one was to close the loopholes that allowed companies to mask what miners were actually breathing. No longer would companies be allowed to scale back production to half of the normal amount when conducting sampling; now, a sample would be deemed valid only if production was at least 80 percent of the normal amount. And no longer could companies turn off the sampling pump after eight hours if the miner worked twelve; the device had to run for the entire shift.

Also gone was the practice of determining a company's compliance with the dust standard by averaging five samples, which could leave individual miners exposed to dangerous concentrations of dust if samples taken from coworkers in less dusty areas canceled out the excess readings. Now MSHA could issue a citation whenever one of its inspectors' samples showed a single miner during a single shift was exposed to concentrations higher than the limit. For samples submitted by companies, MSHA could issue a citation if two of the five samples were too high.

Phase two, effective February 2016, required companies to conduct sampling with a device known as a continuous personal dust monitor, or CPDM. This was the black box Main and others had envisioned in the 1970s. NIOSH had worked with a manufacturer to develop the device and in 2011, after years of testing, had certified it as accurate and reliable. Main himself had tried it out, wearing it underground, and was impressed.

Like the sampling pumps currently in use, the CPDM consisted of a box that fit on the miner's belt with a tube leading to a small inlet that drew in mine air. But the CPDM's box contained a computer, and a digital display showed dust levels in real time. The device also collected information that, at least in theory, would make it easier to detect certain types of tampering, such as hanging the sampler in clean air or covering the piece drawing in air.

The third and final phase, which would take effect in August 2016, was a reduced dust-concentration limit. On this point, the department would receive criticism from some who felt the rule wasn't tough enough. MSHA originally proposed reducing the standard from 2.0 milligrams of dust per cubic meter of air to 1.0, the concentration recommended by NIOSH in 1995. But in the contentious process of finalizing the rule, the agency had compromised, settling on a standard of 1.5. Main understood the concern over this, but to him, the continuous monitor and the plugged loopholes were the critical points; no matter how strict the standard was, it wouldn't matter if companies could game the samples.

Just days after the ceremony in Morgantown, the industry made one last attempt to block the new rule. A coalition led by the National Mining Association sued the Labor Department, alleging that regulators had exceeded their authority and had issued an irrational and flawed rule. In January 2016, the U.S. Court of Appeals for the Eleventh Circuit ruled against the industry and upheld the regulation. Later that year, the Labor Department announced that samples taken since the first two phases went into effect showed 99 percent compliance. To Main, it was proof that companies could control the dust and still operate profitably.

The lingering question, of course, was whether the compliance numbers were real. Rates had been similarly high under the old rule, but that had largely been an illusion. "These numbers are nothing like those; I think there's a greater reliability thanks to all the things we did," Main later said, pointing to the closed loopholes and continuous samplers. "I have far more faith in what we see today."

He was hopeful that, in the coming years, the new requirements would halt and reverse the disturbing upward trend in incidence of severe disease. "If we had those dust controls that we have now twenty years ago," he said, "you wouldn't have the number of claims filed now that we're seeing."

In the meantime, the devastating effects of past exposures became ever more apparent. Over the following years, I kept in touch with NIOSH epidemiologist Scott Laney as he and his colleagues continued to publish studies on the resurgence of black lung, each seemingly more alarming than the last. In a 2018 study, the researchers reported their most dire findings yet: In the 1970s, shortly after preventive efforts began, more than 30 percent of veteran miners (those who had worked at least twenty-five years) had black lung. That number had dropped to 5 percent in the late 1990s, but by 2015, it had doubled to 10 percent. In central Appalachia, the picture was even worse: one in five veteran miners had black lung, and more than 5 percent of them had complicated pneumoconiosis—the highest level ever recorded.

Laney and his colleagues also uncovered new clusters of disease. Amid a coal-industry downturn, more out-of-work miners underwent medical screening in anticipation of filing benefits claims, and many of their X-rays revealed cases of complicated pneumoconiosis that had gone undetected by the surveillance program. Laney remarked, "We're at the epicenter of one of the largest industrial medicine disasters that the United States has ever seen."

It might be decades before the effects of the new rule become clear. If miners today are still breathing dangerous amounts of dust, they might not develop noticeable disease for years. Even then, many may forgo examination and treatment and continue working—a practical obstacle that has confounded efforts to understand the current resurgence of black lung. Though the number of working coal miners in the United States has steadily declined in recent years, tens of thousands still work underground.

The success of disease-prevention efforts also depends on politics. Four years after Main announced the finalization of the dust-control rule, his successor—former coal executive David Zatezalo, appointed by President Donald Trump—announced a "retrospective study" that miners' advocates saw as a pretext for gutting the rule. After bipartisan backlash from members of Congress, Zatezalo said that his agency had no immediate plans to roll back the rule. Still, the rule's protections are only as good as the government's willingness to enforce them, something that remains far from certain under administrations such as Trump's that adopt a more hands-off approach to the coal industry.

The decisions made today by coal companies, policymakers, and regulators will determine whether another generation of miners endures the same needless suffering as those before them or if we will finally make good on the promise to end a preventable, ongoing disaster.

On the third floor of the state courthouse in Beckley, Al Karlin sat alone on a bench at the end of the hall. His neatly folded gray suit jacket lay beside him, and he peered through reading glasses at the stack of papers in his lap, leafing through tabbed documents and scribbling a few final notes on a white legal pad.

John Cline paced nearby. He'd augmented his usual work attire—corduroys and a button-down shirt—with a tie and jacket. "I'm sort of window dressing," he told me, grinning. When it was their turn inside the courtroom, Al would do the heavy lifting, trying to keep the civil suit against Jackson Kelly alive. John knew it was best to leave him alone with his thoughts during these last minutes of preparation. "He gets geared up," John said. "You just hope he gets in the zone."

Just before the scheduled hearing time, Jeffrey Wakefield, the lead attorney for Jackson Kelly's defense team, strode past in a dark suit, carrying a poster board wrapped in plastic. "Visuals," John said with a nervous smile. Wakefield's penchant for props had become a running joke between John and Al.

At three thirty p.m. on July 1, 2014, the courtroom doors swung open, and both sides filed in. "What's your expected time of presentation?" Al asked Wakefield.

"I have no idea," he replied. "Maybe fifteen or twenty minutes. What's yours?"

"That depends on you," Al shot back, prompting grins and nods of approval from the supporters sitting in the rows behind John and Al—a group that included John's son Brooks, his sister Ann and her husband, retired miner Ken Dangerfield, and Dolores Carroll, the only plaintiff able to attend. Wakefield's verbosity was another running joke. His arguments tended to be long and technical. Still, this was the strength of Jackson Kelly's defense. The firm's positions were most convincing when the discourse remained in the realm of abstract legal principles. This noble sheen looked more tarnished in the arena of the case's particular facts.

For Al and John, the hearing was a fight for survival. Jackson Kelly had argued that, based on the Fourth Circuit's decision in Gary Fox's benefits case, Judge H. L. Kirkpatrick III was now legally obligated to throw out the civil suit against the firm.

Wakefield went first, unveiling the poster board and placing it on an easel. It displayed bullet points, each a key quote from the Fourth Circuit's decision. The appeals court, he said, had found that Jackson Kelly had no legal duty to turn over all of its evidence; therefore, not sharing that evidence couldn't possibly amount to fraud. Over and over, he stressed the word *duty,* and, mid-argument, he highlighted it on the poster board for effect. Al and his clients, Wakefield contended, were essentially asking the judge to impose a civil version of the *Brady* rule, which requires prosecutors in criminal cases to turn over exculpatory evidence but does not apply to civil suits.

Thirty minutes later, it was Al's turn. Just as important as what the Fourth Circuit had said, he began, was what it had *not* said. The court had found merely that Jackson Kelly's actions didn't clear the high bar of fraud on the court. The ruling didn't address whether the firm's conduct might meet the standard for common-law fraud, which was at issue in the civil suit. Further, the court had stated explicitly, "We bestow no blessing and place no imprimatur on the company's conduct."

Al walked to the front of the courtroom to get a good look at Wakefield's poster board, which still sat on the easel. The whole duty argument, he said, was a misdirection. The plaintiffs weren't alleging that Jackson Kelly had violated a legal duty; they were alleging that the firm had committed fraud by intentionally misrepresenting evidence to their own experts, the opposing side, and the administrative law judge.

Today, unlike at the oral argument before the Fourth Circuit, Al was on. When the hour-long hearing came to a close, he thought that he'd gotten his points across, that Kirkpatrick understood the issues and wasn't looking for a way to punt the case.

At various points during the hearing, Dolores Carroll had scowled at Wakefield, especially when he'd mentioned her late husband, Clarence. Afterward, outside the courtroom, she was still upset. "I wanted to get up and yell at him," she said. "They talked like he was just nothing. I wanted to get up and tell him, 'You're talking about my husband. You didn't know him.' "

Two months later, Kirkpatrick rejected Jackson Kelly's argument. The civil case and the benefits case were "clearly distinguishable," he wrote. The Fourth Circuit had decided only that the firm's actions in one case—Gary Fox's—did not constitute fraud on the court. What Kirkpatrick had to decide, he wrote, was something different: whether the firm's black lung unit had engaged in a pattern of misrepresentation that amounted to common-law fraud.

Kirkpatrick wrote: "Following the logic of Jackson Kelly's argument through to its inevitable conclusion would permit attorneys to engage in deceitful, and indeed reprehensible, conduct as a matter of course, just as long as such activity is calculated to fall short of outright bribery, jury tampering, or other criminal acts deemed sufficiently egregious to clear the very high bar required to establish fraud on the court."

The motion to dismiss, he concluded, was denied. Now, after five years of litigation over whether the case could even go forward, the battleground shifted to discovery—what, if anything, would Jackson Kelly have to turn over to the plaintiffs? The following months would determine whether John ever got his chance to peek behind the curtain. He and Al knew they were in for a hell of a fight.

The daily bustle of Capitol Hill had not yet started, and hearing room SD-430 of the Dirksen Senate Office Building was silent and unoccupied save for two men: John Cline and retired coal miner Robert Bailey Jr., who had advanced black lung. They sat among the rows of chairs that would soon be filled with spectators in dark suits. In the center of the room was a long table covered with a green cloth facing an elevated platform where leather chairs were arranged behind a wooden desk shaped like a horseshoe. Later that day, July 22, 2014, the two men would sit at that table and speak to members of the U.S. Senate as television cameras rolled.

John and Bailey had met just days earlier, after both had accepted invitations to describe their experiences with the black lung–benefits systems at a public hearing. Bailey had been in the middle of a week of testing at a hospital in Northern Virginia. His wife, who had taken off work to be with him and couldn't afford to miss more time to travel to DC, worried about his making the trip by himself. That was when John called, offered Bailey a ride, and said he'd already booked hotel rooms for both of them.

On the morning of the hearing, both men's nerves had woken them early, so they'd decided to go ahead and make their way to Capitol Hill, arriving before even the event's organizers.

The hearing, titled *Coal Miners' Struggle for Justice: How Unethical Legal and Medical Practices Stack the Deck Against Black Lung Claimants,* was a product of the efforts of the working group of congressional staffers. They hoped it would bring increased public attention to the problems with the black lung–benefits system and create an official record to support the legislation they were still drafting. They knew that, with the current composition of the Senate, they didn't have the votes to pass the bill; there was zero Republican interest in anything that might increase costs for the coal industry. But the bill and the record supporting it would be there, waiting, so if the chamber's makeup did change, members and their staffs would have something they could pick up and push forward. Meanwhile, raising the profile of the issue, they thought, would keep the pressure on the current administration to act. Some of the reforms the staffers had identified would require congressional action, but they believed others were within the existing authority of the Labor Department.

Already, the department had taken some action, starting with the relatively low-hanging fruit—changes it could make through administrative actions alone. In February, the division overseeing black lung–benefits claims had announced a pilot program to raise the quality of the medical reports written by doctors who performed the exams paid for by the department. Often, these doctors didn't clearly address the medical issues that were determinative in benefits claims; the pulmonologists commonly retained by companies, however, were skilled at doing just that. In many cases, the defense experts also evaluated additional evidence that had been developed after the initial exam, adding credibility to their opinions in the eyes of judges.

Under the new initiative, if a miner had worked at least fifteen years underground and the initial exam indicated he was eligible for benefits, the Labor Department would pay for an additional report from the original doctor addressing any new evidence. In a related memo, Solicitor of Labor Patricia Smith directed lawyers in regional offices throughout the country to intervene in instances when miners with seemingly meritorious claims were unrepresented before an administrative law judge. In such cases, government lawyers would step in to ensure that the department-paid physician had an opportunity to review additional evidence and had clearly addressed the key legal issues in a report.

In June, the department issued a bulletin to its district directors, who made initial claims determinations, describing the Center for Public Integrity and ABC News stories about Dr. Paul Wheeler and Johns Hopkins and directing the officials to "(1) take notice of this reporting and (2) not credit Dr. Wheeler's negative readings for pneumoconiosis in the absence of persuasive evidence either challenging the CPI and ABC conclusions or otherwise rehabilitating Dr. Wheeler's readings."

Although the bulletin technically applied only to district directors, Smith instructed department attorneys to watch for cases before administrative law judges that contained readings by Wheeler. In such cases, they were supposed to file a motion asking the judge to take official notice of the articles and the Labor Department bulletin and give no weight to Wheeler's opinions.

The department had also searched its own records for cases in which Wheeler had weighed in and the claim had been denied. It had identified more than a thousand and had begun contacting all of these claimants to tell them how they could seek to reopen their cases or file new ones.

Further reforms—such as the mandatory disclosure of all medical evidence that the miners' advocates sought—would be more time-consuming and difficult, as they required a formal rule-making process. "Regulatory changes are not off the table," Smith had said at the time, "but the department has a full regulatory agenda and we'd have to figure out how to fit something in."

Members of Congress and miners' advocates, however, had pressed the department to do more. Senator Bob Casey, a Pennsylvania Democrat, responded to the administrative changes: "While this is certainly an encouraging development, it's far from what's needed to ensure these miners and their families receive justice." John Cline and other miners' advocates sent Smith's office a detailed letter outlining the numerous problems that remained and what they thought was needed to fix them.

Adding to the momentum of the reform efforts was the announcement on the afternoon of April 14—the fifth anniversary of Gary Fox's death, almost to the exact minute—that the Center for Public Integrity series had been awarded the Pulitzer Prize for Investigative Reporting.

By May, the department appeared to have reconsidered its reluctance to launch a complex rule-making process. A brief public notice said that officials were starting to develop a rule that "would address disclosure of medical evidence."

John had allowed himself a moment of excitement, but he knew that this short announcement was just the beginning of a long process. It was far from certain that the actual contents of the rule would be helpful, or even that it would be issued at all.

Now, in July, Senator Casey called to order the hearing of the subcommittee he chaired, and he made it clear that he intended to keep up the pressure. He and Senator Tom Harkin, an Iowa Democrat whose father had worked more than twenty years in the mines and had contracted black lung, questioned Smith and Labor Department deputy secretary Chris Lu about the steps they'd already taken and urged them to continue their efforts.

After forty minutes, the administration officials rose, and a new panel of witnesses, including John and Bailey, took their spot at the table before the senators. All eyes in the chamber turned to John, as did the video cameras sending out live broadcasts. A photographer crawled up to the table where John sat, pointed a telephoto lens at his face, and began snapping pictures.

This was just the sort of situation John had tried to avoid for most of his life. As much as John relished individual conversations and small gatherings, he'd always been fearful of public speaking. As a kid, he'd gotten nervous about things as quotidian as being introduced by his dad at a father-son Kiwanis lunch. He had spoken at a congressional hearing once before, but it had been a much briefer presentation in a more low-profile setting in eastern Kentucky more than two decades earlier. Over the years, though, he'd grown more comfortable, particularly when it came to presentations about subjects that aroused his passion. Relating the flaws he saw in the black lung–benefits system and telling his clients' stories, he could be engaging.

As Casey peered down at him, John adjusted himself in his seat and began. He thanked the senators and gave a brief overview of the benefits system. "The adversarial process is not only complex, but it has been abused by deceptive tactics by coal companies and their attorneys, or at least some of them," he said. "In my written statement, I provided five examples, and I would like to focus my remarks today on the case of Gary Fox, who I represented near the end of his life."

Telling Gary's story to the senators, John seemed to settle in. His speech slowed to a more comfortable cadence. He told of the denied claim, the withheld pathology reports, Gary's decline and death at age fifty-eight while awaiting a lung transplant. "There could have been a different outcome, in my opinion, if the purposes of the act had not been subverted," John said. "If Mr. Fox had prevailed in his first claim and been able to get out of the dust eight years sooner as the act intended, his life may have been prolonged."

He praised the Labor Department's recent announcement that it would consider a rule about the disclosure of medical evidence. "Requiring the disclosure is the only way, I think, to properly protect the health of miners," he said. "Mr. Fox asked me to pursue this issue. He knew it wouldn't benefit him, but he dearly hoped that it would help others not to have to go through the same experience."

Soon, the cameras turned toward Bailey, and the retired West Virginia coal miner spoke. "What is it like to have black lung? I thought I knew after spending so many years in the mines, but, really, I never knew until I got to this chapter of my life. Some days are a lot better than others. But every day is not a good day."

Bailey had entered the mines after finishing high school, and he'd worked underground for more than thirty-five years. Breathing problems had forced him to retire at age fifty-six, and he'd filed a benefits claim. Almost four years later, a judge had awarded benefits. Now Bailey was undergoing evaluations for a possible lung transplant, but the insurer for his former employer was questioning whether the procedure was necessary and indicating the company might not pay for it.

Still, Bailey told the senators: "I'm fortunate. It took me almost four years. Other miners have been there a lot longer and they haven't received anything, and a lot of it is through negative reading of the records. So I hope to support them with this, what we do today, for those that are still trying to get their benefits and for those who haven't gotten to where we are now. I hope to prevent them from experiencing that."

Bailey eventually did get a double-lung transplant after his former employer agreed to pay for it. Complications kept him hospitalized for weeks. Just when he thought he was well enough to return home, doctors discovered fluid around his lungs and sent him to a specialist at George Washington University Hospital in DC.

I went to see him there. He lay in a hospital bed with his wife, Brenda, seated in the chair beside him. Despite the numerous setbacks, he wore a broad grin. "For the last year," he said, "every breath hurt. Now it doesn't hurt to breathe. Now I can walk across the room without an oxygen tank. That's about all I need."

He had a couple of pictures he wanted to show me. First, the lungs that doctors had removed—a mottled black mess. "Like a burnt steak," he said. Second, the donor lungs that were now inside him. They were pink and bloody and, in his eyes, beautiful.

In early 2019, just over three years after the transplant, doctors told Bailey that he had liver cancer, and they suspected that the arsenal of medicines he was taking to keep his body from rejecting his new lungs had left him vulnerable to the disease. He faced the diagnosis with his usual equanimity. The last time we spoke, he was receiving hospice care as cancer overwhelmed his body. He expressed hope that his advocacy before and since the 2014 Senate hearing had called attention to black lung's ongoing toll and the need to address injustices in the benefits system. He died on March 15, 2019.

Two weeks after Johns Hopkins suspended the work of Wheeler and the Pneumoconiosis Section, Steve Day filed a new benefits claim, this time represented by a new lawyer: John Cline. He selected Dr. Don Rasmussen to perform the exam paid for by the Labor Department. Steve's breathing impairment was "severe," "irreversible," and "totally disabling," Rasmussen found, and Steve's X-ray indicated that his disease had progressed from category B to category C, the most advanced stage of complicated pneumoconiosis.

Eastern Associated Coal Corporation, however, was not ready to concede the case. The lawyer hired by the company, Paul Frampton, sent Steve to be examined by Dr. George Zaldivar in June 2014, and he, too, found that Steve was totally disabled by complicated pneumoconiosis. Frampton sent the X-ray taken during the exam to a radiologist for another reading, and she also saw category C complicated pneumoconiosis.

This time, sending the X-ray to Wheeler and his colleagues was not an option. If Frampton had found another radiologist who read Steve's films as negative, he had not yet produced the report.

Steve's health continued to deteriorate. He developed a gallbladder infection and needed surgery. The procedure was not especially dangerous, but for someone with lungs as compromised as Steve's, even routine medical procedures become risky. The surgery went fine, but coming out of anesthesia, Steve had trouble breathing on his own.

When John and I visited him at Raleigh General Hospital in Beckley in early July, though, he appeared to be on the mend. His sole complaint was the food. If only they would let him have a Dr Pepper and a bag of potato chips, he'd be a happy man, he said, raising empty hands and smiling as if he were clutching his desiderata at that very moment.

When I arrived back in DC, I got an e-mail from John: "Steve is in a nursing home—pulmonary rehab program—and doing well. He finally got a Dr. Pepper and some chips. Another patient brought him some garden cucumbers and tomatoes that also made him very happy."

But the vicissitudes of end-stage black lung are cruel. Days later, nursing-home staff found Steve nonresponsive, in respiratory failure. He was taken back to the intensive care unit of Raleigh General Hospital, where doctors intubated him to help him breathe.

His daughter Patience visited on July 25. It was Steve and Nyoka's forty-fifth wedding anniversary, but Nyoka's rheumatoid arthritis had gotten so bad that she could barely get out of bed. She'd written a long note and given it to Patience, who read it to her father. "She wanted me to wish him a happy anniversary," Patience said later, "said that she loved him, she missed him terribly, and that if her body would have let her, she would have been there."

A mask covered Steve's nose and mouth, and his eyes were closed. Patience sat in a chair and propped her arm on the bed rail. Steve's breathing was so violent that it shook Patience's upper body. Hospital staff had told her they didn't know if Steve would be able to hear her, but she liked to think he'd get the message.

"I let him know that everybody loved him and that everybody missed him," she said. "I told him that, if he felt like he had to go, that nobody would be mad. That we would miss him, but, if he was tired and wanted to go on, that it's okay. I wanted him to know that I wasn't going to be mad, that nobody was, and that I just loved him."

Steve died the next day. A few days later, he lay in an open casket with an American flag draped over it. The Evans Funeral Home Chapel in Oceana, West Virginia, was full of pictures—Steve in his military uniform; a young Steve planting a kiss on Nyoka while giving a wry look to the camera as if he'd been caught in the act; Steve underground in his mining gear; Steve doting on the grandchildren he'd dubbed "Little Buddy" and "Tiny Buddy"; an older Steve and Nyoka walking side by side, Steve's portable oxygen tank in a pack slung over his shoulder.

Steve had made it clear that he wanted an autopsy performed, but there was a practical obstacle: The total bill for the procedure would likely reach almost two thousand dollars, and for the Days, money was tight. Still, they scraped together the funds.

The report from a Charleston pathologist arrived in late August 2014. Steve's lungs, the doctor found, contained large black masses with scarred and dead tissue. It was overwhelming evidence of complicated coal workers' pneumoconiosis. With the permission of Steve's family, I shared the report with three physicians who specialized in the disease: pathologist Francis Green, former NIOSH official and longtime WVU pulmonologist Jack Parker, and the current head of NIOSH's division focused on respiratory diseases, Dr. David Weissman.

"A majority of his lungs had been replaced by scar tissue with coal dust," Green said after reviewing the report. "I think any pathologist seeing those lungs would have to say that this is about as advanced as you can get." I asked where Steve's case fit among others he'd seen over the years. "It has to rank as the very worst," he said.

Weissman said it was "very concerning" that any doctor who had passed the certification exam for X-ray readers that his office administered would fail to see black lung on Steve's films.

"He's at the extreme end of severity," Parker said. "You just can't get any more severe and survive. To not call this radiographically is to put your head in the sand."

Now, given the findings in the autopsy report, it was instructive to compare the earlier reports of Parker and Wheeler on what Steve's X-rays and CT scans indicated. For the most part, the two doctors had seen the same features on the films, but they reached opposite conclusions about what those features represented. Take, for instance, the location and shape of the opacities on film: Wheeler said they indicated another disease, such as TB or histoplasmosis; Parker said they were typical for black lung. Or the areas of dead tissue both doctors saw: Wheeler said they indicated another disease; Parker said they were classic indicators of black lung.

The autopsy showed that the lesions that were "out of the strike zone" for Wheeler were indeed masses of complicated pneumoconiosis. And the areas of dead tissue were cavities within the masses—the result, as Parker had said they were, of a process called liquefaction necrosis, in which portions of the lung become a sort of viscous black liquid that can rupture into the airways and cause miners to cough up black phlegm.

Wheeler's criteria, now emphatically disproven in Steve's case, were the same ones he'd applied to thousands of other miners' X-rays and CT scans for decades.

In October—three months after Steve's death—Eastern Associated Coal conceded the case and agreed to pay benefits.

Patience's son Eli, Steve's Little Buddy, still saw his grandfather in everyday items around the house in Glen Fork. When Patience restocked a refrigerator with Dr Pepper, Eli said, "Paw-Paw?" Nyoka seemed distant, as if she'd been willing herself on while the fight continued and, now that it was over, had let go. She died a month later.*

Steve had always said that his autopsy would prove Wheeler wrong, and Steve had been right. He'd also hoped that his story might help other miners, and on that point, he was also correct.

When retired miner Larry Ramsey showed up for his hearing before a judge in Beckley, he didn't have a lawyer, but he did have a few papers—a

* Wheeler has not faced any disciplinary action by the Maryland Board of Physicians; the board's website indicates that his license is no longer active. Steve's children and the widow of Junior Barr, another miner whose autopsy had proven Wheeler wrong, sued Wheeler and Johns Hopkins in 2016, alleging fraud. A federal judge tossed out the case in the early stages, and the U.S. Court of Appeals for the Fourth Circuit affirmed the dismissal. Judge J. Harvie Wilkinson III—the same judge who derailed the oral argument in Gary Fox's case and wrote the opinion finding that Jackson Kelly had not committed fraud on the court—again dominated the oral argument and again wrote the opinion. He concluded that Wheeler was protected by the witness-litigation privilege, which generally barred lawsuits against experts for their testimony. Judge Robert B. King wrote a blistering dissent, calling the alleged behavior of Wheeler and others at Johns Hopkins an "unabashed fraud scheme" in which they had "purposefully supplied falsehoods and misrepresentations about medical records in a concerted and repugnant effort to skew the adjudicatory outcomes of claims lodged by hundreds of coal miner victims." His colleagues, King wrote, had wrongly expanded the scope of the witness-litigation privilege and allowed "Wheeler and his partners in crime...to escape liability for using their expertise to prevent our hardworking coal miners from receiving black lung benefits that they are legally due."

news article he'd printed out that featured Steve and detailed Wheeler's record. Lawyers for Ramsey's former employer objected. The article didn't specifically mention Ramsey, so it was not relevant to his claim, one of them argued.

"This doctor, this is who you all got," Ramsey said. "I've never seen him before, but he's the one that's got the medical records that says I don't have black lung and everybody else that don't have black lung." Indeed, the other doctors who had evaluated him or seen his chest films had found complicated pneumoconiosis, but Wheeler had graded them as negative for pneumoconiosis, more likely histoplasmosis.

After the company's lawyer finished stating his objection, the judge said she was "well aware" of the article and the swirling criticism of Wheeler's readings. "I'm taking administrative notice that it's out there, okay, but I don't need to take the actual paper because I've already read it," she said.

Ramsey later won his claim.

Challenges to Wheeler's credibility began popping up in a number of cases. Miners' representatives and Labor Department lawyers asked judges to either accept into evidence or take official notice of the news articles and the Labor Department bulletin instructing claims examiners not to credit Wheeler's readings. Lawyers for companies almost always objected, arguing that the articles were irrelevant because they didn't mention the specific miner whose claim was before the judge or that they were one-sided, unscientific, and untrustworthy. Some judges sided with miners' representatives; others didn't.

One judge, refusing to take notice of the documents, wrote that they were merely "opinions" and "not facts capable of accurate and ready determination." Another wrote: "Unlike a criminal conviction or the qualification of B-reader, none of the Investigative Reports constitute facts which are not subject to reasonable dispute."

Other judges, however, found that the documents were clearly relevant to assessing Wheeler's credibility and either admitted them into evidence or took official notice of them, in many cases discounting the doctor's negative readings. One judge described the data on Wheeler's record and his penchant for seeing TB or histoplasmosis when others saw complicated black lung, and she wrote, "It appears that, in the Claimant's case, Dr. Wheeler's interpretation followed this pattern." She found that the miner did have complicated disease and awarded benefits.

In a few cases, judges discounted Wheeler's opinions and awarded benefits in previously denied claims. For instance, a judge had denied one widow's claim after finding that her husband hadn't suffered from complicated pneumoconiosis; the judge had credited Wheeler's negative reading over those who had seen the disease. A procedural finding on appeal kept the case alive, and this time, a judge wrote that there was "reason for questioning the credibility of Dr. Wheeler's x-ray readings." He found that the miner had indeed suffered from complicated disease, and he awarded the claim.

Considering a new claim by a miner who had previously been denied primarily because of Wheeler's readings, a judge wrote: "Dr. Wheeler here continues the narrative of finding no pneumoconiosis in spite of the contrary longitudinal evidence.... I also decline to take Dr. Wheeler's X-ray interpretations at face value, and, accordingly afford them less weight, because they are speculative, and lack support from the general record." The judge found that the miner did have black lung and awarded benefits.

In September 2015, Johns Hopkins announced that it had concluded its internal review of the Pneumoconiosis Section. A spokesperson refused to discuss what the review had uncovered, but the medical center's actions hinted at what its findings might have been: Wheeler had retired. The unit he had led for forty years was terminated.*

Since the explosion at Massey Energy's Upper Big Branch mine that killed twenty-nine miners in April 2010, Don Blankenship had stepped down

* A spokesperson for Johns Hopkins declined my interview requests and chose not to address a list of key points described throughout this book, instead issuing a statement: "We extend our deepest sympathies to those who have lost loved ones to this horrible disease, and we commend all efforts to ensure the Black Lung Benefits claims process is fair and just for all parties involved. Johns Hopkins took swift action when we learned of potential issues with the program. We suspended the program immediately and later decided it would remain closed."

as CEO, and coal giant Alpha Natural Resources had bought Massey for $7.1 billion. Meanwhile, the office of U.S. Attorney Booth Goodwin, with help from the Labor Department, had been slowly working its way up the company's corporate ladder. Some managers had flipped, agreeing to testify against superiors. Coalfield skeptics, however, suspected the ascent would never reach the top rung—a view with ample historical support.

An indication that this time might be different came in November 2014: A grand jury in the Southern District of West Virginia returned an indictment against Blankenship, charging him with conspiring to violate mine safety and health standards, hindering federal inspectors to cover up those violations, and lying to the Securities and Exchange Commission. The indictment did not directly accuse Blankenship of causing the Upper Big Branch blast.

During a trial that spanned most of October and November 2015, government witnesses, including former Massey miners and managers, described for jurors a man who ruled his central Appalachian empire with an iron fist, cut staffing, instructed subordinates not to follow safety and health laws, and created a culture of fear to keep everyone in line, all in the name of "running coal."

Though much of the testimony focused on safety violations such as those that led to the deadly blast and practices such as radioing underground to warn that inspectors were coming, jurors also heard from Massey miners who described dust-sampling fraud. Indeed, the rules governing dust-sampling were among those Blankenship was charged with conspiring to violate.

One miner who had worked at Upper Big Branch testified that sampling pumps sometimes were placed in less dusty areas away from the coalface. Another miner described covering up pumps with clothing to prevent dust from getting in, and a third said that, when it was his turn to wear a dust pump, he was sometimes told to stand back in the clean air and not perform his normal job. At Upper Big Branch, the miner testified, "if you wore these things the way that you should wear them and the way the law required you to wear them, it would have been impossible to bring them into compliance."

These were just the sorts of things that some of Gary Fox's coworkers at Massey's Elk Run complex had described—practices that masked the potentially dangerous levels of dust that filled the air they all breathed.

Jurors learned that a consultant had warned Massey that its obvious cheating on dust-sampling could invite a criminal investigation, and they heard the recorded phone call in which Blankenship said that "the truth of the matter is black lung is not an issue in this industry that's worth the effort they put into it."

During his closing argument, Goodwin said Blankenship was "the kingpin" of a "massive, massive criminal conspiracy." Blankenship's lead defense lawyer, William W. Taylor III, countered that it was wrongheaded to contend, as he said the government had, that "because [Blankenship] believed black lung disease was not as important as other people thought it was, maybe he's guilty of this conspiracy." He insisted the government hadn't shown any willful wrongdoing. "We require the government to prove more than a man was in charge of a company when a terrible tragedy occurred," he argued. "He's not guilty, and we never should have been here in the first place."

After ten days of deliberations, jurors reached their decision. With family members of those killed at Upper Big Branch watching from the front row, some of them in tears, Judge Irene Berger read the verdict aloud to an overflowing courtroom: guilty of the first charge, conspiring to violate mine safety and health standards, but not guilty of the other two, making false statements to the government and committing securities fraud. Though it might seem counterintuitive, the latter two charges—essentially telling investors and the SEC that Massey followed safety rules even though it didn't—were, apparently, more serious in the eyes of lawmakers. The charges of which Blankenship was acquitted were felonies, and if convicted on those counts, he could have faced up to thirty years in prison. The charge of which he was convicted, however, was a misdemeanor with a maximum sentence of one year. That was the penalty the judge imposed in April 2016, along with a $250,000 fine.

Members of Congress before and since Blankenship's conviction have repeatedly introduced legislation to strengthen the criminal penalties for violating mine safety and health laws, but their attempts have always failed. Nonetheless, the fact that Blankenship was convicted of a crime and sentenced to prison was a remarkable departure from the normal course of coalfield justice. "In the long march of history, this is a significant change—a captain of industry facing safety charges and going to jail for it," Davitt McAteer, the longtime mine health and safety advocate who had served as Mine Safety and Health Administration chief under President Bill Clinton, told *Charleston Gazette-Mail* reporter Ken Ward Jr.

The mother of one of the miners killed at Upper Big Branch said, "I wish it was a lot longer, but we will take this."

Blankenship remained defiant and continued to argue that the government had come after him for political reasons. At his sentencing hearing, he said, "My main point is wanting to express sorrow to the families and everyone for what happened." But he added, "It's important to me that everyone knows that I am not guilty of a crime." A federal court later denied his appeal.

In May 2016, Blankenship surrendered to federal authorities and was taken to a low-security prison in California to begin serving his sentence.

The lawyer heading Jackson Kelly's defense in the civil suit, Jeffrey Wakefield, received welcome news in February 2015. Attorneys from the firm's black lung unit again had avoided sanctions from the West Virginia Office of Disciplinary Counsel.

The office had been investigating three lawyers—Bill Mattingly, Doug Smoot, and Ann Rembrandt—over their actions in Gary Fox's benefits claim. The investigation had started years earlier after administrative law judge Thomas Burke had sent the disciplinary office a copy of his decision finding fraud on the court. In a cover letter, Burke wrote that he believed the firm's lawyers had violated provisions of the West Virginia ethical code that barred knowingly offering false evidence and unlawfully altering or concealing evidence.

The disciplinary office's report outlining its conclusions in the Fox case was similar to the one it had issued on the actions of the three attorneys other than Smoot in the Elmer Daugherty case: a stern warning that carried no penalties. Though the investigative panel that evaluated the allegations found them "quite serious and disturbing," there was not enough evidence to prove specific ethical violations, the report said. The panel noted, however, that some of the same conduct was at issue in the civil case unfolding in Raleigh County; it was still an open question before that court as to "whether Jackson Kelly had a long standing pattern and practice of misconduct in black lung cases."

The panel report continued, "If a court makes any finding of misconduct or other violation of the Rules of Professional Conduct, the court may forward any such findings to this office for consideration."

In other words, the office wasn't foreclosing the possibility of sanctions, but it was looking to the state court to supply any factual support for discipline. That was just the sort of information that Al Karlin and John Cline, now locked in a discovery battle with the firm, were trying to get.

Al and John were drowning in paper. Stacks arrived in waves from the files of Jackson Kelly. It wasn't clear, though, whether any of the documents would be useful.

Having survived the defense team's numerous attempts to have the case tossed out, Al and John had filed discovery requests asking Jackson Kelly to produce essentially everything in its files, including internal memos and e-mails and correspondence with experts, related to the Fox, Carroll, and Eller cases.

The defense team argued that the requests were "overly broad and unduly burdensome" and that many of the documents were protected by the attorney-client privilege and work-product doctrine—shields considered sacrosanct by lawyers. A showdown over the privilege issue seemed inevitable, but for the moment, the two sides agreed that Jackson Kelly would search its files, turn over whatever it didn't consider privileged, and create logs listing the items that it was withholding.

Throughout 2015, the defense sent Al and John piles of paper in batches. By the time of the final submission in November, Wakefield said, the firm had scoured thirty boxes of files as well as its computer system and turned over 37,760 pages of documents. It had also sent hundreds of pages of privilege logs, listing more than 7,300 documents it was withholding.

John, Al, and other attorneys at Al's firm spent months trying to impose some order on the reams of paper. When they had finally assembled the documents in chronological order and examined them, they realized that most of them consisted of little more than the official records in the Fox, Carroll, and Eller cases, copied multiple times.

The voluminous privilege logs were spreadsheets that contained a brief description of each document and the reason the firm was withholding it. Some of the entries hinted at potentially valuable evidence, such as "Handwritten notes regarding film readings by different experts." The log for the Fox case contained a few particularly tantalizing entries. One, dated not long after the firm had been forced to turn over the withheld pathology reports, said "Email regarding legal research on false evidence." Another—"Email regarding lawyers handling Fox claim"—was an exchange between three black lung unit lawyers and Jackson Kelly general counsel and ethics adviser Stephen Crislip discussing administrative law judge Thomas Burke's decision finding fraud on the court issued just a few days earlier.

It seemed clear to Al and John that if there were documents that might prove damning for Jackson Kelly, they would be among those in the privilege logs. To get those documents, however, they would have to clear a high bar: invoking a legal doctrine known as the "crime-fraud exception." The principle behind the exception is that, if a party to a legal dispute is engaging in fraud, it cannot hide behind privilege to conceal evidence of that fraud. Al and John would have to convince Kirkpatrick that there was reason to believe that if the judge reviewed the withheld documents in his chambers, some of them likely would contain evidence of fraud.

At a hearing in Beckley on May 18, Al pressed just that argument. What was already known about the firm's actions in the Fox, Carroll, and Eller cases, he asserted, provided reason enough to believe the withheld documents would reveal fraud. Highlighting the Fox case in particular, he said, "If that wasn't an attempt to put forth a fraudulent defense, I don't know what was."

As Al wrapped up his argument, Kirkpatrick offered an observation grounded in personal experience. The judge described his work as a hearing examiner for the West Virginia workers' compensation system, which included state black lung claims. He knew well the species of specialized litigation in which the same doctors appeared again and again, almost always with the same conclusion, and lawyers for both sides sent their clients to the usual suspects, hoping to get favorable opinions. "But what you have characterized," Kirkpatrick continued, "is something that is much more heightened than simply using experts that hopefully would support a certain position because of trends or past indications. Basically, you have more or less set forth a criminal manipulation."

"Yes, Your Honor," Al said. "What happened here is, you know, different and interesting. It's as if that old system, your doctor who you knew would call every case that he could conceivably in good conscience call in your favor, tells you that this one you shouldn't win. And then you go around trying to manipulate the evidence to win it anyways."

This was the heart of the case, the principle—the obligation not to misrepresent—that Al had struggled to articulate before the Fourth Circuit in Gary Fox's benefits claim. By now, he had honed it, whittled it down to a pointed statement.

In response, Wakefield portrayed Al's argument as an assault on the very foundations of the American way of justice, an unprecedented attempt to erode "the bedrock of the adversarial system." The plaintiffs, he said, were basically asking Kirkpatrick to give them carte blanche to scour a law firm's files. This was one of the strongest points of Jackson Kelly's case. What the plaintiffs sought was indeed unusual, the sort of thing that would make attorneys of all stripes uneasy and set their minds racing down a disconcerting slippery slope.

Jackson Kelly attorneys had never misrepresented evidence, Wakefield asserted. In the Fox case, for example, the lawyers' representations to consulting experts and to the judge were true, he said; Koh's *was* the only pathology report in the record, and it didn't contain a finding of black lung. Choosing to submit only reports that were favorable to the defense didn't amount to "manipulating the evidence in any fraudulent way."

Kirkpatrick interrupted: "I think he's going one step further though.... He's saying that Jackson Kelly represented to their own experts and other experts that this is all there is, this is all the medical records that we have or that's pertinent to this case. And by omission, critical reports and examinations have been overlooked or left out."

That argument, Wakefield said, was an attempt to take some sort of duty to disclose everything—something the Fourth Circuit had said didn't exist—and transform it into the basis for a finding of fraud. The lawyers at Jackson Kelly were simply doing what was normal, even expected of them, within the adversarial system of justice, he argued.

When the hearing was over, Al felt encouraged by some of Kirkpatrick's comments, and his reading of the judge proved correct. In July 2016, Kirkpatrick issued an order granting the motion to compel discovery. Wakefield's argument, he wrote, "side-steps the plaintiff's direct claims in all three lawsuits that the defendant violated its duty to not misrepresent evidence."

Kirkpatrick noted the state supreme court of appeals' condemnation of Jackson Kelly attorney Doug Smoot's actions in the case of Elmer Daugherty. "Obvious similarities between *Smoot* and the present case cannot be ignored," the judge wrote. The firm's arguments in the civil suit, the judge noted, were very similar to the arguments that had proved unpersuasive in Smoot's case. "Lastly, the court can properly conclude that the misconduct alleged by the plaintiffs here, which is almost identical to the misdeeds reported in *Smoot,* can be likewise characterized (if proven) as unlawful, dishonest, deceitful, misrepresentative, and prejudicial to the administration of justice."

The plaintiffs, Kirkpatrick wrote, had met their burden, and the crime-fraud exception applied, overriding Jackson Kelly's claims of privilege. He ordered the firm to turn over the withheld documents to him. He would review each one in his chambers and decide whether there was a connection between the document and the alleged fraud. Then he would decide what the plaintiffs were allowed to see.

For both sides, the path ahead was fraught. After reviewing the 7,300 previously withheld documents, Kirkpatrick might decide that Al and John were entitled to see everything, or nothing. There would likely be further litigation over whatever he determined, with Jackson Kelly almost certain to ask the supreme court of appeals to intervene and block disclosure. Even if Al and John got the documents they sought, there likely would be fights over whom they could depose and what they could ask them, further invocations of privilege, motions for protective orders.

The legal combat had already consumed seven years. All three miners were now dead—Gary had died just before the case began and Eller and Carroll during its early stages. Eller's wife, too, had died as the case had dragged on. Mary Fox was now sixty-three years old, and Dolores Carroll was seventy-five.

For its part, Jackson Kelly faced the possibility that its internal documents would be aired publicly, its attorneys would have to defend themselves under oath, and the firm's name would appear in recurring news accounts repeating the allegations of fraud.

It might take years to get to a trial, and going before a jury carried risks for both sides. Perhaps jurors would find Jackson Kelly's behavior appalling and penalize the firm. Or perhaps jurors would view the case as a technical dispute best left to lawyers to sort out among themselves. Even if jurors did side with the plaintiffs, Jackson Kelly was almost certain to appeal, further extending the process. The state supreme court of appeals appeared to be shifting further to the right, with the elevation of one of its conservative members to the role of chief justice and the election of a new justice whose campaign had been backed heavily by conservative and pro-business groups.

There were compelling reasons for both sides to consider settling. In late 2016, both sides agreed to sit down with a mediator, and in June of the following year, they reached a confidential settlement.

Since the filing of the lawsuit in 2009, much had changed, casting the case in a different light. Settling did mean that John would never see whatever Jackson Kelly might have in its files on Edsil Keener and William Harris and LaVerne DeShazo and who knows how many other miners and widows. But exorcising the ghosts of denied claims past or winning money damages had never been the primary goal.

The goal that Gary, Mary, John, and Al had shared, the reason they'd started this legal battle eight years earlier, was to force a change in Jackson Kelly's practices; they had said as much in the original complaint launching the lawsuit. As the other prospects for achieving that goal—the now-vacated fraud-on-the-court finding in the benefits case and the now-closed West Virginia disciplinary investigations—fell away, it had seemed as if the civil suit was their last chance.

As it turned out, there was another way.

Reading the *Federal Register* is often a soporific endeavor, as its tagline, "the Daily Journal of the United States Government," suggests. It is here, within these pages, that the seldom-noticed agencies whose work affects countless lives present proposals, findings, and rules in all their meticulous glory, their entries often brimming with bureaucratese and bereft of humanity.

Page 23746 of the April 29, 2015, edition was different. It told of a man named Gary Fox: "Mr. Fox worked in coal mines for more than thirty years. In 1997, a chest X-ray disclosed a mass in his right lung. A pathologist who reviewed tissue collected from the mass during a 1998 biopsy diagnosed an inflammatory pseudotumor."

The entry, submitted by the Labor Department's office in charge of overseeing the black lung–benefits system, then unfolded the details of Gary's case and said that it "highlights the long-standing problem claimants face in obtaining a full picture of the miner's health from testifying and non-testifying medical experts as well as examining and non-examining physicians."

To address that long-standing problem, the Labor Department announced, it was proposing a rule that would require both sides to disclose all medical reports they received, regardless of whether anyone requested them. When one side received a report, it would have thirty days to send a copy of the entire thing to the opposing side.

This was the proposed rule that the department had announced it was considering shortly before the Senate subcommittee hearing at which John Cline spoke. Of all the department's planned reforms, this was the most ambitious.

The proposal was anathema to Jackson Kelly's approach to defending claims. The backbone of the firm's arguments over the years was that, in the legislation creating the benefits program, Congress had outlined an adversarial system similar to civil litigation.

The Labor Department now made clear that it agreed with the position John and other miners' advocates had advanced for years: Congress had designed a system that, while adversarial, was supposed to function more like a hybrid workers' compensation system; it was a remedial program meant to protect miners.

Congress had specified that officials charged with deciding benefits claims were not constrained by the formal rules and procedures used in many civil cases. It was clear, the Labor Department argued, that Congress had expressed a "strong preference" for "getting to the truth of the matter," rather than "following the technical formalities associated with regular civil litigation." Full disclosure of medical evidence furthered that goal, the department wrote.

The realities of the benefits system, as it was currently functioning, also offered support. Unrepresented claimants were unlikely to know how to request documents, and even when claimants did find help, lay representatives and lawyers rarely pursued discovery, the department noted. Requiring full disclosure "will put all parties on equal footing," the department concluded.

The proposed rule would also give administrative law judges a cudgel to enforce the disclosure requirements. Over the years, Jackson Kelly had repeatedly pointed out to judges that they lacked the sanctioning authority to compel compliance with discovery orders. Under the department's proposal, lawyers could pay a price for flouting a judge's order. The menu of possible sanctions included, among other things, disqualifying the lawyer from participating in the claim going forward. If the withheld evidence was from a miner's previously denied claim, a judge could reopen that claim and award additional benefits. "This sanction removes an incentive for responsible operators to withhold medical information and, by encouraging operators to comply, helps protect miners like Mr. Fox," the department wrote.

The day that the Labor Department released the proposal, John sent an e-mail to Mary Fox: "Mary, I think Gary would be very pleased with this proposed new rule. It's what he wanted to accomplish, and it's beginning to look as though it could actually happen. Keep your fingers crossed!"

The department had roughly twenty months to shepherd the proposal through the rule-making process before a new president succeeded Obama. This would require speed but not haste; cutting corners could place the rule in future legal jeopardy. The first step was to allow members of the public to submit written comments. The department would then have to evaluate and respond to these submissions; ignoring significant issues raised by commenters risked a lawsuit that could invalidate the rule.

When all the comments on the disclosure rule had arrived, any submission from Jackson Kelly was conspicuously absent.

The longest and most substantive opposition came in a joint submission by the National Mining Association and a group of insurers. "The proposed regulations are both disappointing and disturbing," the organizations wrote. "There is no showing that they are necessary, much less a good idea."

The groups' comment did not specifically name Gary Fox, but it alluded to the facts of his claim and characterized it as an isolated incident that "does not support the Department's overreaction." The rule would cause needless disruptions and delays, all to address a perceived problem that didn't really exist, the groups wrote. Further, they argued, the sanctions provisions were "uniformly illegal and not authorized by law."

"In conclusion," the groups wrote, "we find this proposal unfair and unauthorized and it must be abandoned."

Miners' representatives countered with comments of their own. "Lack of information kills people; having medical records can save lives," wrote a group of lawyers that included Tom Johnson and Anne Davis in Chicago. "The proposed rule clearly provides that the right to know trumps litigation strategy every time."

The United Mine Workers and the Appalachian Citizens' Law Center offered further support, as did H. Morgan Griffith, a Republican member of Congress whose district included much of southwestern Virginia coal country. The proposed rule was a much-needed step to protect miners and "level the playing field in claims cases," the congressman wrote, citing "the tragic case of Mr. Gary Fox."

John's submission rebutted the contention that Gary's case was an isolated incident. He outlined similar behavior by Jackson Kelly lawyers in six other claims and wrote, "It also is important to recognize that these examples do not represent the scope of the problem." Most of the time, withholding of key evidence would go undetected because unearthing it required a rare confluence of circumstances: The miner had to find a lawyer or lay representative who was capable of and willing to pursue discovery, then land before a judge who was willing to order discovery. And even then, the defense could just concede liability to make the whole thing go away.

Over the course of seven pages, John made much the same argument he'd been making for twenty years. This time, however, he had the federal government's attention.

At the end of June 2015, the comment period ended, and the months of waiting began. John and the other miners' advocates could only guess what might be happening behind the scenes in Washington and hope that the department would finish the job in time.

In September, a group of lawmakers composed primarily of coal-state Democrats introduced legislation in both houses of Congress titled the "Black Lung Benefits Improvement Act of 2015." The bill had little chance of passing for the moment, but it amounted to a flag planted in the official record that a future Congress could carry forward.

The bill outlined measures intended to lessen miners' legal disadvantages. To encourage more lawyers to represent miners and widows, the legislation allowed attorneys to collect partial fees as the claim progressed rather than having to wait years for an uncertain payday at its conclusion. The bill did not mention Dr. Paul Wheeler by name, but one provision allowed miners and widows whose previously denied claims included a negative reading by a doctor deemed not credible by the Labor Department to file a new claim. Another section established a pilot program in which either side in a benefits claim could request an X-ray reading by an impartial panel of physicians under the watchful eye of an agency quality-assurance initiative.

But it was the first section of the bill that many of the congressional staffers who worked to develop the legislation considered the most important. It was titled "Mandatory Disclosure of Medical Information and Reports" and was essentially the same as the Labor Department's proposed disclosure rule.

For the staffers, the human toll of withheld evidence was real; it had a face. While drafting the legislation, one staffer said, "Gary Fox's case was definitely something we talked about a lot." The disclosure provision in the bill, they thought, would signal their bosses' support for the proposed rule and keep the pressure on the department to finalize it.

Not that the Labor Department needed much encouragement by that point. "The department had gotten their zeal going," one House staffer said.

In the final year of the Obama administration, the rule made it through the White House budget office. On April 26, 2016, the *Federal Register* contained an entry announcing the change John Cline had sought for so long:

> **Black Lung Benefits Act: Disclosure of Medical Information and Payment of Benefits**
>
> **AGENCY:** Office of Workers' Compensation Programs, Labor
> **ACTION:** Final rule

In pages packed with citations of statutory provisions and case law, the department refuted the comments contending that the rule was beyond its authority. The argument that the rule was an overreaction to one or two isolated incidents was "not accurate," the department wrote. "Although the Department illustrated the need for the rule with a detailed summary of miner Gary Fox's claims," it had cited others in the proposal, and it now pointed to additional examples, including those John had described in his written comment and others that the department had identified. What's more, the department continued, it simply wasn't possible to "quantify the volume of undisclosed medical information in cases where parties do not pursue discovery of that information and, in fact, might not even know of its existence."

The rule, the department concluded, "is effective May 26, 2016."

Solicitor of Labor Patricia Smith felt confident that if the industry sued, the department would prevail. To withstand a lawsuit challenging its legality, any rule needed a sturdy legal foundation—a grounding in the statute, a well-reasoned explanation of its details, and so on—and Smith thought this rule had it. But there was another reason for her confidence, an additional layer of insulation against potential attack that this rule, unlike many others, possessed: a clear example for its need, an accounting of the intolerable condition that would otherwise endure in its absence, a glimpse at the intersection of bureaucracy, corporate power, and human suffering. This the department had in the stories of miners such as Charles Caldwell, William Harris, Elmer Daugherty, Clarence Carroll, Norman Eller, and, first and foremost, Gary Fox.

Seven years after his death, Gary Fox and his story endured. He was there in the minds of congressional staffers and Labor Department officials as they decided to act, in their conversations as they met to discuss what needed to be done, in the reforms they drafted with the hopes of rebalancing the scales of justice. He was there in a hearing room on Capitol Hill as senators voiced outrage, in the pages of the regulation designed to ensure no miner suffered the same injustices he had endured.

And when the Labor Department announced the finalization of the rule in a press release, he was there, too, in the department's promise that the new requirements "will prevent the type of tragic outcome for vulnerable coal miners like Gary Fox," and in Smith's accompanying statement: "Under the new provisions, any miner in a situation like Gary Fox's will be able to make an informed decision regarding continuing to work and caring for their lung condition. Making sure that coal miners have all the information regarding their health will put all parties on equal footing regardless of representation and will lead to better, more accurate claims decisions for coal miners across the country."

Epilogue

December 2016

> The last clear definite function of man—muscles aching to work, minds aching to create beyond the single need—this is man. To build a wall, to build a house, a dam, and in the wall and house and dam to put something of Manself, and to Manself take back something of the wall, the house, the dam; to take hard muscles from the lifting, to take the clear lines and form from conceiving. For man, unlike any other thing organic or inorganic in the universe, grows beyond his work, walks up the stairs of his concepts, emerges ahead of his accomplishments.
>
> —John Steinbeck, *The Grapes of Wrath*

Arvin Hanshaw accepted the menu that the waitress handed him, smiled, and handed it right back. No need to read it. "Brown beans and corn bread," he said. His favorite. Here at Mabel's Diner, they made a mean plate of brown beans and corn bread.

Arvin was solidly built, just under six feet tall, with a mustache and full head of neatly parted black hair just starting to show faint streaks of gray. His bear paw of a right hand engulfed the coffee mug as he took a swig. Until recently, that same hand had manipulated the controls of a hulking continuous mining machine. He'd enjoyed running coal, and he'd been good at it. That was before doctors saw masses on his chest X-ray. Now, at age sixty, he was retired, not exactly by choice. Doctors had told him he had complicated pneumoconiosis and needed to get out of the dust immediately.

Eight months had passed since the Labor Department finalized its medical evidence disclosure rule. There hadn't been a lawsuit challenging it, and coal-company lawyers were now turning over reports that almost certainly would have remained unseen in the past. I had returned to West Virginia to see what effect the rule was having, which was why I now sat across from Arvin in a booth at this roadside diner.

Mabel's had a quaint charm. On the front window, a painted white cow with black spots sipped soda through a straw. The interior was a pastiche of faded red, yellow, and orange décor. The diner was part of a travel plaza just off Route 19—gas pumps, a market, a Dollar General. Rows of eighteen-wheelers sat parked out back.

It wasn't far from Arvin's home here in Nicholas County. He lived in the small town of Poe, back in the Panther Mountains where he'd been born and raised. He was one of thirteen children—eight boys and five girls. All but two of the boys had followed the career path of their father, a coal miner who suffered from black lung. To help support the family, Arvin had quit school after the eighth grade and, at age fourteen, gone to work at a sawmill. A few years later, he'd decided to enter the mines, drawn by the better pay. Before long, he'd worked his way up to running a continuous miner, and he spent most of his three decades underground cutting coal at the face.

"Did I like the coal mines? Yeah, I loved it," he said. "To me, seeing the coal moving out, I liked that. But I didn't know that it was really doing so much harm to me."

By the early 1990s, he noticed that breaths seemed harder to come by, and the state workers' compensation fund found that he was 20 percent disabled. The lump-sum payment that came with this finding, however, was peanuts next to his mining salary. He had a wife, three kids, and a mortgage.

He'd spent most of the last fifteen years of his mining career working for Massey Energy subsidiaries. His experience with the company echoed that of other miners, including some of Gary Fox's coworkers—long hours, intense pressure for production, little more than lip service to safety. He described the same sorts of dust-sampling chicanery I'd heard countless times by then—bosses taking the pump off him and putting it in clean air.

But these days, he said, it'd be much harder to pull some of the old tricks, thanks to the 2014 Labor Department rule mandating use of new continuous sampling devices and closing regulatory loopholes. "It's going to cut down on cheating," he said.

The rule arrived too late to benefit Arvin, though. He'd filed his first claim in September 2011. The doctor who had conducted the exam paid for by the Labor Department found complicated pneumoconiosis. But Dr. Paul Wheeler of Johns Hopkins saw something different on Arvin's X-rays; he thought the masses were more likely histoplasmosis or perhaps TB. Dr. George Zaldivar examined Arvin on the company's behalf and ultimately concluded that he didn't have black lung, citing Wheeler's negative reading. A Labor Department claims examiner had credited Wheeler's opinion over those of Arvin's doctors and denied the claim in June 2012.

This was much the same pattern I'd seen unfold in many cases. What was different this time was what had happened next.

Arvin filed a second claim in February 2013 and chose to have his government-paid exam performed by Dr. Don Rasmussen, who found complicated pneumoconiosis. Arvin got in touch with John Cline, who agreed to represent him, squaring off against Jackson Kelly.

By then, part of the lawyers' strategy that had sunk Arvin's first claim—sending the films to Wheeler—was not an option. Johns Hopkins had suspended its Pneumoconiosis Section three months earlier. Soon after the case landed before the firm, the Labor Department issued the memo instructing its officials not to credit Wheeler's readings. Jackson Kelly had Zaldivar examine Arvin, but this time, the pulmonologist reached a different conclusion. Reports by Wheeler no longer seemed to carry as much weight with him. Like Rasmussen, he found that Arvin's breathing had worsened, and he read the X-ray taken at the exam himself and saw complicated pneumoconiosis.

The Labor Department claims examiner awarded benefits, but the firm appealed. As an administrative law judge considered the case, the Labor Department's disclosure rule took effect. Jackson Kelly attorney Ann Rembrandt sent John a packet of documents with a cover letter saying that the company "is exchanging the following medical information under compulsion to comply with the mandatory disclosure requirements." Included were fifteen interpretations of Arvin's X-rays and CT scans by two radiologists—each one graded as positive for complicated pneumoconiosis. In July 2016, the judge awarded benefits.

Without Wheeler, Jackson Kelly had no case, and because of the disclosure rule, everyone knew it.

Arvin attributed much of his victory to the fact that Jackson Kelly was required to show its hand. I asked if he knew how the new rule had come to be. Had John ever mentioned a man named Gary Fox? He shook his head. I began to tell Gary's story, and Arvin put down his spoon and leaned slightly forward, his brow furrowed. When I got to the part about the two pathology reports Jackson Kelly had obtained, he anticipated my next words. "Withheld them," he said.

There were numerous similarities between Arvin's case and Gary's. Both men had worked in high-exposure jobs and spent the last portion of their careers with a Massey Energy subsidiary. Both had contracted complicated pneumoconiosis at an unusually young age. The size of the masses on Arvin's chest films and the moderate severity of his breathing impairment in 2016 were similar to Gary's at the time of his first claim in 1999. That was where their stories diverged. Gary had lost and endured years of additional dust exposure, the masses in his lungs had grown, and his breathing impairment had become severe. Arvin had won his case and gotten out.

"It not only helped me," Arvin said of the disclosure rule, "that there has helped different people where, hey, if you got any evidence, you got to show it, whether it's good or bad. That's fair. Withholding evidence is not fair." He laughed as if the proposition was axiomatic and having to say it aloud was absurd.

Of course, Arvin's life would not be easy going forward. Doctors had told him that his breathing would progressively worsen and that he might need a lung transplant one day. But now that he was receiving benefits, he had access to treatment that would otherwise have been prohibitively expensive. No longer did he have to choose between paying monthly bills and paying for the costly medicines and medical tests that his insurance plan didn't cover. He planned to enroll in a pulmonary rehabilitation program at the New River clinic in Scarbro.

He had what Gary had been denied: hope that his disease might not reach the worst stage and, perhaps more important, time. And he had an idea of how he wanted to spend a good chunk of it. A few years earlier, he'd become president of the Nicholas County chapter of the Black Lung Association. The group held monthly meetings, and members helped fellow miners and widows file benefits claims and get in touch with clinic workers and lawyers who might provide further aid. Not long after Arvin had won his benefits claim, he'd invited John to speak at one of the association's meetings. Afterward, Arvin said, "I told him, I said, 'John, anything that you need me for, just call me.' He said, 'What I want you to do, keep doing what you're doing right here for these people.'"

Arvin and I cleaned our plates, drained our coffee mugs, and paid the bill, then stood outside Mabel's talking for a few more minutes as the rigs roared past on Route 19. Arvin asked if I was planning to attend the West Virginia Association of Black Lung Clinics' annual conference in Pipestem. Arvin had become a regular at these gatherings. He sometimes spoke about his experience with the disease, and he came back equipped with new knowledge to share with his fellow Black Lung Association members. He had taken up the mantle of the original black lung insurgents, the wildcatters, the legends of the coalfields.

I told Arvin I hoped to make it to Pipestem again this year. If I did, he said, I should let him know. He knew of a place near there where we could grab a good meal. "They make some mean brown beans and corn bread," he said, grinning widely.

We shook hands, and I stood in the parking lot for a few moments, smiling as I watched his stone-gray F-350 diesel merge with the landscape, bound for the Panther Mountains.

The Smith home in Morgantown was a flurry of activity. Just four days remained until Christmas. The Amazon boxes were piling up in the basement, waiting to be wrapped and placed beneath the tree. The household coursed with the energy attendant on the wide-eyed anticipation of children.

Gary Fox's daughter, Terri—now Terri Smith—was curled up in a love seat in the living room, keeping tabs on the comings and goings of her two children and issuing a mixture of gentle guidance and, when appropriate, stern admonitions. "Santa Claus is watching you," she told her son, Luke, pointing two fingers at her eyes, then at his, like Robert De Niro in *Meet the Parents*. At the same time, she was managing to field yet another round of questions from me. When we had met four years earlier, Luke had been a shy one-year-old. Now he proudly presented himself in a series of costumes, disappearing and reappearing moments later, transformed; first in a police uniform, then in various combinations of jeans, shirts, and shiny hats.

Luke was in kindergarten at a local Montessori school and had developed a love of reading, Terri said. Across the room, Luke chimed in, "Read, read, read!" The next year, he would start first grade at the local elementary school, which was ranked among the best in the state.

He now had a sister, two-year-old Parker. She was more sedate, crawling into the love seat with Terri and looking at me with a mix of curiosity and trepidation.

These were the grandchildren Gary had so desperately wanted. Terri and Rick had married five months after his death, and Luke had arrived almost two years later. "He would love them," Terri said of her father. "I don't know if we would ever even see them. My mom and my dad would have them."

Mary's sister, Gloria Roles, and her husband, Richard, tried to help plug the hole that had been left by Gary's death. Richard had bought a vintage Volkswagen Beetle and fixed it up, just as he and Gary had always talked about doing. Luke loved riding in it, and they'd dubbed it "Bugsy." Richard doted on Terri's children. When I'd talked with him, he'd scrolled through picture after picture of them on his phone, recounting birthdays, parades, and picnics. "Trying to make up for them not having Gary," Richard had said, "it's what I try to do."

Terri's husband, Rick, made time for his family despite a demanding job, much the same way Gary had. A cardiologist at Mon Health Medical Center who was often on call, Rick nonetheless managed to carve out opportunities to be with his children. As Terri and I talked that afternoon, he appeared at the house in blue scrubs. He had more work to finish up that evening, but for the moment, priority number one was teaching Luke to play guitar. From a nearby room came the opening riff of "Jessie's Girl" on repeat.

After sixteen years as a nurse, Terri had stopped working to take care of Luke and Parker. Her father's death still felt fresh in her mind, especially around Christmas. If not for the kids, she said, she probably wouldn't have put up the decorations that now adorned the house—garlands on the balcony railing, a wreath, a tree. That was something she and Gary had always done.

Only recently had she learned the details of her father's fight for benefits. Her parents had kept much of it from her, trying to shield her from worry. Now, that knowledge added further reason for the pride she'd always felt in her father. "You would have had to drag the story out of him," she told me. "He wouldn't have really wanted to talk about it. He would have thought that he was just trying to provide for his family."

On that account, he had succeeded. She and Rick were raising their children in a three-story home near the fifteenth hole of a golf course in a gated community. The kids could go to movies and football games, take golf lessons, and attend the best schools.

"I worked for sixteen years in these amazing jobs because my dad sent me to college," Terri said. "Everything else that we have is part of that."

* * *

"You're not going to put that in the book, are you?" Mary Fox asked.

I had brought up how she and Gary met—how she'd been dared to go steal the pitcher of beer from the guy across the bar, after which the two teenagers were virtually inseparable. I'd first heard the story from Terri, then Mary's sister, Gloria, but I was curious to hear Mary's telling of it.

"Sure, why not?" I said. "It's a cute story."

Mary grinned and launched into the tale, which included a pantomime of her reaching for the pitcher and Gary stopping her. We had talked on numerous occasions over the past few years as I'd tried to learn more about Gary and his life, and each time, she'd apologize and say something like "I feel like I'm not helping you; I think I've told you all I know." But each time, she'd softly utter at least a few things that offered a further glimpse of the life she and Gary had shared and the quiet persistence both had shown.

This time, one of those utterances came when she finished telling the beer-pitcher dare story: "He always said he loved me from the minute he first saw me."

We were sitting in the living room of the house in Beckley where Terri had grown up, where Gary had returned at all hours of the night and begun the laborious process of scrubbing the coal dust from his clothes and body. Pictures of the trio lined the walls.

Parked in the driveway was Gary's black Toyota Tacoma, the one that had arrived home day after day coated in coal dust from the Elk Run mine complex, the one that Gary had asked his brother-in-law, Richard Roles, to drive once a week while he was in Pittsburgh awaiting a lung transplant. "I don't want it to sit down and rust," he'd told Richard. "Then it won't be in good condition when I get back."

Now Mary drove it occasionally. "I'll keep it until it falls apart, because he liked it," she'd told me once. "He played with it all the time, working on it. I've had plenty of offers, wanting to buy it. Forget it. That's about the only thing I got left."

For a while after Gary's death, Richard would drive Mary to the mall or the grocery store, keep her company, take her mind off the loss. Time had brought acceptance and a new type of happiness. She relished being with her grandchildren, Luke and Parker. She sometimes joked about finding a nice, handsome man, but in truth, she had no interest in dating. To her, there was only Gary.

Mary had found additional purpose in carrying forward the legal fights that she and Gary considered so important, and she had found further support in John Cline. "If it wasn't for John Cline," she said, "I wouldn't have known what was in and what was out because all that was new to me."

"When you were in the middle of the process," I said, "did you ever think that you would get to the point where— "

"We are now?" she said, finishing my question. "No, never. And I don't think John Cline ever thought that, taking Gary on as a client."

I brought up the Labor Department's disclosure rule that had taken effect a few months earlier.

"That is what all this was about, was getting that changed so other miners wouldn't go through it," she said. "I am glad we finally got it." Gary, she said, "would have been very happy. *Very*."

I mentioned how unusual it was for a federal agency to devote so much space in a formal regulatory document to the telling of one person's story.

Mary smiled and said, "They ought to call it 'Gary's Law.' "

John Cline and I sat in the kitchen of the home he and his sons had built in Piney View. Slivers of light from an overcast sky filtered through the line of bare trees bounding John's land and fell on the papers spread out before us on the smooth maple tabletop.

We had spent the past few days driving around the coalfields. We had followed Route 3 to the tiny hollow of Toney's Branch, where John lived when he'd arrived in West Virginia as a VISTA. We had driven south, tracing the path of the Big Coal River past the Elk Run complex where Gary Fox had worked. We had seen the handiwork of Cline Building Corporation—homes spread across the Beckley area.

One evening, we arrived back at the place where I'd parked my car for the day: the New River clinic in Scarbro. I gathered my notes and tape recorder and retrieved my backpack from the rear seat of John's car, and when I looked up, John was stooped over the front of my car, towel in hand. "Got salt on your headlights," he said, wiping away the grit that city workers had spread on the roads to combat the recent snowfall.

The next day, we visited the cemetery where some of the ashes of John's oldest son, Jesse, were buried. In the wooded area behind the other gravestones, John crouched and cleared the leaves and brush from the stone marker. A year earlier, I had seen him do the same thing in East Aurora, where a similar marker for Jesse sat next to one for John's parents. That day, he had knelt in the grass as a heavy rain pounded his hooded black jacket.

From Jesse's marker overlooking Piney Creek, it was a short drive to Blue Ridge Memorial Gardens, where Gary Fox was buried. We crossed the frosty grass to a gravestone that said GARY N. FOX, SP4 US ARMY, VIETNAM. A stone bench by the roadside was engraved with the words FOX, IN LOVING MEMORY OF GARY NELSON.

Now, John and I sat where he and Gary had met a decade earlier and gone through Gary's case file. It was staggering to think how much had changed in the interim. The long-sought rule to strengthen measures to prevent black lung was now fully in effect. The archetypal coal baron accused of flouting health and safety laws was in prison. A critical legal presumption that helped miners in benefits cases had been restored after thirty years. X-ray readings by Wheeler and his colleagues at Johns Hopkins—work that had affected the lives of thousands of miners over the course of four decades—had come to an abrupt end. As far as John and other miners' advocates could tell, no similarly credentialed radiologists had stepped up to take their place, and the federal government had recently announced a quality-assurance program designed to identify physicians who consistently overdiagnosed or underdiagnosed black lung in films for benefits claims. These doctors now could face sanctions from a suspension to a permanent ban from participating in the benefits system.

And disclosure of all medical evidence developed by both sides was now mandatory.*

When the Labor Department had announced this last reform, John sent me an e-mail: "Haven't had a chance to look—mired in discovery brief." It seemed appropriate. One more good scrap over documents before the disclosure rule took effect and, in large part, obviated the need for such tussles. A few days later, he'd carved out time to sit down and read the rule. There, in the pages of the *Federal Register,* were the arguments he'd been making for so long. At first, he had felt like one man shouting into the void. As evidence had amassed over the years, he'd slowly gained supporters. Now his proposal was the law of the land. Names and faces of miners and widows flashed through his mind.

"There it was, after all those years," he said. "I was overwhelmed."

In the five years I'd known John, I had regarded him with both admiration and a faint, lingering skepticism. The narrative of his life—idealistic

---

* Two sitting administrative law judges told me that, in the years since the disclosure rule took effect, they have seen a growing number of cases in which companies have conceded after having to turn over evidence favorable to the miner or surviving family member. "I think sunshine is a wonderful thing in so many ways," one said, "and this has added to the sunshine in the medical records." I also discussed the disclosure rule with retired judge Edward Terhune Miller, who presided over Gary Fox's first claim. "That's a splendid outcome—long, long overdue," he said.

Overall, the award rate at the initial level—before Labor Department district directors—has risen in recent years. From the 2008 to 2014 fiscal years, miners and surviving family members won at this stage roughly 20 percent of the time. That rate jumped in 2015 and reached just under 33 percent in 2019. Department statistics also indicate that the proportion of claimants who have a lawyer or lay representative at the initial level has increased. In fiscal year 2011, 46 percent of claimants had representation. By fiscal year 2019, that number had reached 68 percent.

There are numerous potential explanations: increasing severity of disease, reflecting the recent resurgence; newly out-of-work miners filing claims amid an industry downturn; the restoration of the fifteen-year presumption in 2010; and the administrative and regulatory reforms enacted by the Labor Department from 2014 to 2016. A department spokesperson said award rates for cases before administrative law judges were not available.

As significant as these changes have been, the black lung–benefits system is hardly a fine-tuned engine of justice today. In many cases, miners and widows still face a daunting and lengthy fight against an opponent with vastly superior resources. They must meet high standards of proof and endure rounds of appeals. Significant backlogs remain, and the trust fund that pays benefits in many cases has substantial financial challenges. Some of these difficulties were built into the benefits system by its congressional creators decades ago. Others could be addressed, at least in part, by increased funding from the federal government.

New Yorker leaves middle-class comfort to serve the common man in Appalachia—seemed too neat, too much like the stuff of a cable-TV Christmas special. I watched for traces of hypocrisy or ulterior motives, not because I hoped to find them but because I felt it was my responsibility to look. For years, I had pried into every corner of his private life. John had not only endured this but welcomed it. He had spoken candidly about what he considered his shortcomings, and he only reluctantly acknowledged his accomplishments after I learned of them from other people. Assuming he hadn't spent a lifetime orchestrating an elaborate deception that would yield no obvious benefit to him personally, the only plausible explanation was that his story was true.

In his writing, John liked to quote the American anthropologist Margaret Mead: "Never doubt that a small group of thoughtful, committed citizens can change the world; indeed, it's the only thing that ever has." I had thought it was a noble but quixotic sentiment, one that offered useful inspiration but usually failed to find support in the course of human events. Now I wasn't so sure.

Whenever I asked John about his achievements, he would deflect attention to others he insisted had done just as much. He was merely trying "to be of use," he often said. He'd once written to me of what he'd learned managing construction jobs: "When *I* did less, *we* did more." He didn't hold any formal position in the Black Lung Association. He didn't like to speak publicly; he preferred to assemble a panel of miners to speak for themselves. He chose to be the face of nothing but the catalyst of countless things.

"I think he's injected into the whole black lung advocacy and black lung health community a spirit of activism, a kind of quiet activism that really very subtly encourages people to do the right thing," John's former boss at the New River clinic and longtime friend Craig Robinson had told me. "He's always been an organizer, really. In addition to being technically astute, he knows how to connect people, and so those connections and the networks that he's strengthened I think will also carry on."

John sometimes spoke of a periodic "itch to build" even though he'd left the home-building business, a desire he'd indulged on a couple of occasions with the help of sons Jesse and Matthew. It was a fundamental impulse that seemed to extend well beyond carpentry. It had dwelled within Gary Fox too. Now, on this cold December afternoon, some of the fruits of this shared impulse were spread across John's kitchen table.

John had brought three case files down from his office, and we now went through them. Each man had spent decades in the mines of southern West Virginia, and each of their files contained a letter from Jackson Kelly or another major defense firm that began in similar fashion: "This letter constitutes the disclosure of medical information..." or "Under compulsion to comply with the mandatory disclosure requirements..." And each one contained readings of X-rays and CT scans on which experts of the firms' own choosing had found complicated pneumoconiosis.

Miners and their representatives across the coalfields were starting to receive similar letters and reports. Debbie Wills, who helped miners pursue benefits from a clinic near Charleston, had seen a number of them; miners brought them in with other materials from their claims. "Until then, I had no clue what they were hiding," she said. "It has really, really made a difference in a number of claims that I've seen with my own eyes.... They bring stuff in and those reports will be part of it, and I'll just say, 'Oh, hallelujah!'"

As John and I flipped through files, I asked how he thought these cases would have unfolded in years past. Most likely, he said, lawyers would have withheld these reports, shopped for more evidence, and tried to build a case around any more favorable opinions they could find. The claims could have dragged on for years. "Now," he said, "these look like easy cases."

We discussed the battles he'd fought over withheld evidence over the course of twenty years, from the curiously incomplete report Charles Caldwell had shown him to the conspicuous absence of pathology reports in Gary Fox's file. He allowed himself a moment of reflection, the closest John came to overt pride. "All the things we've fought for over the years," he said, "I think what we have now is pretty good, pretty fair."

When we finished going through the papers, John returned them to their manila folders, carried them up the stairs to his office, and put them back in their proper places. At age seventy-one, he was as busy and vibrant as ever. The shelves lining the walls were as full as ever, with the excess documents finding homes on the couch or the stairs to the attic. Next to the phone, the folder containing handwritten logs of calls from potential clients was as thick as ever.

At his desk, John stood for a moment, looking out the window toward the mountains in the distance and the gorge below.

The sculptor of this gorge, Piney Creek, is a stubborn contrarian. It flows north, gathering strength from Packs Branch and Batoff Creek. At a junction once called Coalgate, it joins the New River and continues meandering defiantly north, its sharp curves yielding spectacular views and its frothing flight over rapids delighting rafters and kayakers. With the input of Ephraim Creek and Butcher Branch and many more, it reaches the dam at Hawks Nest coursing with power enough to generate 100 megawatts of electricity. On it flows, becoming the Kanawha River, then the Ohio, then the Mississippi, until it empties into the Gulf of Mexico.

It is each of these and all of them. It is recreation and comfort and strength. It is of great use.

Acknowledgments

Working on this book has been the most meaningful experience of my career. I have gotten to meet remarkable people, spend time in a beautiful and culturally rich region, and test myself as a writer in ways I dreamed of as a child.

A project of this scale requires an enormous investment of time and resources—luxuries that are increasingly rare in this age of slashed budgets and staff cuts throughout the media industry. I am incredibly grateful that, over the course of the roughly six years I worked on this book, numerous people believed in the project and afforded me ample amounts of both luxuries so I could complete it.

My editors at the Center for Public Integrity, Jim Morris and Ronnie Greene, supported and guided my reporting on black lung from the first brief piece I wrote on the subject in early 2011 through the twenty-eight-thousand-word series we published in late 2013. In retrospect, I realize how extraordinary it was for a twenty-seven-year-old with limited investigative experience to be allowed to spend so much time on a story. Jim and Ronnie shaped the reporting, kept me going when I thought I'd hit a dead end, and gracefully improved my writing while leaving my voice intact. Both of them are terrific journalists and true friends.

Some of the center's stories were done in partnership with other news outlets. For the story on the causes of black lung's resurgence, we teamed up with NPR, and I got to spend time in the coalfields with reporter Howard Berkes and producer Sandra Bartlett. For the story on the radiology unit at Johns Hopkins, I worked with correspondent Brian

Ross and producer Matthew Mosk of ABC News. The broadcast pieces that resulted from both partnerships were wrenching and powerful.

Even after the publication of the Center for Public Integrity's series in late 2013, I felt there was a larger story to tell, but I knew next to nothing about writing a book. Numerous fellow journalists offered helpful advice then and later. Ronnie, who is the author of two books, put me in touch with his agent, Esmond Harmsworth, who agreed to represent me. Since then, Esmond has provided invaluable guidance.

In 2014, I accepted a job as a reporter on the investigative team at BuzzFeed News. My editor there, Mark Schoofs, offered unwavering support for my work on this book. I am extremely grateful to Mark and BuzzFeed editor in chief Ben Smith for allowing me to take months of leave as I hammered out a first draft.

In 2018, the Columbia University Graduate School of Journalism and the Nieman Foundation provided me with a much-needed infusion of resources and a psychological boost by selecting this book as a recipient of the J. Anthony Lukas Work-in-Progress Award.

In 2019, I joined the investigations desk at the *New York Times,* and I have enjoyed the full backing of my editors there, particularly Dean Murphy and Rebecca Corbett.

Multiple colleagues, including Ronnie, Mark, and Jessica Garrison, read some or all of a draft manuscript and offered insightful feedback.

The fact-checking process took more than a year. Over that time, Lu Fong and Hilary Elkins picked through a heavily annotated version of the manuscript and many thousands of source documents. When time constraints arose, Mary Marge Locker and Jamie Fisher pitched in to check a few chapters. All of them saved me from embarrassing errors; any that remain are my own. Copyeditor Tracy Roe likewise caught mistakes and made much-appreciated suggestions on issues of context, tone, and language. My thanks to production editor Pamela Marshall, who kept the whole process moving smoothly, and to book designer Sean Ford. My editor at Little, Brown, Phil Marino, showed great patience as I worked through multiple drafts and has been a champion of this book.

While reporting and writing this book, I frequently found myself leaning on skills I'd picked up as an undergraduate at the University of Richmond and as a graduate student at the University of Missouri. Richmond journalism professors Steve Nash, Tom and Betsy Mullen, and Mike Spear ensured I had a solid foundation in the craft and nurtured my growing love for it. Numerous Missouri professors taught me valuable techniques and principles, and a dream team of advisers—Mark Horvit, Steve Weinberg, and David Herzog—helped me execute my first major investigative project.

One lesson I have learned is that no matter how passionate you are about the subject of your book, there will be moments when you feel that your work stinks, that the process will never end, and that you're wasting your time. I have been fortunate to have family and friends who listened and talked me down when those moments inevitably arose. They also provided much-needed emotional support as I reported and wrote about the difficult events in the final chapters, including the deaths of miners I had come to know.

As I slogged through the years of reporting and writing, my parents, Elizabeth and Roger, were an unflagging source of strength, just as they have been my entire life. When I was growing up, we read books together every night. They encouraged me to write, even praising my adolescent attempts at short stories and my half-baked teenage novel outlines, and they still have copies of an issue of a children's literary magazine containing a silly bug-related story I wrote in elementary school. More recently, they offered feedback on draft sections of this book. I cannot recall the last holiday or family trip that didn't involve me disappearing for hours to make calls, write a new section, or fact-check a passage, yet they never once complained. They have given me nothing but unconditional love and support. I owe them more than I can ever repay.

Finally, I want to thank the many people who allowed me to tell their stories in this book. These miners and their families and advocates welcomed me into their homes and hospital rooms. They shared intimate details about their personal struggles and gave me access to nonpublic legal files containing sensitive medical and financial information. Without their trust and unflinching honesty, this book would not have been possible.

I especially want to thank Gary Fox's family, including Mary Fox and Terri Smith, for spending so much time answering my questions and providing records. Though I never got to meet Gary, I feel as if I knew him. Likewise, John Cline and his family opened up their lives to me to a remarkable degree.

We reporters tend to be a cynical bunch. We typically write about injustices and abuses of power; we highlight humanity's thoughtlessness, greed, and cruelty. It is easy for us to become convinced that such traits are the norm among our fellow citizens. Occasionally, though, we get to work on a story that reminds us that this isn't true, that people possess an enormous capacity for empathy, generosity, and self-sacrifice in the interest of the commonweal. Working on this book was such a reminder for me, an affirmation that, despite the immense challenges we all face, hope remains.

Notes

A Note on Sources

This book is based on roughly eight years of reporting that included numerous trips to southern West Virginia, hundreds of interviews, and the review of tens of thousands of documents. Most scenes and dialogue throughout are based on either my firsthand observations or contemporaneous documentation such as transcripts, notes, and video or audio recordings. In the less common instances in which I have relied on the memories of sources, I have attempted to interview all participants, and I asked sources to repeat their accounts in multiple interviews so I could evaluate their recollections for internal consistency and check them against verifiable facts. Statements about individuals' thoughts and feelings are based on interviews with the individuals themselves or, in a few instances, people with whom they discussed their thoughts and feelings; documents such as notes and e-mails sometimes provided further corroboration.

To describe specific federal black lung–benefits cases, I relied heavily on case files obtained from the U.S. Department of Labor. These documents are not publicly available. The department publishes final decisions issued by administrative law judges, but the files themselves are protected by privacy laws and are not releasable under the Freedom of Information Act. To obtain the full files, I identified cases of potential interest, then contacted miners or their surviving family members and asked if they would be willing to fill out and sign a form granting me access. Most agreed. I sent these forms to the department, which then provided me with the full record from each case. Individual claim files varied in length from a few hundred pages to many thousands of pages. I also relied, to a lesser degree, on case documents obtained from other sources. Because these documents are so numerous—a single paragraph in this book is often based on information from a handful of documents—I have generally not cited each one individually.

The already public final decisions since 2000, which are posted online, also proved helpful in many respects. I used them to create a database documenting the roughly fifteen hundred cases in which Dr. Paul Wheeler of Johns Hopkins was involved, and they pointed to other key issues and interesting cases that warranted further scrutiny.

I obtained some documents, such as the filings in criminal cases involving dust-sampling fraud as described in chapter 14 and the e-mails of administrative law judges quoted in chapter 22, through public records requests.

I sought interviews or comments from the people and companies featured in this book; in some cases, I have noted their responses (or lack thereof) in footnotes. I attempted to contact the Jackson Kelly attorneys mentioned in this book by e-mail, certified letter, and, when possible, phone. I provided a list of specific points and general themes that I hoped to discuss. With one exception—Mary Rich Maloy, who agreed to discuss only her work outside of Jackson Kelly and whose comments are reflected in a footnote in chapter 12—the attorneys either refused to talk with me or did not respond to my requests.

The firm's general counsel, Michael Victorson, sent me an e-mail that I have quoted in a footnote on page 139 in chapter 11.

A spokesperson for Johns Hopkins declined my interview requests and chose not to address a list of key points described throughout this book, instead issuing a statement that I have quoted in a footnote on page 335 in chapter 24.

I repeatedly attempted to reach Dr. Paul Wheeler by phone and certified letter to discuss the material about him in this book, but I received no reply. As described in chapter 13, I spoke briefly by phone with Dr. William Scott, who worked alongside Wheeler for many years. Scott insisted he had always followed the black lung–benefits system's rules for classifying X-rays, and he hung up the phone when I cited documents that indicated he had not. I tried to contact other doctors who worked in the medical center's Pneumoconiosis Section over the years, but they did not respond to calls and messages.

Former Massey Energy CEO Don Blankenship declined to be interviewed at length. I spoke with him by phone, and he said: "No one ever did more for mine safety than Don Blankenship did, and people that worked at Massey know that. The press has lied about that for thirty years." As described in further detail in a footnote on page 177 in chapter 14, he answered a couple of my questions related to allegations of dust-sampling fraud but declined to answer follow-up questions.

Though some people declined to address my questions, I have tried to accurately represent their views to the extent possible based on other documentation, such as legal filings and public statements.

This book—especially the historical descriptions in earlier chapters—benefited greatly from previous work by scholars and journalists. For anyone interested in learning more about black lung and the fight that culminated in the Federal Coal Mine Health and Safety Act of 1969, I strongly recommend two books: *Black Lung: Anatomy of a Public Health Disaster,* by labor historian Alan Derickson, and *Digging Our Own Graves: Coal Miners and the Struggle Over Black Lung Disease,* by Barbara Ellen Smith, a professor emerita at Virginia Tech and chronicler of Appalachian economic justice movements.

Longtime *Charleston Gazette-Mail* reporter Ken Ward Jr.'s courageous coverage of the coal industry, among other subjects, enlightened me; I know of perhaps no other reporter whose work better represents the ideal of journalism as a public good—a service that is becoming increasingly rare as local and regional newspapers across the country cut staff or close entirely.

Other journalists have also documented the toll taken by black lung. A 1998 series titled "Dust, Deception and Death" by the *Louisville Courier-Journal* detailed widespread fraud that undermined efforts to eradicate the disease. Howard Berkes of NPR, with whom I worked on a 2012 story about black lung's resurgence, has continued reporting on the subject. In 2018, he led an NPR team that unearthed new details about the severity of the resurgence and helped pinpoint potential reforms (see, for example, https://www.npr.org/2018/12/18/675253856/an-epidemic-is-killing-thousands-of-coal-miners-regulators-could-have-stopped-it), and they collaborated with PBS's *Frontline* to produce a powerful documentary (see https://www.pbs.org/wgbh/frontline/film/coals-deadly-dust/).

The historical details in this book also benefited from the willingness of a few people to share with me their personal collections of notes, news clippings, meeting fliers, letters, and similar materials. John Cline in particular is a meticulous note-taker, and he has kept supporting documents dating back decades, everything from letters he wrote to his family in 1968 to meeting minutes from 1988 to logistical details of a lobbying trip to Washington, DC, in 2019.

In the endnotes, I have attempted to provide the reader with an understanding of the factual support for statements in the text and, in some instances, additional context and suggestions for further reading. The source materials and lists of interviewees contained in the endnotes are not necessarily exhaustive. Often I relied on additional source materials, but I have not included a lengthier list because of space limitations and a desire not to overwhelm the reader with technical references. In a few instances, I have not named an interviewee because the person spoke only under the condition of anonymity, but such cases are extremely rare. Virtually every interview conducted for this book was on the record.

PROLOGUE: JANUARY 10, 2007

3 **As each man appraised the other:** Much of this prologue is drawn from documents within the files for Gary Fox's two federal black lung–benefits claims, John Cline's detailed notes of his interactions with Gary, and numerous interviews in which I discussed these materials with John. To the extent possible, I have corroborated facts with additional sources of information.

5 **peculiar legal gauntlet that miners and widows had to traverse:** For an overview of the federal black lung–benefits system, see the website of the Division of Coal Mine Workers' Compensation, an office within the U.S. Department of Labor's Office of Workers' Compensation Programs: https://www.dol.gov/owcp/dcmwc/.

6 **turned down about 85 percent of claims:** Division of Coal Mine Workers' Compensation,

"Black Lung PDO Claim Decisions at the District Director Level FY 2006-2017," https://www.dol.gov/owcp/dcmwc/statistics/PartCClaimsDecisions.htm.

7 **monthly payments set by law:** Division of Coal Mine Workers' Compensation, "Black Lung Monthly Benefit Rates," https://www.dol.gov/owcp/dcmwc/regs/compliance/blbene.htm.

7 **$876.50 for a miner and his wife:** Division of Coal Mine Workers' Compensation, "Benefit Rates Under Part C, 1973–2018," https://www.dol.gov/owcp/dcmwc/benefits_part_c.htm.

7 **fifteen thousand dollars for five years of work:** This is a rough estimate based on my communications with a handful of lawyers who routinely represent claimants and my review of documents from individual case files, including petitions in which attorneys who had won a case asked an administrative law judge to order the coal company to pay their accumulated legal fees.

8 **an extremely sick man:** Case files obtained from the Labor Department contain detailed medical records. In Gary Fox's case, as in many others, these include not only the results from the medical examinations performed as part of the benefits claim but also years of records from treating physicians. In addition, Gary's wife, Mary, provided me with records from Gary's treatment at facilities run by the U.S. Department of Veterans Affairs.

8 **advanced black lung:** The Federal Coal Mine Health and Safety Act, 30 U.S.C. § 921(c)(3), specifies that miners with this advanced stage of disease, known as complicated pneumoconiosis or progressive massive fibrosis, automatically qualify for benefits.

PART ONE: INTRODUCTION

9 **the worst coal-mine disaster in forty years:** Governor's Independent Investigation Panel, *Upper Big Branch: The April 5, 2010, Explosion: A Failure of Basic Coal Mine Safety Practices* (May 2011), https://web.archive.org/web/20110601060618/http://www.nttc.edu/programs&projects/minesafety/disasterinvestigations/upperbigbranch/UpperBigBranchReport.pdf.

13 **Government researchers were tracking the resurgence:** National Institute for Occupational Safety and Health, *Coal Mine Dust Exposures and Associated Health Outcomes: A Review of Information Published Since 1995* (April 2011), https://www.cdc.gov/niosh/docs/2011-172/pdfs/2011-172.pdf?id=10.26616/NIOSHPUB2011172.

13 **more than three hundred coal workers died in accidents:** Mine Safety and Health Administration, "Coal Fatalities for 1900 Through 2018," https://arlweb.msha.gov/stats/centurystats/coalstats.asp.

13 **black lung claimed about ten thousand:** National Institute for Occupational Safety and Health, "Coal Workers' Pneumoconiosis: Number of Deaths by State, U.S. Residents Age

15 and Over, 1995–2004," https://wwwn.cdc.gov/eworld/Data/Coal_Workers_Pneumoconiosis_Number_of_deaths_by_state_US_residents_age_15_and_over_20012010/596.

13 **article that we published:** Chris Hamby, "Persistent Black Lung, Old Scourge of Coal, Found in Autopsies of Most Massey Miners," Center for Public Integrity, May 19, 2011, https://publicintegrity.org/environment/persistent-black-lung-old-scourge-of-coal-found-in-autopsies-of-most-massey-miners/.

14 **"pleasing tho' dreadful":** John Alexander Williams, *West Virginia: A History* (Morgantown: West Virginia University Press, 2001), 22.

CHAPTER 1: GARY

Interviews: Frank Ratcliff, Jerry Bailey, Raymond Keffer, Gene Stewart, Dale Birchfield, Harold Ford, Marshall Litton, Henry Bradley, Carter Stump, Mary Fox, Carnes Shrewsbury, Gary Hairston, Ronnie Lilly, Ronald Acord, Gloria Roles, Richard Roles, Ernie Fox, Jerry Fox, Freddie Fox, Earl Waddell, John Smallwood

17 **Gary and his coworkers spent more time here:** Sections about the conditions in which Gary worked are based largely on the accounts of his coworkers and documents from his two federal black lung–benefits claims, including his testimony during a September 19, 2000, hearing at the Robert C. Byrd Courthouse in Beckley, West Virginia.

17 **would-be coal baron named H. Paul Kizer:** Kizer's life and the scandals in which he was involved received significant media coverage; see, e.g., Paul Nyden, "Kizer Sees Himself as Victim," *Charleston Gazette,* April 22, 1990; Paul Nyden, "Kizer to Run for Governor," *Charleston Gazette,* January 5, 2003; "AMCI Takes Over East Gulf Sales," *International Coal Report,* November 11, 1994; "Key Witness Against Moore Thinks Moore Should Get Law License," Associated Press, December 10, 2002; Barry Bearak, "Column One: Scandals as Thick as Coal Dust," *Los Angeles Times,* July 8, 1990; Brian Farkas, "Moore Pleads Guilty to Federal Extortion, Tax Charges," United Press International, April 13, 1990.

19 **might not work right if a man had a beard:** The Mine Safety and Health Administration includes in its guidance on proper respirator use this admonition: "Shave your face. Facial hair . . . must be removed prior to wearing your respirator"; see https://web.archive.org/web/20160729231845/ https://arlweb.msha.gov/illness_prevention/ideas/respirator.htm.

19 **an imperfect last line of defense:** A key concept in industrial hygiene is the "Hierarchy of Controls," which states that employers should rely on personal protective equipment, such as respirators, only after more direct attempts to eliminate or control the hazardous substance at the source have failed. See, for example, the Mine Safety and

Health Administration's guidance: https://arlweb.msha.gov/Readroom/HANDBOOK/MNMInspChapters/Chapter1.pdf. Miners have also filed dozens of lawsuits against respirator manufacturers, alleging that they contracted black lung at least in part because the devices failed to protect them. See, for example, Peter Hayes, "As Black Lung Rises, Juries Weigh Respirator Suits," Bloomberg Law, May 25, 2017.

20 **Coal River Valley:** "Journey Up Coal River: A Project of Aurora Lights," http://auroralights.org.215-3.purplecat.net/map_project/about.php. William H. Dean, "Coal River," *West Virginia Encyclopedia,* https://www.wvencyclopedia.org/articles/1365.

20 **For generations, people here had worked the mines:** John Alexander Williams, *West Virginia: A History* (Morgantown: West Virginia University Press, 2001), 102–5; Library of Congress, "Tending the Commons: Folklife and Landscape in Southern West Virginia," https://www.loc.gov/collections/folklife-and-landscape-in-southern-west-virginia/.

21 **eighteenth-century land speculators staked their claims:** Williams, *West Virginia,* 106–9, 139–41; Philip Sturm, "The Frontier," *West Virginia Encyclopedia,* https://www.wvencyclopedia.org/articles/2074.

21 **one nineteenth-century prophet:** Barbara Freese, *Coal: A Human History* (New York: Penguin, 2004), 105–6.

21 **coal bosses sent thousands to their deaths:** Mine Safety and Health Administration, "Coal Fatalities for 1900 Through 2015," http://arlweb.msha.gov/stats/centurystats/coalstats.asp; Mine Safety and Health Administration, "MSHA Fatality Statistics," http://arlweb.msha.gov/stats/charts/chartshome.htm; National Institute for Occupational Safety and Health, "Coal Mining Disasters: 1839 to Present," http://www.cdc.gov/niosh/mining/statistics/content/coaldisasters.html.

22 **"the Battle of Blair Mountain":** Robert Shogan, *The Battle of Blair Mountain: The Story of America's Largest Labor Uprising* (New York: Perseus, 2006); "The Mine Wars," PBS, January 26, 2016, http://www.pbs.org/wgbh/americanexperience/films/theminewars/; Clayton D. Laurie, "The United States Army and the Return to Normalcy in Labor Dispute Interventions: The Case of the West Virginia Coal Mine Wars, 1920–1921," West Virginia Archives and History, www.wvculture.org/history/journal_wvh/wvh50-1.html.

22 **beginning to feel the sting of the Appalachian coal industry's decline:** Rory McIlmol and Evan Hansen, "The Decline of Central Appalachian Coal and the Need for Economic Diversification," Downstream Strategies, January 19, 2010, https://www.downstreamstrategies.com/documents/reports_publication/DownstreamStrategies-DeclineOfCentralAppalachianCoal-FINAL-1-19-10.pdf. The U.S. Energy Information Administration provides a wealth of data and analysis online at https://www.eia.gov/coal/.

22 **Poverty rates would rise:** Appalachian Regional Commission, "An Economic Analysis of the Appalachian Coal Industry Ecosystem," summary report by West Virginia

University and the University of Tennessee, January 2018, https://www.arc.gov/assets/research_reports/CIESummary-AppalachianCoalIndustryEcosystemAnalysis.pdf.

22 **ravaged by a flood of prescription opioids:** National Institute on Drug Abuse, "Opioid Summaries by State," revised May 2019, https://www.drugabuse.gov/drugs-abuse/opioids/opioid-summaries-by-state.

24 **Skelton was the creation of coal baron Samuel Dixon:** Robert W. Craigo, ed., *The New River Company: Seventy Years of West Virginia Coal History* (Mount Hope, WV: New River Company, 1976). This is a detailed company history, including maps and photographs, commissioned by the New River Company. The book begins with a letter from company president E. V. Bowman to "New River Company Employees, Past and Present," that says, among other things: "This booklet presents the story in both words and pictures of a time when a cooperative effort existed between miners and their employer to secure better living conditions for themselves and their families."

24 **the region was undergoing a transformation:** Williams, *West Virginia,* 102–14; Barbara Ellen Smith, *Digging Our Own Graves: Coal Miners and the Struggle over Black Lung Disease* (Philadelphia: Temple University Press, 1987), 12–13.

25 **The region was soon teeming with coal camps:** Rhonda Janney Coleman, "Coal Miners and Their Communities in Southern Appalachia, 1925–1941," part 2, *West Virginia Historical Society Quarterly* 15, no. 3 (July 2001); Williams, *West Virginia*, 139–42; Smith, *Digging Our Own Graves*, 12–16; Alan Derickson, *Black Lung: Anatomy of a Public Health Disaster* (Ithaca, NY: Cornell University Press, 1998), 26–33.

26 **the scene of the competition that spawned the legend of John Henry:** National Park Service, "John Henry and the Coming of the Railroad," https://www.nps.gov/neri/learn/historyculture/john-henry-and-the-coming-of-the-railroad.htm.

26 **a three-story mansion on a hill overlooking Skelton:** Visitors to Beckley can tour a restored version of this mansion; for information, see https://beckley.org/superintendent-house/.

26 **the state lost 13 percent of its population:** Williams, *West Virginia,* 178–79.

26 **The hard years came to Skelton:** Craigo, *The New River Company,* 60, 82, 101.

26 **little acknowledgment among doctors:** Derickson's *Black Lung* and Smith's *Digging Our Own Graves* provide detailed accounts of the history of denial and misdiagnosis.

29 **Gary was drafted:** Details about Gary's service in the army and related medical care are drawn from his military records.

30 **Gary had gone back to work:** Records from Gary's two federal black lung–benefits claims, including his itemized statement of earnings from the Social Security Administration, comport with the memories of his family members.

CHAPTER 2: THE MINES

Interviews: Frank Ratcliff, Gene Stewart, Jerry Bailey, Terri Smith, Mary Fox

31 **a coalfield adage:** Various forms appear in publications and can be heard on occasion in conversations in the coalfields today. This particular version of the saying is from Barbara Ellen Smith, *Digging Our Own Graves: Coal Miners and the Struggle over Black Lung Disease* (Philadelphia: Temple University Press, 1987), 32.

32 **greater share of production:** Mine Safety and Health Administration, "Statistics: Historical Key Indicator Charts," https://www.msha.gov/data-reports/statistics; U.S. Energy Information Administration, "Annual Coal Report," https://www.eia.gov/coal/annual/.

32 **the middle of a layer cake:** Barbara Freese, *Coal: A Human History* (New York: Penguin, 2004), 17–20; West Virginia Geological and Economic Survey, "Physiographic Provinces of West Virginia," http://www.wvgs.wvnet.edu/www/geology/geolphyp.htm," Ohio Geological Survey, "Coal: Educational Leaflet No. 8," https://minerals.ohiodnr.gov/portals/minerals/pdf/coal/el08.pdf.

32 **The techniques people developed to extract this precious mineral:** Medieval Technology and American History, "The Medieval Roots of Colonial Iron Manufacturing Technology," Center for Medieval Studies at the Pennsylvania State University, http://www.engr.psu.edu/mtah/articles/roots_colonial_iron_technology.htm; Keith Dix, *What's a Coal Miner to Do?: The Mechanization of Coal Mining* (Pittsburgh: University of Pittsburgh Press, 1988).

32 **As mining boomed in Appalachia:** Information about mining practices around the start of the twentieth century comes from various sources, including Smith, *Digging Our Own Graves*, 35–37; Dix, *What's a Coal Miner to Do?*, 1–14; Paul H. Rakes, "Mining Methods in the Hand-Loading Era," *West Virginia Encyclopedia*, https://www.wvencyclopedia.org/articles/1835; James Green, "A Day in the Life of a West Virginia Coal Miner," Literary Hub, https://lithub.com/a-day-in-the-life-of-a-west-virginia-coal-miner/; William Graebner, *Coal-Mining Safety in the Progressive Period: The Political Economy of Reform* (Lexington: University Press of Kentucky, 1976), 112–13.

34 **many mines had upgraded their equipment:** Dix, *What's a Coal Miner to Do?*; Smith, *Digging Our Own Graves*, 47–55; Alan Derickson, *Black Lung: Anatomy of a Public Health Disaster* (Ithaca, NY: Cornell University Press, 1998), 139–40, 172–73; Paul H. Rakes, "Coal Mine Mechanization," *West Virginia Encyclopedia*, http://www.wvencyclopedia.org/articles/1364.

34 **the three deadly gases:** Freese, *Coal*, 47–51; Utah Labor Commission, "Test Preparation Study Guide for Underground Mine Foreman Certification," https://digitallibrary.utah.gov/awweb/awarchive?type-file&item-82229. National Coal Mining Museum for England, "Mine Gases," https://www.ncm.org.uk/downloads/27/Mine_gases.pdf.

35 **there were ten deaths:** Accident reports for Itmann and Birchfield retrieved from the Mine Safety and Health Administration's fatality archive database; see https://arlweb.msha.gov/training/library/FatalRewrdsSearch.asp. Many reports were available only upon request from the agency's Technical Information Center and Library.

35 **intuit what the mountain was telling him:** Interviews with coworkers, including Frank Ratcliff, Gene Stewart, and Jerry Bailey; Christopher Mark, "The Introduction of Roof Bolting to U.S. Underground Coal Mines (1948–1960): A Cautionary Tale," Pittsburgh Research Laboratory, National Institute for Occupational Safety and Health, 2002; Kentucky Geological Survey, "Kettlebottoms in Mine Roofs," http://www.uky.edu/KGS/coal/coal-mining-geology-Kettlebottoms.php; Shae Davidson, "Kettle Bottom," *West Virginia Encyclopedia,* https://www.wvencyclopedia.org/articles/1196.

36 **More than one hundred thousand miners have died:** Mine Safety and Health Administration, "Coal Fatalities for 1900 Through 2019," http://arlweb.msha.gov/stats/centurystats/coalstats.asp; Mine Safety and Health Administration, "MSHA Fatality Statistics," http://arlweb.msha.gov/stats/charts/chartshome.htm; National Institute for Occupational Safety and Health, "Coal Mining Disasters: 1839 to Present," http://www.cdc.gov/niosh/mining/statistics/content/coaldisasters.html.

37 **In 1831 a doctor in Scotland:** James Craufurd Gregory, "Case of Peculiar Black Infiltration of the Whole Lungs, Resembling Melanosis," *Edinburgh Medical and Surgical Journal* 36 (1831): 390.

37 **Across the Atlantic, in Pennsylvania coal country:** "Report of the Schuylkill County Medical Society," *Transactions of the Medical Society of Pennsylvania,* 5th ser., pt. 2 (June 1869): 489.

37 **At the 1881 annual meeting of the Colorado State Medical Society:** Derickson, *Black Lung,* 1.

37 **there was broad recognition:** Smith, *Digging Our Own Graves,* 3–11; Derickson, *Black Lung,* 1–12.

38 **government agencies in coal-mining states:** Derickson, *Black Lung,* 12–20.

38 **Many started as breaker boys:** Mine Safety and Health Administration, "A Pictorial Walk Through the 20th Century: Little Miners," https://web.archive.org/web/20180618175914/ https://arlweb.msha.gov/century/little/page1.asp; Ohio State University Department of History, "The Boys in the Breakers," https://ehistory.osu.edu/exhibitions/gildedage/content/breakerboys.

38 **An 1884 folk song captured the indignity:** Jack Johnson, "The Miner's Life."

38 **Health benefits and pensions were almost nonexistent:** Smith, *Digging Our Own Graves,* 40–43; Derickson, *Black Lung,* 22–33.

38 **Tradition held that the best way for a man to clear the dust out of his system:** One popular concoction among miners was "morning bitters," which George Korson described as "whiskey with a mixture of snake-root, gold seal, and/or calamus root,

sweetened with rock candy"; see George Korson, *Black Rock: Mining Folklore of the Pennsylvania Dutch* (Baltimore: Johns Hopkins University Press, 1960), 258.

39 **despite what appeared to be a growing consensus:** Derickson, *Black Lung,* 43–59; Smith, *Digging Our Own Graves,* 5–22.

39 **coal dust was "relatively harmless":** Radiologists Henry K. Pancoast and Eugene P. Pendergrass, quoted by Derickson, gave this monograph a title that exemplified the tendency to blame silica—not coal dust—for any lung impairment suffered by miners: *Pneumoconiosis (Silicosis): A Roentgenological Study with Notes on Pathology.*

39 **coal miners were "practically immune from tubercular infection":** A bulletin put out by the U.S. Bureau of Labor Statistics quoted Carr's assertion, which he made in a 1905 letter to the *Journal of the American Medical Association,* along with similar assertions by others, but officials ultimately concluded that these were yet unproven generalizations that warranted further inquiry; see Frederick L. Hoffman, U.S. Bureau of Labor Statistics, *Mortality from Respiratory Diseases in Dusty Trades (Inorganic Dusts)* (Washington, DC: Government Printing Office, 1918).

39 **If any miners really were sick, there was some other cause:** Smith, *Digging Our Own Graves,* 17–18; Derickson, *Black Lung,* 17.

39 **blame everything on a different type of dust:** Derickson, *Black Lung,* 43–54, 58–59; Smith, *Digging Our Own Graves,* 16–21.

40 **the federal government did little to correct the record:** Derickson, *Black Lung,* 69–86; Smith, *Digging Our Own Graves,* 19–22.

40 **a reinvigorated United Mine Workers union:** Smith, *Digging Our Own Graves,* 55–66; Alan Derickson, "The United Mine Workers of America and the Recognition of Occupational Respiratory Diseases, 1902–1968," *American Journal of Public Health* 81, no. 6 (June 1991): 3–4.

40 **major advancements were happening in Great Britain:** Andrew Meiklejohn, "The Development of Compensation for Occupational Diseases of the Lungs in Great Britain," *British Journal of Industrial Medicine* 11 (1954); Derickson, *Black Lung,* 119–22; Smith, *Digging Our Own Graves,* 30.

CHAPTER 3: JOHN

Interviews: George Bragg, Melody Bragg, Dan Doyle, Craig Robinson, Judy Robinson, John Cline, Brenda Halsey Marion, Susie Criss, Cathy Stover, Ken Dangerfield, Tim Burke, Ann Cline, Cathi Cline, Andy Cline

41 **The man in the photograph:** Melody Bragg, "Fayette Folks: John Cline—the Miner's Friend," *Fayette Tribune.* The exact date of the article is lost, but both its author and a librarian for the newspaper confirmed the rough time frame.

41 **the domain of the New River Company:** Robert W. Craigo, ed., *The New River Company: Seventy Years of West Virginia Coal History* (Mount Hope, WV: New River Company, 1976), 105. The company commissioned this publication to commemorate its history.

42 **a federal grant to diagnose and treat working and retired miners:** Health Resources and Services Administration, "Black Lung Clinics Program," https://www.hrsa.gov/get-health-care/conditions/black-lung/index.html.

42 **breathing tests used to detect impaired lung function:** National Institute for Occupational Safety and Health, "Spirometry," https://www.cdc.gov/niosh/topics/spirometry/default.html.

43 **West Virginia's workers' compensation system:** "Workers' Compensation: Death and Disability Benefits," West Virginia Code, chapter 23, article 4, http://www.wvlegislature.gov/wvcode/code.cfm?chap=23&art=4. The state black lung–benefits system is run by the Occupational Pneumoconiosis Board; see https://www.wvinsurance.gov/Workers-Compensation/Claims-Services-Workers-Compensation/OPBoard.

43 **federal black lung benefits–program:** Information about the program is available at the website of the Division of Coal Mine Workers' Compensation, part of the U.S. Department of Labor's Office of Workers' Compensation Programs; see https://www.dol.gov/owcp/dcmwc/.

45 **a quintessential slice of small-town America:** Observations based on my reporting trip to East Aurora in May 2015.

45 **John's paternal great-grandfather:** Details about John's family history are drawn largely from a letter dated December 26, 2004, that John wrote to his three sons. In it, he synthesized numerous sources—such as relatives' genealogy research and personal written accounts and archival documents—into a narrative. I have supplemented this information with my own interviews and, to the extent possible, examination of public documents.

48 **John enrolled at Allegheny College:** Official records confirm John's recollection of his education.

CHAPTER 4: WILDCAT

Interviews: Craig Robinson, Judy Robinson, Donald Rasmussen, John Cline, Pat Richards, Tim Burke

50 **explosions ripped through the underground tunnels:** Details about the Farmington disaster and its aftermath are drawn primarily from the following: Bonnie E. Stewart, *No. 9: The 1968 Farmington Mine Disaster* (Morgantown: West Virginia University Press, 2012); Jeffrey B. Cook, "Farmington Mine Disaster," *West Virginia Encyclopedia,*

https://www.wvencyclopedia.org/articles/2241; Barbara Ellen Smith, *Digging Our Own Graves: Coal Miners and the Struggle over Black Lung Disease* (Philadelphia: Temple University Press, 1987), 101–4; Ken Hechler, *The Fight for Coal Mine Health and Safety: A Documented History* (Charleston, WV: Pictorial Histories Publishing, 2011), 75; Alan Derickson, *Black Lung: Anatomy of a Public Health Disaster* (Ithaca, NY: Cornell University Press, 1998), 149–50; Richard Fry, "Making Amends: Coal Miners, the Black Lung Association, and Federal Compensation Reform, 1969–1972," *Federal History* 5 (January 2013): 5.

52 **Buff began advocating:** Brit Hume, *Death and the Mines: Rebellion and Murder in the UMW* (New York: Grossman, 1971), 102; Ken Hechler, "I. E. Buff," *West Virginia Encyclopedia,* https://www.wvencyclopedia.org/articles/689.

52 **frustration swelled among rank-and-file miners:** Smith, *Digging Our Own Graves,* 108–14; Derickson, *Black Lung,* 140–51; Hume, *Death and the Mines,* 66–72, 104–5; Barbara Ellen Smith, "Black Lung: The Social Production of Disease," *International Journal of Health Services* 11, no. 3 (1981), http://journals.sagepub.com/doi/pdf/10.2190/LMPT-4G1J-15VQ-KWEK.

52 **"You all have black lung":** Derickson, *Black Lung,* 148; Hechler, *The Fight for Coal Mine Health and Safety,* 22, 104–5; Joseph A. Loftus, "Doctors Leading Fight to Curb Miners' Disease," *New York Times,* January 7, 1969; Robert G. Sherrill, "West Virginia Miracle: The Black Lung Rebellion," *Nation,* April 28, 1969.

53 **Rasmussen grew up in Colorado:** Details of Rasmussen's life are drawn from my interviews with him as well as from the following: Betty Dotson-Lewis, "Oral History Interview with Dr. Donald L. Rasmussen," interviews in July 2002, http://dotson-lewis.blogspot.com/2011/04/black-lung-dr-donald-rasmussen.html; Paul J. Nyden, "Rasmussen, Pioneering Black Lung Researcher, Dies at 87," *Charleston Gazette,* July 23, 2015, https://www.wvgazettemail.com/news/rasmussen-pioneering-black-lung-researcher-dies-at/article_bcaff185-dad0-563d-8f78-93ddada4a976.html; Sam Roberts, "Dr. Donald L. Rasmussen, Crusader for Coal Miners' Health, Dies at 87," *New York Times,* August 2, 2015, https://www.nytimes.com/2015/08/03/health/research/dr-donald-l-rasmussen-crusader-for-coal-miners-health-dies-at-87.html.

54 **Coal companies and their allies in the medical establishment:** Smith, *Digging Our Own Graves,* 119–20; Derickson, *Black Lung,* 154–58.

54 **The scientific evidence:** Derickson, *Black Lung,* 170–71; Richard Fry, "Fighting for Survival: Coal Miners and the Struggle Over Health and Safety in the United States, 1968–1988" (PhD diss., Wayne State University, 2010), 120.

55 **mocked the label *black lung:*** Derickson, *Black Lung,* 154–56, 177–78.

55 **It is now widely accepted:** A couple of examples from the growing body of scientific literature: Edward L. Petsonk, Cecile Rose, and Robert Cohen, "Coal Mine Dust Lung Disease. New Lessons from an Old Exposure," *American Journal of Respiratory and*

Critical Care Medicine 187, no. 11 (June 1, 2013); A. Scott Laney and David N. Weissman, "Respiratory Diseases Caused by Coal Mine Dust," *Journal of Occupational and Environmental Medicine* 56, no. 10 (September 2014).

55 **rank-and-file miners created the Black Lung Association:** Smith, *Digging Our Own Graves,* 111; Derickson, *Black Lung,* 151–52; Hume, *Death and the Mines,* 112–13.

56 **The breaking point came on February 18, 1969:** Smith, *Digging Our Own Graves,* 115; Derickson, *Black Lung,* 15; Hume, *Death and the Mines,* 134.

56 **two thousand miners and their allies:** Edward Peeks, "Coal Miners Sound Call of 'No Law, No Work,'" *Charleston Gazette,* February 27, 1969; Ben A. Franklin, "Coal Miners Win in West Virginia," *New York Times,* February 27, 1969.

56 **John wrote of the experience:** John retained and shared with me copies of letters he wrote to family members between 1967 and 1974.

57 **John had been drafted:** John retained and shared with me his Selective Service System records, related communications, and a news article concerning his indictment.

57 **The bulk of the coal-mining jobs:** John Alexander Williams, *West Virginia: A History* (Morgantown: West Virginia University Press, 2001), 178–82; Barbara Freese, *Coal: A Human History* (New York: Penguin, 2004), 160–61; Richard Cartwright Austin, *Moral Imagination in Industrial Culture, 1966–1975* (Dungannon, VA: Creekside Press, 2010), 22–24.

58 **"rediscovery" of Appalachian poverty:** Guy Carawan and Candie Carawan, *Voices from the Mountains: The People of Appalachia—Their Faces, Their Words, Their Songs* (Athens: University of Georgia Press, 1996).

58 **a foundation for future successes:** Smith, *Digging Our Own Graves,* 91–92; Derickson, *Black Lung,* 144–45; Jefferson Cowie, *Stayin' Alive: The 1970s and the Last Days of the Working Class* (New York: New Press, 2010), 32–33.

59 **more than forty thousand miners had walked off the job:** Smith, *Digging Our Own Graves,* 122; Derickson, *Black Lung,* 161; Franklin, "Coal Miners Win in West Virginia."

59 **The final legislation:** Fry, "Fighting for Survival," 90–91; "Miners Vote to Continue Strike Until 'Black Lung' Bill Is Signed," Associated Press, March 9, 1969; Derickson, *Black Lung,* 161–62; Smith, *Digging Our Own Graves,* 123–26.

59 **The rebellion spread:** Fry, "Fighting for Survival," 90–94; Derickson, *Black Lung,* 162–66.

60 **"The wave of wildcat strikes":** "Coal Miners' Revolt," *New York Times,* February 25, 1969.

60 **"The crusade to clean up the mines":** "The Black Lungers," *New York Times,* February 3, 1969.

60 **Congressman Ken Hechler:** Niel M. Johnson, "Oral History Interview with Ken Hechler," Harry S. Truman Presidential Library and Museum, November 29, 1985, https://www.trumanlibrary.gov/library/oral-histories/hechler; Richard Goldstein,

"Ken Hechler, West Virginia Populist and Coal Miners' Champion, Dies at 102," *New York Times,* December 11, 2016; "Hechler, Kenneth William," U.S. House of Representatives History, Art and Archives, https://history.house.gov/People/Detail?id=14821&cid=38165&cs=1&ce=115&f=All; Hechler, *The Fight for Coal Mine Health and Safety*.

61 **the largest companies and the national trade association took a different tack:** Smith, *Digging Our Own Graves,* 131–33; Hume, *Death and the Mines,* 154–55; Fry, "Fighting for Survival," 117; Derickson, *Black Lung,* 173.

61 **the Department of Health, Education, and Welfare recommended:** Alan Derickson, "'Nuisance Dust': Unprotective Limits for Exposure to Coal Mine Dust in the United States, 1934–1969," *American Journal of Public Health* 103, no. 2 (February 2013), https://www.ncbi.nlm.nih.gov/pmc/articles/PMC3558784/.

62 **The United Mine Workers' stance:** Hechler, *The Fight for Coal Mine Health and Safety,* 92, 95; Hume, *Death and the Mines,* 82; Fry, "Fighting for Survival," 112; Derickson, *Black Lung,* 171.

62 **Buff brought a toned-down version:** Ben A. Franklin, "Lack of Care Is Laid to Coal Industry," *New York Times,* March 14, 1969.

62 **the debate broadened:** Peter S. Barth, *The Tragedy of Black Lung: Federal Compensation for Occupational Disease* (Kalamazoo, MI: W. E. Upjohn Institute for Employment Research, 1987), 23–27; Derickson, *Black Lung,* 175–77; Fry, "Fighting for Survival," 113–15.

62 **The bill that made it through both houses of Congress:** Federal Coal Mine Health and Safety Act of 1969, Public Law 91-173, *U.S. Statutes at Large* 83 (1969): 742–804.

63 **He was threatening to veto the entire thing:** Fry, "Fighting for Survival," 150–52; Barth, *Tragedy of Black Lung,* 26–27; "Senate Approves Mine Safety Bill," Associated Press, December 18, 1969.

63 **"Given the epochal character of the law":** "Lost in a Coal Shaft," *New York Times,* December 22, 1969.

63 **Hechler chartered two small planes:** Hechler, *The Fight for Coal Mine Health and Safety,* 277; "Mines Are Picketed Over Bill on Safety," Associated Press, December 29, 1969; Ben A. Franklin, "President to Sign Mine Safety Bill Despite Doubts," *New York Times,* December 30, 1969.

64 **He issued a statement**: Richard Nixon, "Statement on Signing the Federal Coal Mine Health and Safety Act of 1969," December 30, 1969, American Presidency Project, https://www.presidency.ucsb.edu/documents/statement-signing-the-federal-coal-mine-health-and-safety-act-1969.

64 **disillusioned with the law's implementation:** Smith, *Digging Our Own Graves,* 146–69; Fry, "Fighting for Survival," 248–69.

65 **a newsletter called the *Black Lung Bulletin:*** One article complained: "Many bureaucrats also tend to look down on some miners and widows because of their lack of education.

They seem to think that it takes a degree to know even the simplest things." The newsletters also urged miners to write the Black Lung Association headquarters to receive a free copy of "Black Lung Bill Battles Social Security," a sixteen-page illustrated booklet tracing the exploits of breathless coal miner "Black Lung Bill." Bill's buddies at the Black Lung Association teach him the intricacies of the claims process and help him overcome an initial denial. In the end, Bill prevails and dispenses his own brand of justice (in cartoon format) by turning a Social Security stooge upside down and shaking loose some cash.

66 **debate in Congress over proposed reform legislation:** William M. Buckley, "Rocky Road Seen for Miners' Aid," *Charleston Gazette,* October 28, 1971; William Greider, "Black Lung Program: It Isn't Working," *Washington Post,* December 20, 1971; Jerry Landauer, "Costly Black Lung Assistance Bill Causes Dilemma for Budget-Minded Republicans," *Wall Street Journal,* March 10, 1972; Jerry Landauer, "Who Will Pay for Black Lung?" *Wall Street Journal,* June 21, 1972; Hechler, *The Fight for Coal Mine Health and Safety,* 79; Barth, *Tragedy of Black Lung,* 25–28; Smith, *Digging Our Own Graves,* 171–72.

66 **His concerns were much the same:** Landauer, "Costly Black Lung Assistance Bill"; Fry, "Fighting for Survival," 282; Landauer, "Who Will Pay for Black Lung?"; "Nixon Approves Extension of Aid to Lung Victims," Associated Press, May 20, 1972.

66 **Nixon signed the legislation:** Richard Nixon, "Statement About Signing the Black Lung Benefits Act of 1972," May 20, 1972, American Presidency Project, https://www.presidency.ucsb.edu/documents/statement-about-signing-the-black-lung-benefits-act-1972.

66 **The law was another resounding victory:** Black Lung Benefits Act of 1972, Public Law 92-303, *U.S. Statutes at Large* 86 (1972): 150–57.

66 **the wall of denial and neglect that had surrounded workplace illness:** Cowie, *Stayin' Alive,* 254, 259–60; Lorin E. Kerr, "The United Mine Workers of America Look at Occupational Health," *American Journal of Public Health* 61, no. 5 (May 1971); "Danger on the Job," *New York Times,* September 1, 1970.

67 **In 1970, Congress passed the Occupational Safety and Health Act:** Occupational Safety and Health Act of 1970, Public Law 91-596, *U.S. Statutes at Large* 84 (1970): 1590–1620.

67 **the agency has been able to address only a small fraction of the toxic substances:** Remarkably, the Occupational Safety and Health Administration itself has acknowledged that its exposure limits for many substances "are outdated and inadequate for ensuring protection of worker health." See Occupational Safety and Health Administration, "Permissible Exposure Limits—Annotated Tables," https://www.osha.gov/dsg/annotated-pels/.

CHAPTER 5: FOOTER

Interviews: John Cline, Grant Crandall, Debbie Wills, Brenda Halsey Marion, Susie Criss, Craig Robinson, Judy Robinson, Dan Doyle, Ken Dangerfield, Daniel Richmond, Arland Griffith, Donna Meadows, Misty Martin

68 **eagerly joined the effort**: Most of this chapter's details about the activities of the Black Lung Association are drawn from interviews with members and from documents they retained over the years and shared with me, including meeting minutes, sign-in sheets, newsletters, photographs, and fliers.

68 **one of the more than fifty sites:** Office of Technology Assessment, *Health Care in Rural America, OTA-H-434* (Washington, DC: U.S. Government Printing Office, September 1990).

69 **the Robert Wood Johnson Foundation:** The Grants Explorer page on the foundation's website shows that the New River clinic was awarded $412,331 in 1975 as part of the Rural Practice Project; see https://www.rwjf.org/en/how-we-work/grants-explorer.html#k=New%20River&end=1985.

70 **federal government program offering grants for black lung clinics:** New River Health, "Black Lung," https://www.newriverhealthwv.com/breathing-center-black-lung; Health Resources and Services Administration, "Black Lung Clinics Program," https://www.hrsa.gov/get-health-care/conditions/black-lung/index.html.

71 **Cline Building Corporation:** The website for the West Virginia Secretary of State indicates that Cline Building Corporation was established in 1975 by Crawford and John Cline, with its office in Piney View; see https://apps.wv.gov/SOS/BusinessEntitySearch/Details.aspx?Id=OY3l6m+pK9MKodIWBsp4hA==&Search=8Gg%20hYtBYOonYaoGphwaZw==&Page=0.

71 **Clivus Multrum:** For more on these composting toilets, see https://clivusmultrum.com/.

74 **initial claims approval rate was hovering around 4 percent:** U.S. General Accounting Office, *Black Lung Program: Further Improvements Can Be Made in Claims Adjudication,* GAO/HRD-90-75 (Washington, DC, 1990), 19, https://www.gao.gov/assets/220/212290.pdf.

PART TWO: INTRODUCTION

Interviews: Don Rasmussen, Thomas Marcum, Edward Petsonk, Scott Laney, Anita Wolfe, Ray Marcum, Donald Marcum, James Marcum

77 **a small medical clinic in Beckley:** I visited Rasmussen's clinic with Howard Berkes and Sandra Bartlett, both of NPR, on April 27, 2012.

78 **X-ray surveillance program that tracked the disease:** National Institute for Occupational

Safety and Health, "Coal Workers' Health Surveillance Program," https://www.cdc.gov/niosh/topics/cwhsp/default.html.

78 **The surveillance data had indicated:** The National Institute for Occupational Safety and Health has published a series of papers about black lung's resurgence. A few examples: A. Scott Laney et al., "Potential Determinants of Coal Workers' Pneumoconiosis, Advanced Pneumoconiosis, and Progressive Massive Fibrosis Among Underground Coal Miners in the United States, 2005–2009," *American Journal of Public Health* 102, suppl. 2 (May 2012), https://www.ncbi.nlm.nih.gov/pmc/articles/PMC3477901/; A. Scott Laney, Edward L. Petsonk, Michael D. Attfield, "Pneumoconiosis Among Underground Bituminous Coal Miners in the United States: Is Silicosis Becoming More Frequent?," *Occupational and Environmental Medicine* 67, no. 10 (2010), https://oem.bmj.com/content/67/10/652; Eva Suarthana et al., "Coal Workers' Pneumoconiosis in the United States: Regional Differences 40 Years After Implementation of the 1969 Federal Coal Mine Health and Safety Act," *Occupational and Environmental Medicine* 68, no. 12 (2011): 908–13, https://oem.bmj.com/content/68/12/908; David J. Blackley, Cara N. Halldin, and A. Scott Laney, "Continued Increase in Prevalence of Coal Workers' Pneumoconiosis in the United States, 1970–2017," *American Journal of Public Health* 108, no. 9 (September 2018), https://www.ncbi.nlm.nih.gov/pmc/articles/PMC6085042/.

81 **finalized those articles:** Chris Hamby, "Black Lung Surges Back in Coal Country," Center for Public Integrity, July 8, 2012, https://publicintegrity.org/workers-rights/black-lung-surges-back-in-coal-country/; Howard Berkes, "Special Series: Black Lung Returns to Coal Country," NPR, July 2012, https://www.npr.org/series/156453033/black-lung-returns-to-coal-country.

CHAPTER 6: "GO JUMP"

Interviews: John Cline, Grant Crandall

83 **holding a hearing:** U.S. House of Representatives, Committee on Education and Labor, Subcommittee on Labor Standards, *Field Hearings on Black Lung,* 101st Congress, 2nd sess., 1990.

83 **He had previously filed a benefits claim and lost:** Details throughout this chapter about Ralph Manning's health and legal case are drawn primarily from documents in his federal black lung–benefits claim file.

85 **more stringent eligibility criteria:** U.S. General Accounting Office, *Program to Pay Black Lung Benefits to Coal Miners and Their Survivors—Improvements Are Needed* (Washington, DC, 1977), 43–45, https://www.gao.gov/assets/120/119419.pdf.

85 **The legislation that President Jimmy Carter signed into law in 1978:** Black Lung Benefits Reform Act of 1977, Public Law 95-239, *U.S. Statutes at Large* 92 (1978): 95–106.

85 **so liberalized the program:** U.S. General Accounting Office, "Statement of Morton E. Henig, Senior Associate Director, Human Resources Division, Before the Subcommittee on Oversight, Committee on Ways and Means, United States House of Representatives, on the Black Lung Program and Black Lung Disability Trust Fund" (Washington, DC, 1981), https://www.gao.gov/assets/100/99873.pdf; U.S. Social Security Administration, *Social Security Bulletin: Black Lung Benefits Revision* 45, no. 11 (November 1982), https://www.ssa.gov/policy/docs/ssb/v45n11/v45n11p26.pdf; U.S. General Accounting Office, *Black Lung Program: Further Improvements Can Be Made in Claims Adjudication,* GAO/HRD-90-75 (Washington, DC, 1990), 19, https://www.gao.gov/assets/220/212290.pdf.

85 **General Accounting Office reviewed claims:** U.S. GAO, "Statement of Morton E. Henig," 5.

86 **A companion law:** Black Lung Benefits Revenue Act of 1977, Public Law 95-227, *U.S. Statutes at Large* 92 (1978): 11–24.

86 **the fund spent $1.9 billion on benefits payments:** U.S. GAO, "Statement of Morton E. Henig," 4.

86 **"automatic pension":** "UMW Combats Reagan on Black Lung Cuts," *Coal Age,* April 1981, 13–14.

86 **"we will close down every coal mine":** Lavinia Edmunds, "Strike Threatened Over Black Lung Plans," States News Service, February 20, 1981.

87 **The union staged a rally in Washington:** Ben A. Franklin, "Miners Protest Planned Reduction in U.S. Aid to Black Lung Program," *New York Times,* March 10, 1981; Jim Ragsdale, "Marching Miners Hear Pledges of Support," *Charleston Gazette,* March 11, 1981; "Administration: Black Lung Victims Won't Lose Benefits," Associated Press, March 10, 1981; "UMW Official Says Reagan Got Message," Associated Press, March 15, 1981.

87 **approved a bill by large margins:** Black Lung Benefits Revenue Act of 1981, Public Law 97-119, *U.S. Statutes at Large* 95 (1981): 1635–45. The Senate voted 63–30 and the House of Representatives 363–47 in favor of the legislation; see https://www.congress.gov/bill/97th-congress/house-bill/5159/all-actions?overview=closed#tabs.

87 **Reagan signed it into law:** Ronald Reagan, "Statement on Signing Black Lung Program Reform Legislation," December 29, 1981, American Presidency Project, https://www.presidency.ucsb.edu/documents/statement-signing-black-lung-program-reform-legislation.

87 **even the miners' usual allies grudgingly supported the legislation:** Bob Stiegel, "Future for Black Lung Claims Not Totally Black—Rahall," *Beckley Post-Herald and Register,* March 28, 1982; Bob Stiegel, "Financial Aspects of Black Lung Act Draw Attention," *Beckley Post-Herald,* March 30, 1982; Bob Stiegel, "Senate Campaign Opponents Agree on Black Lung Measure," *Beckley Post-Herald,* March 31, 1982.

88 **had cut the approval rate:** U.S. GAO, *Further Improvements Can Be Made,* 17–19.

88 **companies appealed 97 percent of initial approvals:** U.S. GAO, *Program to Pay Black Lung Benefits,* 11.

88 **the system was plagued by "excessive litigation":** Office of Workers' Compensation Programs, *Report of the OWCP Black Lung Task Force: Program Description and Recommendations* (Washington, DC, 1977), viii–xviii.

89 **"We've fought some claims we shouldn't have":** Bob Arnold, "Clumsy Assist? Government Attack on 'Black Lung' Ills Has Mixed Results," *Wall Street Journal,* August 28, 1975.

89 **backlogs were increasing at the appeals levels:** U.S. GAO, *Further Improvements Can Be Made,* 25–26.

89 **"I quit taking them":** Bob Geiger, "Miners Have Less Chance of Getting Benefits Today," *Charleston Gazette,* December 22, 1987.

89 **"I hate the federal black-lung act":** Joseph S. Stroud, "Black Hole for Miners," Knight-Ridder Newspapers, October 16, 1993.

89 **Harold Hayden, an official for the United Mine Workers:** "Statement of Harold Hayden, Representative of United Mine Workers, District 29," U.S. House of Representatives, Committee on Education and Labor, Subcommittee on Labor Standards, *Oversight Hearing on the Administration of the Black Lung Program,* 100th Cong., 2nd sess., 1988.

90 **an assessment of the dismal state of the benefits program:** U.S. House of Representatives, *Field Hearings on Black Lung.*

90 **the law made them eligible for benefits:** Black Lung Benefits Act, 30 U.S.C. § 902(d), https://uscode.house.gov/view.xhtml?req=granuleid:USC-prelim-title30-section902&num=0&edition=prelim: "The term 'miner' means any individual who works or has worked in or around a coal mine or coal preparation facility in the extraction or preparation of coal."

91 **A miner could apply for a waiver:** U.S. GAO, *Further Improvements Can Be Made,* 30–32.

92 **the distinct complexities of black lung:** Division of Coal Mine Workers' Compensation, "Compliance Guide to the Black Lung Benefits Act," https://www.dol.gov/owcp/dcmwc/regs/compliance/blbenact.htm.

CHAPTER 7: "THE GREAT COAL DUST SCAM"

96 **The Labor Department had promised a bombshell announcement:** Department of Labor, "Department of Labor Press Conference with Labor Secretary Lynn Martin; Bill Tattersall, Assistant Secretary of Labor for Mine Safety and Health; Ed Claire, Associate Solicitor, Division of Mine Safety and Health," April 4, 1991.

96 **the system the government had devised to protect miners:** For more on the Mine Safety and Health Administration's respirable-dust-sampling program, see

Mine Safety and Health Administration, "End Black Lung, Act Now!," https://arlweb.msha.gov/S&HINFO/BlackLung/homepage2009.asp; "Report of the Secretary of Labor's Advisory Committee on the Elimination of Pneumoconiosis Among Coal Mine Workers," October 1996, https://arlweb.msha.gov/S&HINFO/BlackLung/1996Dust%20AdvisoryReport.pdf; National Institute for Occupational Safety and Health, *Criteria for a Recommended Standard: Occupational Exposure to Respirable Coal Mine Dust* (Cincinnati, OH, September 1995), https://www.cdc.gov/niosh/docs/95-106/pdfs/95-106.pdf?id=10.26616/NIOSHPUB95106.

96 **It was a cumbersome apparatus:** To watch a Mine Safety and Health Administration industrial hygienist demonstrate proper use of a respirable-dust-sampling device, visit https://www.youtube.com/watch?v=CHTJ8i55HUk.

97 **the average dust levels in samples submitted by coal companies:** This is based on my analysis of data obtained from the Mine Safety and Health Administration.

97 **"It should be impossible to get black lung disease":** Peter T. Kilborn, "Coal Miners Contend Their Plight Is Worsening," *New York Times*, March 17, 1991, https://www.nytimes.com/1991/03/17/weekinreview/the-nation-coal-miners-contend-their-plight-is-worsening.html?searchResultPosition=1.

99 **about 80 percent were in West Virginia, Kentucky, or Virginia:** Karen Ball, "Government Finds Widespread Tampering in Coal Industry," Associated Press, April 4, 1991.

99 **"This contention they're making of an industry-wide conspiracy":** "Coal Companies Seek Independent Study of Tampering," *U.S. Coal Review*, April 9, 1991.

99 **the nation's largest coal producer, Peabody Coal:** Guilty plea, information, and related documents in *United States of America v. Peabody Coal Company*, U.S. District Court, Southern District of West Virginia, 2:91-00022.

99 **Some in the coalfields called it:** Mary Jordan, "Miners' Lives and Livelihood at Issue; Workers Fearful, Angry as Hearing on Coal Dust Testing Begins," *Washington Post*, April 15, 1991; "Peabody Bon-Bons Give Industry Bad Name," *International Coal Report*, April 8, 1991.

99 **Newspaper editorial boards cheered Martin:** "An Addiction to Cheat," *St. Louis Post-Dispatch*, April 10, 1991; "Black Mark Against Coal Industry Cheaters," *Atlanta Journal and Constitution*, April 21, 1991; "Keep the Pressure on Coal Companies," *St. Louis Post-Dispatch*, April 23, 1991.

100 **revoked the licenses:** Karen Ball, "Labor Department Stiffens Fines for Some Coal Dust Tampering Violations," Associated Press, April 15, 1991; Philip Dine, "U.S. Increases Fines for Dust-Tampering," *St. Louis Post-Dispatch*, April 16, 1991.

100 **just 2 percent of its $213 million in profits:** "Large Coal Companies Doing Pretty Well," *U.S. Coal Review*, March 13, 1991.

100 **"Our industry has been tried and found guilty":** "Labor Department Increases Fines for Sample Tampering," *U.S. Coal Review*, April 17, 1991.

100 **hearings before an administrative law judge:** Details about the proceedings before

the Federal Mine Safety and Health Review Commission (FMSHRC) are drawn from three decisions and one dissenting opinion: "common-issues decision": *In Re: Contests of Respirable Dust Sample Alteration Citations,* 15 FMSHRC 1456 (July 1993), https://www.fmshrc.gov/decisions/alj/93071456.PDF; "mine-specific decision": *Keystone Coal Mining Corp.,* 16 FMSHRC 857 (April 1994), https://www.fmshrc.gov/decisions/alj/94040857.pdf; decision on appeal: *In Re: Contests of Respirable Dust Sample Alteration Citations, Keystone Coal Mining Corp. v. Secretary of Labor,* 17 FMSHRC 1819 (November 1995), https://arlweb.msha.gov/SOLICITOR/FMSHRC/freqdecs/In%20re%20Contests%20of%20Respirable%20Dust%20Sample%20Alteration%20Citations,%2017%20FMSHRC%201819%20(Nov.%201995)%20(opinion%20by%20Commissioners%20Doyle%20and%20Holen).pdf; dissent on appeal: *In Re: Contests of Respirable Dust Sample Alteration Citations, Keystone Coal Mining Corp. v. Secretary of Labor,* dissent by Marc Lincoln Marks (December 1995), https://www.fmshrc.gov/decisions/commission/dustdis.pdf.

100 **mortally wounded the government's case:** *In Re: Contests of Respirable Dust Sample Alteration Citations,* 1835–36.

100 **"playing Noah Webster":** *In Re: Contests of Respirable Dust Sample Alteration Citations,* dissent, 3–5.

101 **Broderick was convinced that AWCs:** *In Re: Contests of Respirable Dust Sample Alteration Citations,* 15 FMSHRC 1456, 1513–23.

101 **"impossible to shoulder":** *In Re: Contests of Respirable Dust Sample Alteration Citations,* dissent, 4–5.

101 **Labor Department tried to introduce testimony:** *In Re: Contests of Respirable Dust Sample Alteration Citations,* 1873–75.

102 **almost 43 percent of samples:** *Keystone Coal Mining Corp.,* 16 FMSHRC 857, 878–81.

102 **"to ask us to believe that elephants fly":** *In Re: Contests of Respirable Dust Sample Alteration Citations,* dissent, 29–31.

102 **Another key piece of statistical evidence:** Ibid., 28–29.

103 **refused to even consider the silica-dust samples:** *Keystone Coal Mining Corp.,* 16 FMSHRC 857, 887–88.

103 **a panel of commissioners voted:** *In Re: Contests of Respirable Dust Sample Alteration Citations,* 17 FMSHRC 1819.

103 **appointed by George H. W. Bush:** *PN1523-2—Arlene Holen—Federal Mine Safety and Health Review Commission,* 101st Cong., 2nd sess., 1990, https://www.congress.gov/nomination/101st-congress/1523/2/actions?r=1274&overview=closed.

103 **appointed by Ronald Reagan:** "Nominations & Appointments, July 12, 1985: Nomination of Joyce A. Doyle to Be a Member of the Federal Mine Safety and Health Review Commission," Ronald Reagan Presidential Library and Museum, https://www.reaganlibrary.gov/research/speeches/71285d.

103 **nominated by Bill Clinton:** *PN1675—Marc Lincoln Marks—Federal Mine Safety and Health Review Commission,* 103rd Cong., 2nd sess., 1994, https://www.congress.gov/nomination/103rd-congress/1675.

103 **represented Pennsylvania in Congress as a Republican:** James Coates, "Disaffected Republicans Signing On with Clinton," *Chicago Tribune,* October 27, 1992, https://www.chicagotribune.com/news/ct-xpm-1992-10-27-9204070477-story.html; Al Kamen, "Mining for Trouble at Coal Convention," *Washington Post,* October 14, 1994, https://www.washingtonpost.com/archive/politics/1994/10/14/mining-for-trouble-at-coal-convention/990a616d-c406-4a4a-8aa9-29859e0332f2/.

103 **concluded his dissent with a pointed rebuke:** *In Re: Contests of Respirable Dust Sample Alteration Citations,* dissent, 35.

103 **A federal appeals court affirmed Broderick's decision:** *Secretary of Labor, Petitioner, v. Keystone Coal Mining Corporation and Federal Mine Safety and Health Review Commission, Respondents, Southern Ohio Coal Company, et al., Intervenors,* 151 F.3d 1096 (D.C. Cir. 1998).

103 **the Labor Department vacated the thousands of citations:** *In Re: Contests of Respirable Dust Sample Alteration Citations,* order of dismissal (December 1998).

104 **criminal cases that unfolded:** This is based on data provided by the Mine Safety and Health Administration. For more on this subject, see chapter 14 of this book.

CHAPTER 8: THE NEW COALITION

Interviews: John Cline, Craig Robinson, Brenda Halsey Marion, Grant Crandall, Tom Johnson, Robert A. Cohen, Steve Sanders, Paul Siegel, Cecile Rose, Debbie Wills, Mary Natkin, Timothy MacDonnell, Tammy Cline, Tim Burke, Matthew Cline

105 **multiple iterations of them that might apply:** Division of Coal Mine Workers' Compensation, "Historical Information—Selected Statutory, Regulatory and Judicial History," https://www.dol.gov/owcp/dcmwc/HistoricalInformation.htm.

106 **a good idea to have more of these get-togethers:** John Cline and other miners' advocates retained and shared with me documents such as letters, memos, meeting minutes, presentations, and notes from this period.

106 **migrated north in search of factory jobs:** John Alexander Williams, *West Virginia: A History* (Morgantown: West Virginia University Press, 2001), 177–79; Guy Carawan and Candie Carawan, *Voices from the Mountains: The People of Appalachia—Their Faces, Their Words, Their Songs* (Athens: University of Georgia Press, 1996), 71.

108 **doctors and scientists whose pioneering research:** These included researchers from the National Institute for Occupational Safety and Health whose work underpinned much of the landmark 1995 publication "Criteria for a Recommended Standard: Occupational

Exposure to Respirable Coal Mine Dust" (https://www.cdc.gov/niosh/docs/95-106/pdfs/95-106.pdf?id=10.26616/NIOSHPUB95106), which recommended cutting in half the existing standard for the concentration of dust allowed in mine air and synthesized a large body of scientific evidence to show that conditions such as emphysema and chronic bronchitis could be caused by coal dust.

109 **established a black lung legal clinic:** "Advanced Administrative Litigation Clinic (Black Lung)," Washington and Lee University School of Law, https://law.wlu.edu/clinics/advanced-administrative-litigation-clinic-(black-lung); Wendy Lovell, "David v. Goliath," *W&L Law Alumni Magazine* (Fall 2005), http://law2.wlu.edu/faculty/facultydocuments/murchisonb/blacklungstory.pdf.

111 **Terry, a former coal camp:** Rhonda Janney Coleman, "Coal Miners and Their Communities in Southern Appalachia, 1925–1941," part 1, *West Virginia Historical Society Quarterly* 15, no. 2 (April 2001), http://www.wvculture.org/history/wvhs/wvhs1502.html; "Terry, WV," *Coalfields of the Appalachian Mountains: A Scrapbook of Appalachian Coal Towns,* http://www.coalcampusa.com/sowv/river/terry-wv-coal-mine/terry-wv-coal-mine.htm.

CHAPTER 9: "OUTLAW"

Interviews: Frank Ratcliff, Harold Ford, Henry Bradley, Marshall Litton, Mary Fox, Terri Smith, Gary Hairston, Jerry Bailey, Raymond Keffer, Gene Stewart, Carter Stump, Dwarfus Dale Permelia, Ronnie Lilly, Dewey Keiper, John Smallwood, Gloria Roles, Richard Roles, John Cline

114 **Blankenship was a true son of West Virginia:** "Meet Don Blankenship," Don Blankenship: American Competitionist (his personal website), https://www.donblankenship.com/meet-don; Laurence Leamer, *The Price of Justice: A True Story of Greed and Corruption* (New York: St. Martin's, 2013); Peter A. Galuszka, *Thunder on the Mountain: Death at Massey and the Dirty Secrets Behind Big Coal* (New York: St. Martin's, 2012); Strat Douthat, "'Most Hated Man in Mingo' Is Determined Union Foe," *Charleston Gazette,* March 31, 1985; Ian Urbina, "Wealthy Coal Executive Hopes to Turn Democratic West Virginia Republican," *New York Times*, October 22, 2006, https://www.nytimes.com/2006/10/22/us/22blankenship.html.

115 **In 1982, he started at a Massey subsidiary:** "Meet Don Blankenship"; Galuszka, *Thunder on the Mountain,* 37–39; Ian Urbina and John Leland, "A Mine Boss Inspires Fear, but Pride, Too," *New York Times,* April 7, 2010, https://www.nytimes.com/2010/04/08/us/08blankenship.html.

115 **The first clash:** Bryan T. McNeil, *Combating Mountaintop Removal: New Directions in the Fight Against Big Coal* (Urbana: University of Illinois Press, 2011), 80–82; Michael Shnayerson, *Coal River: How a Few Brave Americans Took On a Powerful Company—and*

the Federal Government—to Save the Land They Love (New York: Farrar, Straus and Giroux, 2008), 31–32; Peter Slavin, "Razing Appalachia," *Los Angeles Times Magazine*, May 5, 2002.

115 **a larger contract dispute erupted:** Don Blankenship, "An American Political Prisoner," https://www.donblankenship.com/american-political-prisoner; Douthat, "'Most Hated Man in Mingo'"; Shnayerson, *Coal River*, 32–34; Galuszka, *Thunder on the Mountain*, 38–39; Urbina, "Wealthy Coal Executive"; Leamer, *The Price of Justice*, 66–68.

116 **Blankenship became E. Morgan Massey's handpicked successor:** "Morgan Massey Retires," *Coal Week*, February 17, 1992; "Massey Retires as Chairman and CEO of Company," *U.S. Coal Review*, February 19, 1992; Leamer, *The Price of Justice*, 49, 159; David Segal, "The People v. the Coal Baron," *New York Times*, June 20, 2015, https://www.nytimes.com/2015/06/21/business/energy-environment/the-people-v-the-coal-baron.html.

116 **Blankenship staked the company's future on central Appalachia:** "Massey Makes Smart Buys," *FT Energy Newsletters—International Coal Report*, August 21, 1992, 7; Leamer, *The Price of Justice*, 49; "A.T. Massey Coal Co. Acquires West Virginia Coal Properties," PR Newswire, August 12, 1992; "Eastern Producers Cut Back," *Coal Week*, March 27, 1995; Galuszka, *Thunder on the Mountain*, 29; "Massey Builds New Coking Complex," *Coal Week International*, August 2, 1994; "Competitors Watch Massey," *FT Energy Newsletters—International Coal Report*, January 22, 1996, 9.

117 **a fifty-million-dollar judgment:** *Caperton v. A. T. Massey Coal Co.*, Supreme Court of Appeals of West Virginia, No. 33350 (2008), http://www.courtswv.gov/supreme-court/docs/spring2008/33350R.pdf; *Caperton v. A. T. Massey Coal Co.*, 556 U.S. 868 (2009); Adam Liptak, "Judicial Races in Several States Become Partisan Battlegrounds," *New York Times*, October 24, 2004, https://www.nytimes.com/2004/10/24/politics/campaign/judicial-races-in-several-states-become-partisan.html; John Gibeaut, "Caperton's Coal," *ABA Journal*, February 2, 2009, http://www.abajournal.com/magazine/article/capertons_coal.

117 **"What you have to accept in a capitalist society generally":** Leamer, *The Price of Justice*, 14; "Don Blankenship: 'Survival of the Most Productive,'" YouTube, https://www.youtube.com/watch?v=S9lBWdK37VM.

117 **controversial mining practice known as mountaintop removal:** Michael Shnayerson, "The Rape of Appalachia," *Vanity Fair*, November 20, 2006, https://www.vanityfair.com/news/2006/05/appalachia200605; Jeff Goodell, "Don Blankenship: The Dark Lord of Coal Country," *Rolling Stone*, December 9, 2010, https://www.rollingstone.com/culture/culture-news/don-blankenship-the-dark-lord-of-coal-country-184288/; David Roberts, "Robert F. Kennedy Jr. Takes on Mountaintop-Mining Magnate Don Blankenship," *Grist*, January 22, 2010, https://grist.org/article/2010-01-22-robert-kennedy-debate-mountaintop-mining-magnate-don-blankenship/.

117 **The people living in the hollows below felt the brunt of the blasting:** A large body of scientific literature on mountaintop removal exists, and entire books have been written on the subject. The following two federal government reports provide a survey of some of the primary concerns that have been evaluated: Claudia Copeland, *Mountaintop Mining: Background on Current Controversies,* U.S. Library of Congress, Congressional Research Service, RS21421 (2015), https://fas.org/sgp/crs/misc/RS21421.pdf; U.S. Environmental Protection Agency, *The Effects of Mountaintop Mines and Valley Fills on Aquatic Ecosystems of the Central Appalachian Coalfields* (March 2011).

117 **vast amounts of waste:** David Kohn, "The 300-Million-Gallon Warning," *Mother Jones,* March/April 2002, https://www.motherjones.com/environment/2002/03/300-million-gallon-warning/; Shnayerson, "The Rape of Appalachia"; Goodell, "Dark Lord of Coal Country"; "Massey's Sidney Coal Unit in Kentucky Suffered Another Spill," *U.S. Coal Review,* March 17, 2003; "WVDEP Sues Massey Over Tug Fork Spill," *Platts Coal Outlook,* April 29, 2002; U.S. Environmental Protection Agency, *Martin County Coal Slurry Spill Site: Order on Consent for Compliance,* March 9, 2001, https://www.epa.gov/sites/production/files/2014-03/documents/martin_county_coal_89.pdf; Ken Ward Jr., "Massey Sludge Spill Probes Still Unfinished," *Charleston Gazette,* October 11, 2001.

118 **he became more politically active:** Urbina, "Wealthy Coal Executive"; Blankenship, "An American Political Prisoner"; Galuszka, *Thunder on the Mountain,* 43.

118 **His message resonated among many in the coalfields:** Urbina and Leland, "A Mine Boss Inspires Fear"; Leamer, *The Price of Justice,* 159–60, 229; Evan Osnos, "Don Blankenship, West Virginia's 'King of Coal,' Is Guilty," *New Yorker,* December 3, 2015, https://www.newyorker.com/business/currency/don-blankenship-west-virginias-king-of-coal-is-guilty; Michael Shnayerson, "The Truth About Don Blankenship," *Vanity Fair,* April 16, 2010, https://www.vanityfair.com/news/2010/04/the-truth-about-don-blankenship; Lawrence Messina, "Justice Says Vacation Meeting with Massey CEO Didn't Influence Decision," *Beckley Register-Herald,* January 15, 2008, https://www.register-herald.com/archives/justice-says-vacation-meeting-with-massey-ceo-didn-t-influence/article_9c7f3ea2-7f37-51b4-906e-7b7c61bc736b.html; Howard Berkes, "Massey CEO's Pay Soared as Mine Concerns Grew," NPR, April 17, 2010, https://www.npr.org/templates/story/story.php?storyId=126072828; Len Boselovic, "Calls Grow for Ouster of CEO at Massey," *Pittsburgh Post-Gazette,* May 2, 2010.

118 **he couldn't escape the creditors:** "Westmoreland's Low Vol Faces Peril," *FT Energy Newsletters—International Coal Report,* December 14, 1992; "Bankrupt WV Companies Continue to Produce for Westmoreland," *U.S. Coal Review,* January 6, 1993; "Kizer Coal Companies File Reorganization Plan," *U.S. Coal Review,* December 15, 1993; Andy Yale, "The Closing of the American Mine," *Harper's,* December 1, 1994; "Former WV Governor Pleads Guilty to Extortion of Coal Executive," *U.S. Coal Review,* April 18, 1990.

119 **Union mines were hamstrung:** Douthat, "'Most Hated Man in Mingo'"; Blankenship,

"An American Political Prisoner," 12–13; Leamer, *The Price of Justice,* 14, 20, 67, 120, 135–36; "With Prices Rising, Massey Ready to Reap Benefits of Selling Short," *Platts Coal Outlook,* November 6, 2000.

119 **Blankenship micromanaged his operations:** Superseding indictment, *United States of America v. Donald L. Blankenship,* Criminal No. 5:14-cr-00244, U.S. District Court for the Southern District of West Virginia, March 10, 2015; Consolidated Amended Class Action Complaint for Violations of the Federal Securities Laws, *In Re: Massey Energy Co. Securities Litigation,* Civil Action No. 5:10-cv-00689, U.S. District Court for the Southern District of West Virginia, March 11, 2011; Governor's Independent Investigation Panel, *Upper Big Branch: The April 5, 2010, Explosion: A Failure of Basic Coal Mine Safety Practices* (May 2011), https://web.archive.org/web/20110601060618/http://www.nttc.edu/programs&projects/minesafety/disasterinvestigations/upperbigbranch/UpperBigBranchReport.pdf; Galuszka, *Thunder on the Mountain,* 156–58.

120 **a mine at the Elk Run complex was having problems:** Videotaped deposition of Stan Edwards, *Ricky Tucker and Deborah Tucker v. Consolidation Coal Company, et al.,* Civil Action No. 11-C-78, Circuit Court of Wyoming County, WV.

120 **"I'm looking to make an example out of somebody":** Superseding indictment, *U.S. v. Blankenship,* 29–30.

120 ***Running Coal:*** Memorandum from Don Blankenship to all deep mine superintendents, October 19, 2005.

121 **"S-1, P-2, M-3":** "Our Formula for Success," Massey Energy, https://web.archive.org/web/20110126145057/http:/www.masseyenergyco.com:80/company/success.shtml; Ken Ward Jr., "Massey Chief Testifies at Trial," *Charleston Gazette,* November 15, 2008; Governor's Independent Investigation Panel, *Upper Big Branch,* 94–96.

121 **exercises in semantics:** Mine Safety and Health Administration, *Internal Review of MSHA's Actions at the Upper Big Branch Mine–South, Performance Coal Company, Montcoal, Raleigh County, West Virginia* (March 2012), D1–D4, https://arlweb.msha.gov/PerformanceCoal/UBBInternalReview/UBBInternalReviewReport.pdf; Mine Safety and Health Administration, *Report of Investigation: Fatal Underground Mine Explosion April 5, 2010, Upper Big Branch Mine–South, Performance Coal Company, Montcoal, Raleigh County, West Virginia, ID No. 46-08436* (December 2011), 61–62, https://arlweb.msha.gov/Fatals/2010/UBB/FTL10c0331noappx.pdf; Senate Committee on Appropriations, Subcommittee on Departments of Labor, Health and Human Services, and Education, and Related Activities, *Investing in Mine Safety: Preventing Another Disaster,* 111th Cong., 2nd sess., 2010, 56–57; U.S. House of Representatives, Committee on Education and Labor, *H.R. 5663, Miner Safety and Health Act of 2010,* 111th Cong., 2nd sess., 2010, 59; Senate Committee on Health, Education, Labor, and Pensions, *Putting Safety First: Strengthening Enforcement and Creating a Culture of Compliance at Mines and Other Dangerous Workplaces,* 111th Cong., 2nd sess., 2010, 42.

121 **prodigious number of safety and health violations:** Giovanni Russonello, "Massey Had Worst Mine Fatality Record Even Before April Disaster," Investigative Reporting Workshop, American University School of Communication, November 23, 2010, https://web.archive.org/web/20110504001348/http://www.investigativereportingworkshop.org/investigations/coal-truth/story/massey-had-worst-mine-fatality-record-even-april-d/; Governor's Independent Investigation Panel, *Upper Big Branch,* 93; Mine Safety and Health Administration, *Report of Investigation: Upper Big Branch.*

121 **the meager fines regulators could impose:** "Criteria and Procedure for Proposed Assessment of Civil Penalties," *Code of Federal Regulations,* title 30, part 100, https://www.govinfo.gov/content/pkg/CFR-2009-title30-vol1/pdf/CFR-2009-title30-vol1-sec100-3.pdf.

122 **"'if you can get the footage, we can pay the fines'":** Memorandum from Stephanie Ojeda, senior corporate counsel of Massey Coal Services, Inc., to Don Blankenship, subject line "Report of June 17, 2009 Meeting with Bill Ross Regarding MSHA Violations," June 25, 2009.

125 **Gary had been keeping track of his lung issues:** Gary's treatment records later became part of his federal black lung–benefits case file. Mary also shared with me additional medical records from Gary's treatment at facilities run by the U.S. Department of Veterans Affairs.

CHAPTER 10: COTTON

Interviews: Patsy Caldwell, John Cline, Grant Crandall, Robert F. Cohen Jr., Gregory Fino, Mary Natkin, Robert A. Cohen, Tom Johnson

126 **Caldwell had selected Dr. Don Rasmussen:** Many of the details throughout this chapter are drawn from documents contained in Charles Caldwell's federal black lung–benefits case file.

127 **supposed to lead to an automatic award of benefits for miners:** "Irrebuttable Presumption of Total Disability or Death Due to Pneumoconiosis," *Code of Federal Regulations,* title 20, section 718.304, https://www.govinfo.gov/content/pkg/CFR-2010-title20-vol3/pdf/CFR-2010-title20-vol3-sec718-304.pdf.

129 **required to submit to another round of testing:** Division of Coal Mine Workers' Compensation, "Guide to Filing for Black Lung Benefits (Miner)," https://www.dol.gov/owcp/dcmwc/filing_guide_miner.htm:

> *If we find that a coal company is liable for your claim, that company has the right to have you examined by a physician of its choice. If you refuse to be examined, your claim may be denied without any further consideration of your entitlement to benefits.*

130 **an academic journal published an article he had written:** N. L. Lapp, W.K.C. Morgan, and G. Zaldivar, "Airways Obstruction, Coal Mining, and Disability," *Occupational and Environmental Medicine* 51, no. 4 (April 1994): 234–38, https://oem.bmj.com/content/oemed/51/4/234.full.pdf.

CHAPTER 11: THE FIRM

131 **The firm traces its roots:** "History," Jackson Kelly PLLC, https://www.jacksonkelly.com/firm/history.

131 **formed in 1822 by Benjamin H. Smith:** Elizabeth Jill Wilson, *The First 150: Spilman Thomas and Battle's History of Service,* 9–13, https://www.spilmanlaw.com/Spilman/media/Dev/Spilman-History.pdf.

131 **Entrepreneurs, many of them located outside the state:** John Alexander Williams, *West Virginia: A History* (Morgantown: West Virginia University Press, 2001), 42–45; Barbara Rasmussen, "Land Ownership," *West Virginia Encyclopedia,* https://www.wvencyclopedia.org/articles/1293.

131 **Smith made it his mission:** Brooks F. McCabe Jr., "Benjamin Harrison Smith, Land Titles, and the West Virginia Constitution," *West Virginia History: A Journal of Regional Studies* 6, no. 1 (Spring 2012), https://muse.jhu.edu/article/474947; Wilson, *The First 150,* 13–18.

132 **"The state's economic development":** McCabe, "Benjamin Harrison Smith," 1.

132 **Edward B. Knight:** Wilson, *The First 150,* 21–24; McCabe, "Benjamin Harrison Smith."

132 **"the Lawyers' Constitution":** Williams, *West Virginia,* 92.

133 **Brown, Jackson, and Knight**: Wilson, *The First 150,* 31; "James Brown," West Virginia Department of Arts, Culture, and History, http://www.wvculture.org/history/statehood/images/brownjames.html; James F. Brown III Collection, West Virginia Department of Arts, Culture, and History, http://www.wvculture.org/history/collections/manuscripts/ms2016-071.html; "History," Jackson Kelly PLLC; George Wesley Atkinson, ed., *Bench and Bar of West Virginia* (Charleston, WV: Virginian Law Book Company, 1919), 248.

133 **The formation in 1890 of the United Mine Workers:** Robert Shogan, *The Battle of Blair Mountain: The Story of America's Largest Labor Uprising* (New York: Perseus, 2006), 35–38; Jerry Bruce Thomas, "United Mine Workers of America," *West Virginia Encyclopedia,* https://www.wvencyclopedia.org/articles/835.

133 **standing with the coal bosses:** Williams, *West Virginia,* 156–57; Edward M. Steel, ed., *The Court-Martial of Mother Jones* (Lexington: University Press of Kentucky, 1995), 142; Roger Shattuck, "A Talk with Mary Lee Settle," *New York Times,* October 26, 1980, https://archive.nytimes.com/www.nytimes.com/books/98/10/25/specials/settle-talk.html.

134 **A measure of the elite status the firm had attained:** Williams, *West Virginia,* 156; "James F. Brown III Collection," West Virginia Department of Arts, Culture, and History.

134 **former West Virginia governor Homer A. Holt:** "History," Jackson Kelly PLLC; Williams, *West Virginia,* 160–68; Jerry B. Thomas, " 'The Nearly Perfect State': Governor Homer Adams Holt, the WPA Writers' Project and the Making of *West Virginia: A Guide to the Mountain State,*" *West Virginia History* 52 (1993): 91–108, http://www.wvculture.org/history/journal_wvh/wvh52-7.html.

135 **lobbying on behalf of the coal industry:** Barbara Ellen Smith, *Digging Our Own Graves: Coal Miners and the Struggle Over Black Lung Disease* (Philadelphia: Temple University Press, 1987), 124–25; Brit Hume, *Death and the Mines: Rebellion and Murder in the UMW* (New York: Grossman, 1971), 125–26.

135 **"West Virginia's oldest and most prestigious law firm":** Williams, *West Virginia,* 156–57.

136 **just a few of its many services:** For example: Written comments on behalf of the Midwest Ozone Group opposing a 2012 Environmental Protection Agency proposal to strengthen air pollution standards for particulate matter; comments during a 2015 public hearing on behalf of the West Virginia Coal Association opposing an Interior Department proposal to require mining companies to do more to avoid polluting streams; written comments on behalf of the National Construction Association, National Stone, Sand, and Gravel Association, Portland Cement Association, American Iron and Steel Institute, Georgia Mining Association, and Georgia Construction Aggregate Association opposing a Mine Safety and Health Administration rule requiring mining companies to conduct enhanced safety examinations; participation on behalf of International Coal Group in a Mine Safety and Health Administration investigation of a fatal accident at one of the company's mines in West Virginia; lawsuit on behalf of numerous trade organizations challenging an Environmental Protection Agency rule regulating toxic emissions from power plants.

136 **"We understand the coal industry from the inside":** "Industries: Coal," Jackson Kelly PLLC, https://www.jacksonkelly.com/industries/coal.

137 **During a 1988 congressional committee hearing in Beckley:** "Statement of Harold Hayden, Representative of United Mine Workers, District 29," U.S. House of Representatives, Committee on Education and Labor, Subcommittee on Labor Standards, *Oversight Hearing on the Administration of the Black Lung Program,* 100th Cong., 2nd sess., 1988.

137 **William S. Mattingly:** "William S. Mattingly," Jackson Kelly PLLC, https://www.jacksonkelly.com/professionals/william-s-mattingly; "Transcript of Hearing Before the Investigative Panel of the Lawyer Disciplinary Board, State of West Virginia, *In Re: Douglas A. Smoot, Esquire,*" June 18, 2009 (388, 415), and June 19, 2009 (97–101); written comments on Proposed Revisions of the Rules of Practice and Procedure for Hearings Before the Office of Administrative Law Judges, RIN 1290-AA26, February 4, 2013; written comments on Proposed Regulations Implementing the Byrd Amendments

to the Black Lung Benefits Act, RIN 1240-AA04, May 29, 2012; written comments on Specifications for Medical Examinations of Underground Coal Miners, RIN 0920-AA21, March 9, 2012; "Sixty-Five Jackson Kelly PLLC Attorneys Recognized by the Best Lawyers in America," *West Virginia Executive,* April 19, 2014, http://www.wvexecutive.com/sixty-five-jackson-kelly-pllc-attorneys-recognized-best-lawyers-america/; "Jackson Kelly PLLC Attorneys Recognized by the Best Lawyers in America," Jackson Kelly PLLC, August 16, 2015, https://www.jacksonkelly.com/insights/jackson-kelly-pllc-attorneys-recognized-by-the-best-lawyers-in-america; "William S. Mattingly—Awards," Jackson Kelly PLLC, https://www.jacksonkelly.com/professionals/william-s-mattingly#awards.

137 **Mattingly offered his take:** William S. Mattingly, "If Due Process Is a Big Tent, Why Do Some Feel Excluded from the Big Top?," *West Virginia Law Review* 105 (2003): 791–826.

139 **a U.S. Senate committee had held a hearing:** Senate Committee on Labor and Human Resources, *Black Lung Benefits Restoration Act,* 103rd Cong., 2nd sess., 1994. The quoted passage is from a letter to Senator Paul Simon from Virgil E. Young of Rockwood, Tennessee, dated March 21, 1994, that is included in the hearing record on pages 61–62.

CHAPTER 12: DISCOVERY

Interviews: Robert F. Cohen Jr., John Cline, Tom Johnson, Grant Crandall

142 **John sent a letter to Maloy:** Details about Charles Caldwell's case are drawn largely from the documents contained in his federal black lung–benefits claim as well as from John Cline's contemporaneous notes and other documentation.

143 **Zaldivar's report was one of the rare exceptions**: Jackson Kelly's position was that if the opposing side asked, it was required to turn over a report by an "examining physician," such as Dr. Zaldivar, under Rule 35(b) of the Federal Rules of Civil Procedure, which says, "The party who moved for the examination must, on request, deliver to the requester a copy of the examiner's report."

143 **If asked for pretty much any other evidence:** Jackson Kelly's position was that if the firm didn't plan to submit a physician's report for the record, that physician was a "non-testifying expert." Under Rule 26 of the Federal Rules of Civil Procedure, the firm argued, reports by non-testifying experts are exempt from disclosure unless the requestor demonstrates "exceptional circumstances on which it is impracticable for the party seeking discovery to obtain facts or opinions on the same subject by other means."

144 **rebelling against what they saw as the corrupt United Mine Workers hierarchy:** Intertwined with the black lung uprising that began in 1968 was the rank-and-file campaign to oust UMW president Tony Boyle and restore democratic rule to the

union. The saga included the murders of opposition leader Jock Yablonski and his wife and daughter, a federally supervised election in which a Black Lung Association leader unseated Boyle, and Boyle's conviction on charges of ordering the Yablonski killings. For more, see, among other sources, Brit Hume's *Death and the Mines: Rebellion and Murder in the UMW,* Ken Hechler's *The Fight for Coal Mine Health and Safety: A Documented History,* and Jefferson Cowie's *Stayin' Alive: The 1970s and the Last Days of the Working Class.*

146 **the lawyers who made up the black lung unit:** As noted previously, all of the Jackson Kelly black lung attorneys—with the exception of Mary Rich Maloy, as described in a footnote—either refused to speak with me or did not respond to my requests for interviews. The information about their community service, personal interests, and pro bono work is drawn from publicly available sources, such as online biographies and social media postings.

CHAPTER 13: "SUPERIOR CREDENTIALS"

Interviews: Daniel A. Henry, David Weissman, Jack Parker, Benjamin Park, Robert A. Cohen, Cecile Rose, Tom Johnson, Steve Sanders, Edward Terhune Miller

149 **Coal companies paid a premium for this prestigious imprimatur:** Many of the details about Dr. Paul Wheeler's views throughout this chapter are drawn from his depositions in individual benefits cases and his statements, as referenced or quoted directly by administrative law judges in written decisions, in hundreds of additional cases. Wheeler later reiterated many of his views in an interview at Johns Hopkins on July 24, 2013. (By then I was working with ABC News on a story about Wheeler and his colleagues. Johns Hopkins made Wheeler available for an on-camera interview conducted by ABC News' Brian Ross at which I and ABC producer Matthew Mosk were present.) Among the depositions on which I have relied most heavily are the following: *Milton D. Kincaid v. Eastern Associated Coal LLC,* Claim No. 2012-BLA-05822, July 10, 2013; *Cubert Spence v. West Virginia Solid Energy, Inc.,* Claim No. 2005-BLA-00018, December 6, 2004; *Gary Fox v. Elk Run Coal Co., Inc.,* Claim No. 2000-BLA-00598, September 7, 2000; *Reaford D. Syck v. Sidney Coal Co., Inc./A. T. Massey,* Claim No. 2010-BLA-05179, September 17, 2010; *William Harris v. Westmoreland Coal Company,* Claim No. 90-BLA-02730, July 8, 1994; *Catherine D. Bartley v. Union Carbide Corporation,* Claim No. 2001-BLA-01008, May 14, 2003; *Michael Day v. Eastern Associated Coal Corp.,* Claim No. 2006-BLA-05411, February 10, 2009.

149 **judges usually relied heavily on physicians' credentials:** This approach has met with the approval of the Benefits Review Board in many cases, and it is grounded in the regulations governing the benefits program; 20 CFR § 718.202 states that "where two or

more X-ray reports are in conflict, in evaluating such X-ray reports consideration must be given to the radiological qualifications of the physicians interpreting such X-rays."

150 **Pneumoconiosis Section:** In depositions, Wheeler often used this name to describe the unit he headed, and some, though not all, reports from the unit have this title on the letterhead. Officials in Johns Hopkins' media relations department generally called it the "B reader program" in our communications, a reference to the National Institute for Occupational Safety and Health's designation for radiologists who had passed the government exam demonstrating proficiency in classifying pneumoconiosis on X-rays.

150 **reading X-rays and CT scans for outside clients:** *Kincaid,* 48; *Syck,* 7.

150 **long list of publications, presentations, and plaudits to his name:** Curriculum vitae of Paul S. Wheeler, MD; Otha W. Linton and Bob W. Gayler, *Johns Hopkins Radiology, 1896–2010* (Baltimore: Johns Hopkins Hospital, 2011), 112–13.

150 **"The newer ones were trained by me":** *Bartley,* 8.

151 **he voiced it over and over:** *Kincaid,* 16–18, 37–38; *Spence,* 21–25; *Fox,* 27–28; *Syck,* 9–20; *Harris,* 34–54; *Bartley,* 19–48; *Day,* 13–29.

151 **numerous published studies dating as far back as 1974:** H. E. Amandus et al., "Pulmonary Zonal Involvement in Coal Workers' Pneumoconiosis," *Journal of Occupational Medicine* 16, no. 4 (1974): 245–47; A. Scott Laney and Edward L. Petsonk, "Small Pneumoconiotic Opacities on U.S. Coal Worker Surveillance Chest Radiographs Are Not Predominantly in the Upper Lung Zones," *American Journal of Industrial Medicine* 55 (2012): 793–98; A. Cockcroft et al., "Prevalence and Relation to Underground Exposure of Radiological Irregular Opacities in South Wales Coal Workers with Pneumoconiosis," *British Journal of Industrial Medicine* 40 (1983): 169–72; A. Cockcroft et al., "Shape of Small Opacities and Lung Function in Coalworkers," *Thorax* 37 (1982): 765–69; J. P. Lyons et al., "Significance of Irregular Opacities in the Radiology of Coalworkers' Pneumoconiosis," *British Journal of Industrial Medicine* 31 (1974): 36–44; Edward L. Petsonk, Cecile Rose, and Robert Cohen, "Coal Mine Dust Lung Disease: New Lessons from an Old Exposure," *American Journal of Respiratory and Critical Care Medicine* 187, no. 11 (2013): 1178–85.

151 **A study published in a prominent radiology journal:** Hiroaki Arakawa et al., "Pleural Disease in Silicosis: Pleural Thickening, Effusion, and Invagination," *Radiology* 236, no. 2 (2005): 685–93.

151 **one of Wheeler's colleagues in the Johns Hopkins radiology department:** R. Nick Bryan, "Farewell to Radiology Editor: Stanley S. Siegelman, MD," *Radiology* 205, no. 3 (1997): 878–79; "Radiology and Radiological Science: Why Johns Hopkins?," Johns Hopkins Medicine, https://www.hopkinsmedicine.org/radiology/education/residency/diagnostic-radiology-residency-program/why-hopkins.html (the website noted that "mentorship is readily available and perhaps best embodied by Dr. Stanley S. Siegelman, who spent 12 years as the editor in chief of Radiology and served as program director at our institution for 35 years").

152 **"I don't think I need medical literature":** *M. F. A., widow of and o/b/o L. S. A. (D) v. Peerless Eagle Coal Company,* Claims No. 2006-BLA-05465 and 2006-BLA-05466, October 30, 2006 (as quoted in a decision by administrative law judge Linda S. Chapman, https://www.oalj.dol.gov/DECISIONS/ALJ/BLA/2006/MA_v_PEERLESS_EAGLE_COAL__2006BLA05465_(MAR_06_2007)_090834_CADEC_SD.PDF, 26).

152 **his mentor, Dr. Russell Morgan:** M. W. Donner and O. M. Gatewood, "Memorial Tribute to Russell H. Morgan, 1911–1986," *American Journal of Roentgenology* 148, no. 1 (1987): 16–17, https://www.ajronline.org/doi/pdf/10.2214/ajr.148.1.16; "Russell H. Morgan, 75, Dead; Pioneer in Field of Radiology," *New York Times,* February 27, 1986, https://www.nytimes.com/1986/02/27/obituaries/russell-h-morgan-75-dead-pioneer-in-field-of-radiology.html.

152 **"certified genius":** *Syck,* 8.

152 **NIOSH asked him to devise an exam:** Linton and Gayler, *Johns Hopkins Radiology,* 128–29; Russell H. Morgan, "Proficiency Examination of Physicians for Classifying Pneumoconiosis Chest Films," *American Journal of Roentgenology* 132, no. 5 (1979): 803–8, https://www.ajronline.org/doi/pdf/10.2214/ajr.132.5.803.

152 **Morgan trained him to read X-rays for pneumoconiosis:** Linton and Gayler, *Johns Hopkins Radiology,* 113; *Kincaid,* 5–9; *Bartley,* 7; *Fox,* 11–12; *Day,* 4–6.

152 **The regulations governing the benefits system specify:** "Chest Roentgenograms (X-Rays)," *Code of Federal Regulations,* title 20, section 718.102(b), https://www.govinfo.gov/content/pkg/CFR-2010-title20-vol3/pdf/CFR-2010-title20-vol3-sec718-102.pdf.

153 **system established in 1950 by the International Labour Organization:** International Labour Office, "Guidelines for the Use of the ILO International Classification of Radiographs of Pneumoconioses," rev. ed., 2011, https://www.ilo.org/wcmsp5/groups/public/---ed_protect/---protrav/---safework/documents/publication/wcms_168260.pdf; National Institute for Occupational Safety and Health, "Chest Radiography: ILO Classification," https://www.cdc.gov/niosh/topics/chestradiography/ilo.html.

153 **automatically resulted in an award of benefits:** "Irrebuttable Presumption of Total Disability or Death Due to Pneumoconiosis," *Code of Federal Regulations,* title 20, section 718.304, https://www.govinfo.gov/content/pkg/CFR-2010-title20-vol3/pdf/CFR-2010-title20-vol3-sec718-304.pdf.

153 **neglected to quote the next paragraph of the guidelines:** "Guidelines for the Use of the ILO International Classification," 2. National Institute for Occupational Safety and Health director John Howard would later rebut Wheeler's approach in a statement to the U.S. Senate; see https://www.help.senate.gov/imo/media/doc/Howard8.pdf, 4.

154 **Wheeler occasionally derided the system:** *Fox,* 24; *Syck,* 39–40; *Bartley,* 27.

154 **he maintained his certification as a B reader:** Letters and certificates, obtained under the Freedom of Information Act from the National Institute for Occupational Safety and Health, confirm Wheeler's B-reader status dating to 1973.

155 **"histoplasmosis is far more common":** *Kincaid,* 59.

156 **he had performed an autopsy:** *Kincaid,* 50–52; *Bartley,* 13–14; *Day,* 25–26; *Syck,* 35–36; *Harris,* 42–43.

156 **"one night in an abandoned chicken coop":** *Kincaid,* 57.

156 **His own father, he said, had suffered from both ailments:** *Bartley,* 14, 21; *Kincaid,* 57; *Day,* 43–44.

156 **Wheeler had counterarguments:** *Bartley,* 23; *Day,* 27–29; *Harris,* 43–46.

156 **assumption that the disease had become rare:** *Kincaid,* 54–55; *Bartley,* 20; *Syck,* 23; *Spence,* 25–26.

157 **"cabins that are air conditioned":** *Syck,* 29–30.

157 **Wheeler dismissed the NIOSH data:** *Kincaid,* 55; *Syck,* 35–37; *Spence,* 26–27.

157 **"My father felt that if I was to be a radiologist":** *Bartley,* 39.

158 **"Except apparently in your part of the country":** *Kincaid,* 62.

158 **a letter that the coal company's lawyers submitted:** Letter from Wheeler to attorney Michael F. Blair of the law firm PennStuart, dated September 8, 2009, regarding the case *Orville Blankenship v. Harman Development Corp.*

158 **"As long as we're dealing with X-ray patterns":** *Kincaid,* 49.

159 **it's not necessary in most cases:** "Criteria for the Development of Medical Evidence-General," *Code of Federal Regulations,* title 20, section 718.101; Division of Coal Mine Workers' Compensation, "Guide to Filing for Black Lung Benefits (Miner)," https://www.dol.gov/owcp/dcmwc/filing_guide_miner.htm; "Pneumoconiosis," Chest Foundation, American College of Chest Physicians, https://foundation.chestnet.org/patient-education-resources/pneumoconiosis/; "Diagnosing and Treating Pneumoconiosis," American Lung Association, https://www.lung.org/lung-health-and-diseases/lung-disease-lookup/pneumoconiosis/diagnosing-treating-pneumoconiosis.html.

159 **These dangers were documented in the medical literature:** See, for example, Renda Soylemez Wiener, Daniel C. Wiener, and Michael K. Gould, "Risks of Transthoracic Needle Biopsy: How High?," *Clinical Pulmonary Medicine* 20, no. 1 (2013): 29–35; John P. Hutchinson et al., "In-Hospital Mortality After Surgical Lung Biopsy for Interstitial Lung Disease in the United States. 2000 to 2011," *American Journal of Respiratory and Critical Care Medicine* 193, no. 10 (2015): 1161–67.

159 **"That's temporary":** *Kincaid,* 52–53.

159 **hefty fee for the services:** *Kincaid,* 46–49; *Spence,* 32–33; Decision and Order of Stuart A. Levin, *T.V. v. Consolidation Coal Co./Consol Energy, Inc.,* Claim No. 2007-BLA-05607, August 21, 2009, https://www.oalj.dol.gov/DECISIONS/ALJ/BLA/2007/TV_v_CONSOLIDATION_COAL_C_2007BLA05607_(AUG_21_2009)_111321_CADEC_SD.PDF, 5 (footnote 2); statement from Johns Hopkins media relations on October 30, 2013.

161 **Wheeler told a U.S. Senate subcommittee:** Senate Committee on Labor and Human

Resources, Subcommittee on Labor, *Investigation as to Whether or Not There Is a National Asbestos Crisis; And If So, What Should Be Done About It,* 98th Cong., 2nd sess., 1984, 21–39.

161 **some questionable claims being awarded under the liberalized standards:** U.S. General Accounting Office, "Statement of Morton E. Henig, Senior Associate Director, Human Resources Division, Before the Subcommittee on Oversight, Committee on Ways and Means, United States House of Representatives, on the Black Lung Program and Black Lung Disability Trust Fund" (Washington, DC, 1981), https://www.gao.gov/assets/100/99873.pdf.

161 **plummeted to around 4 percent:** U.S. General Accounting Office, *Black Lung Program: Further Improvements Can Be Made in Claims Adjudication,* GAO/HRD-90-75 (Washington, DC, 1990), 19, https://www.gao.gov/assets/220/212290.pdf.

161 **an additional 132 years to collect that same amount:** Monthly benefits rates are set by law at 37.5 percent of the base salary of a federal employee at level GS-2, step 1—a pay grade that government guidelines say would be appropriate for a person with a high-school education and no experience—and increase modestly for miners with dependents. Historical rates are available at https://www.dol.gov/owcp/dcmwc/benefits_part_c.htm.

162 **"outstanding qualifications":** Decision and Order of Jeffrey Tureck, *W. N. v. Eastern Associated Coal Corp.,* Claim No. 2005-BLA-05823, August 6, 2007, 5, https://www.oalj.dol.gov/DECISIONS/ALJ/BLA/2005/WN_v_EASTERN_ASSOC_COAL_C_2005BLA05823_(AUG_06_2007)_133638_CADEC_SD.PDF.

162 **"superior knowledge of radiology":** Decision and Order of Clement J. Kichuk, *Draper L. Woodard v. Dominion Coal Company,* Claim No. 1997-BLA-01611, January 30, 2001, 10, https://www.oalj.dol.gov/DECISIONS/ALJ/BLA/1997/woodard_noble_hdec_v_dominion_coal_corp_d_1997bla01611_(jan_30_2001)_162536_cadec_sd.PDF.

162 **"highly distinguished record":** Decision and Order of William S. Colwell, *Carnish H. Delp v. Eastern Associated Coal,* Claim No. 2004-BLA-05403, October 31, 2005, 5, https://www.oalj.dol.gov/DECISIONS/ALJ/BLA/2004/DELP_CARNIS_H_v_EASTERN_ASSOC_COAL_D_2004BLA05403_(OCT_31_2005)_142058_CADEC_SD.PDF.

162 **"the prestigious Johns Hopkins University Medical Institute":** Decision and Order of Daniel F. Solomon, *J. B. for B. G. B. v. Performance Coal Company,* Claim No. 2005-BLA-05675, September 21, 2006, 6, https://www.oalj.dol.gov/DECISIONS/ALJ/BLA/2005/BB_v_PERFORMANCE_COAL_CO__2005BLA05675_(SEP_21_2006)_110502_CADEC_SD.PDF.

163 **judges sometimes tried to work with those:** For example, Decision and Order of Robert B. Rae, *Donald L. Lane v. East Star Mining, Inc.,* Claim No. 2008-BLA-05560, February 28, 2012, 5–6, https://www.oalj.dol.gov/DECISIONS/ALJ/BLA/2008

/LANE_DONALD_L_v_EAST_STAR_MINING_INC_2008BLA05560_(FEB_28_2012)_120606_CADEC_SD.PDF; Decision and Order of Alice M. Craft, *Randall L. Varney v. McCoy Elkhorn Coal Corp.,* Claim No. 2009-BLA-05025, January 28, 2010, 21, https://www.oalj.dol.gov/DECISIONS/ALJ/BLA/2009/VARNEY_RANDALL_L_v_MCCOY_ELKHORN_COAL_C_2009BLA05025_(JAN_28_2010)_111119_CADEC_SD.PDF; Decision and Order of Linda S. Chapman, *M. F. M. v. Sewell Coal Co.,* Claim No. 2004-BLA-06395, April 10, 2007, 15, https://www.oalj.dol.gov/DECISIONS/ALJ/BLA/2004/MM_v_SEWELL_COAL_CO_DIR-O_2004BLA06395_(APR_10_2007)_135757_CADEC_SD.PDF.

163 **the third award stuck:** Decision and Order of Linda S. Chapman, *Mary Frances Adkins, widow of and o/b/o Samuel L. Adkins v. Peerless Eagle Coal Company,* Claims No. 2006-BLA-05465 and 2006-BLA-05466, August 5, 2010, https://www.oalj.dol.gov/Decisions/ALJ/BLA/2006/ADKINS_MARY_FRANCESW_v_PEERLESS_EAGLE_COAL__2006BLA05465_(AUG_05_2010)_151230_CADEC_SD.PDF.

CHAPTER 14: "NO BAD SAMPLES"

Interviews: Carter Stump, Henry Bradley, Dewey Keiper, Debbie Wills, Grant Crandall, Brenda Halsey Marion, Frank Ratcliff, Tim Bailey, Gene Stewart, Harold Ford, Raymond Keffer, Ronald Acord, Jerry Bailey, Marshall Litton, Bill Harvit, Carnes Shrewsbury, Dwarfus Dale Permelia, Gary Hairston, Ronnie Lilly

165 **a section foreman at Elk Run's White Knight mine:** Details of Duncan's case (*USA v. Duncan,* U.S. District Court for the Southern District of West Virginia, Case No. 2:00-cr-00263) are drawn from the criminal information filed December 21, 2000; written plea agreement filed January 10, 2001; and transcripts of his plea and sentencing hearings, which took place on January 10, 2001, and March 21, 2001, respectively.

167 **94 percent of underground mine sections:** From the Secretary of the Interior's 1973 Annual Report on Federal Coal Mine Health and Safety, as summarized in U.S. General Accounting Office, *Improvements Still Needed in Coal Mine Dust-Sampling Program and Penalty Assessments and Collections,* December 31, 1975, 15, https://www.gao.gov/assets/120/114174.pdf.

167 **a 1975 General Accounting Office report:** Ibid.

168 **the company submitted dozens of samples:** I obtained dust-sampling data from the Mine Safety and Health Administration under the Freedom of Information Act and analyzed it using Microsoft Access.

168 **The first scandal broke in 1975:** Morton Mintz, "Are Coal Mine Operators Cheating on Black Lung?," *Washington Post,* March 18, 1979, https://www.washington

post.com/archive/opinions/1979/03/18/are-coal-mine-operators-cheating-on-black-lung/171eac3b-1a84-4057-9710-d88010ebf387/; "Lung Disease Is an Unspoken Issue in Trial of Coal Company Officials," *New York Times,* March 26, 1979; Morton Mintz, "U.S. Black-Lung Prosecutors Thwarted," *Washington Post,* April 7, 1979, https://www.washingtonpost.com/archive/politics/1979/04/07/us-black-lung-prosecutors-thwarted/9ee0a687-6fc2-4fb9-9aad-5e1e8fec8940/?noredirect=on; "Coal Company, 2 Aides Cleared of Black-Lung Case Conspiracy," *New York Times,* April 9, 1979, https://www.nytimes.com/1979/04/09/archives/coal-company-2-aides-cleared-of-blacklung-case-conspiracy-causes.html?_r=0.

169 **In 1980, the government tried again:** Estes Thompson, "Two Coal Companies Indicted on Safety Charges," Associated Press, July 10, 1980; "Coal Company Officials Charged with Mine Safety Violations," Associated Press, July 21, 1980; "Settlement Reached in Mine Safety Case," Associated Press, November 26, 1980; Ben A. Franklin, "Mine Company Gets a Fine of $100,000," *New York Times,* December 5, 1980; Glenn Frankel, "Pilgrim Mine: A Legacy of Fear, Doubt; Workers Bitter About Outcome of Mine-Safety Case," *Washington Post,* December 20, 1980.

169 **just four dust-related cases, exacting a total of about $145,000 in fines:** I calculated these totals using a list of cases provided by the Department of Labor.

170 **published in 1984 in the *American Journal of Industrial Medicine:*** Leslie I. Boden and Morris Gold, "The Accuracy of Self-Reported Regulatory Data: The Case of Coal Mine Dust," *American Journal of Industrial Medicine* 6, no. 6 (1984): 427–40.

170 **A 1990 study published in the *American Industrial Hygiene Association Journal:*** Noah S. Seixas et al., "Assessment of Potential Biases in the Application of MSHA Respirable Coal Mine Dust Data to an Epidemiologic Study," *American Industrial Hygiene Association Journal* 51, no. 10 (1990): 534–40.

170 **"There was already a record of tampering in 1978":** Martha Bryson Hodel, "Coal Dust Sample Cheating Not New," Associated Press, April 21, 1991.

170 **efforts stalled during the 1980s:** U.S. General Accounting Office, "Mine Safety and Health: Tampering Scandal Led to Improved Sampling Devices," February 25, 1993, https://www.gao.gov/assets/220/217479.pdf.

171 **In depositions and interviews, miners who worked at Elk Run:** Particularly relevant were depositions taken as part of a lawsuit filed by Ricky Tucker (*Ricky Tucker and Deborah Tucker v. Consolidation Coal Company, et al.,* Civil Action No. 11-C-78, Circuit Court of Wyoming County, WV), who was diagnosed with black lung after working about thirty years in the mines for subsidiaries of Consolidation Coal, Maben Energy, and Massey Energy. He worked at the Elk Run complex during roughly the same period that Gary Fox did. Numerous Elk Run miners and managers were questioned during depositions about the conditions at Elk Run at the time.

174 **"Do not cut any overcasts"**: Superseding indictment, *United States of America v.*

Donald L. Blankenship, U.S. District Court for the Southern District of West Virginia, 5:14-cr-00244, March 10, 2015, 25.

174 **"We'll worry about ventilation or other issues":** Ibid., 23.

174 **Blankenship recorded some of his phone conversations:** Ken Ward Jr., "Defense Resumes Fight Over Blankenship Phone Recordings," *Charleston Gazette-Mail,* October 8, 2015, https://www.wvgazettemail.com/news/defense-resumes-fight-over-blankenship-phone-recordings/article_7ae60dd3-8ff5-5b2f-a509-d081782a9514.html; Wendy Holdren, "Jurors Hear More Recorded Phone Calls from Blankenship," *Beckley Register-Herald,* October 14, 2015, https://www.register-herald.com/news/jurors-hear-more-recorded-phone-calls-from-blankenship/article_faf2ba20-c976-505b-98e7-ed42219c11b9.html; Ken Ward Jr., "Black Lung Sampling at Issue in Blankenship Trial," *Charleston Gazette-Mail,* November 28, 2015, https://www.wvgazettemail.com/news/black-lung-sampling-at-issue-in-blankenship-trial/article_c5b7b1a6-f741-5edb-a3f6-81dd9644e4e4.html. The recording itself is available on YouTube: https://www.youtube.com/watch?v=8dBlQCX2XSM&feature=youtu.be.

175 **"weakness, lack of concern, high non-compliance":** Memorandum from Bill Ross to Stephanie Ojeda, senior corporate counsel for Massey Coal Services, subject line "Recommended Measures for Reduction of MSHA Violations," July 2, 2009. On July 6, 2009, Ojeda forwarded the memo by e-mail to Don Blankenship and other Massey officials.

175 **"still cheating on the Respirable Dust Sampling":** E-mail from Bill Ross to Gary Frampton, subject line "Hazard Elimination," January 7, 2010.

175 **memo to Blankenship in 2009:** Memorandum from Stephanie Ojeda to Don Blankenship (with two other Massey officials copied), subject line "Report of June 17, 2009 Meeting with Bill Ross Regarding MSHA Violations," June 25, 2009.

176 **knew the rules well and used them to their advantage:** For an explanation of the respirable dust-sampling rules as they existed prior to changes in 2014, see "Report of the Secretary of Labor's Advisory Committee on the Elimination of Pneumoconiosis Among Coal Mine Workers," October 1996, https://arlweb.msha.gov/S&HINFO/BlackLung/1996Dust%20AdvisoryReport.pdf; "Review of the Program to Control Respirable Coal Mine Dust in the United States," Report of the Coal Mine Respirable Dust Task Group, June 1992, https://arlweb.msha.gov/s&hinfo/blacklung/Reports/1992ReviewCoalMineDust.pdf.

177 **Elk Run mines were frequently on reduced standards:** My analysis of MSHA dust-sampling data obtained under FOIA.

177 **federal investigators discovered it at one Massey mine:** Mine Safety and Health Administration, *Internal Review of MSHA's Actions at the Upper Big Branch Mine–South, Performance Coal Company, Montcoal, Raleigh County, West Virginia* (March 2012), https://arlweb.msha.gov/PerformanceCoal/UBBInternalReview/UBBInternalReviewReport.pdf.

178 **"Very consistent themes emerged":** Laura E. Reynolds et al., "Work Practices and

Respiratory Health Status of Appalachian Coal Miners with Progressive Massive Fibrosis," *Journal of Occupational and Environmental Medicine* 60, no. 11 (2018): 575–81, https://www.ncbi.nlm.nih.gov/pmc/articles/PMC6607434/.

CHAPTER 15: WINDOW

Interviews: Kathryn South, Robert A. Cohen, Cecile Rose, Grant Crandall, Steve Sanders, Tom Johnson, Brenda Halsey Marion

181 **the hang of lobbying on a budget:** Most of the details in this chapter about the activities of the Black Lung Association are drawn from interviews with members and from documents they retained over the years and shared with me, including meeting minutes, sign-in sheets, newsletters, photographs, and fliers.

183 **The legislation Congress was considering:** U.S. House of Representatives, *Black Lung Benefits Restoration Act of 1994,* HR 2108, 103rd Cong., introduced in House May 12, 1993, https://www.congress.gov/bill/103rd-congress/house-bill/2108; U.S. Senate, *Black Lung Benefits Restoration Act,* S 1781, 103rd Cong., introduced in Senate November 23, 1993, https://www.congress.gov/bill/103rd-congress/senate-bill/1781.

183 **"coal miners' lottery":** Representative Richard Armey (R-TX) speaking on HR 2108, 103rd Cong., 2nd sess., *Congressional Record,* May 19, 1994, https://webarchive.loc.gov/congressional-record/20160315164745/http://thomas.loc.gov/cgi-bin/query/C?r103:./temp/~r103C7kSgk.

183 **"extra pension":** Representative John Boehner (R-OH) speaking on HR 2108, 103rd Cong., 2nd sess., *Congressional Record,* May 19, 1994, https://webarchive.loc.gov/congressional-record/20160315164716/http://thomas.loc.gov/cgi-bin/query/C?r103:./temp/~r103R1reZD.

183 **"new and excessive costs":** Representative Craig Thomas (R-WY) speaking on HR 2108, 103rd Cong., 2nd sess., *Congressional Record,* May 19, 1994, https://webarchive.loc.gov/congressional-record/20160315164716/http://thomas.loc.gov/cgi-bin/query/C?r103:./temp/~r103R1reZD.

185 **the task force had issued a report:** "Review of the Program to Control Respirable Coal Mine Dust in the United States," Report of the Coal Mine Respirable Dust Task Group, June 1992, https://arlweb.msha.gov/s&hinfo/blacklung/Reports/1992ReviewCoalMineDust.pdf.

185 **similar recommendations in a 1995 document:** National Institute for Occupational Safety and Health, *Criteria for a Recommended Standard: Occupational Exposure to Respirable Coal Mine Dust* (Cincinnati, OH, September 1995), https://www.cdc.gov/niosh/docs/95-106/pdfs/95-106.pdf?id=10.26616/NIOSHPUB95106.

185 **report from an advisory committee:** Mine Safety and Health Administration, "Report

of the Secretary of Labor's Advisory Committee on the Elimination of Pneumoconiosis Among Coal Mine Workers," October 1996, https://arlweb.msha.gov/S&HINFO/BlackLung/1996Dust%20AdvisoryReport.pdf.

186 **"a massive undertaking":** Memorandum from Donald S. Shire, associate solicitor for black lung benefits, to J. Davitt McAteer, assistant secretary for Mine Safety and Health, subject line "Proposed Revision of the Black Lung Program Regulations," March 25, 1996.

186 **proposed a rule that incorporated many of the Black Lung Association's suggestions:** Employment Standards Administration, "Proposed Rule: Regulations Implementing the Federal Coal Mine Health and Safety Act of 1969, as Amended," *Federal Register* 62, no. 14 (January 22, 1997): 3338–435, https://www.govinfo.gov/content/pkg/FR-1997-01-22/pdf/97-44.pdf.

187 **final rule implementing the reforms to the benefits program:** Employment Standards Administration, "Final Rule: Regulations Implementing the Federal Coal Mine Health and Safety Act of 1969, as Amended," *Federal Register* 65, no. 245 (December 20, 2000): 79920-80107, https://www.govinfo.gov/content/pkg/FR-2000-12-20/pdf/00-31166.pdf.

187 **The National Mining Association sued:** *National Mining Association v. Secretary of Labor,* 153 F.3d 1264 (11th Cir. 1998).

187 **proposing two prevention-related rules:** Mine Safety and Health Administration, "Proposed Rule: Verification of Underground Coal Mine Operators' Dust Control Plans and Compliance Sampling for Respirable Dust," *Federal Register* 65, no. 131 (July 7, 2000): 42122–85, https://www.govinfo.gov/content/pkg/FR-2000-07-07/pdf/00-16149.pdf#page=1; Mine Safety and Health Administration and Centers for Disease Control and Prevention, "Proposed Rule: Determination of Concentration of Respirable Coal Mine Dust," *Federal Register* 65, no. 131 (July 7, 2000): 42068–122, https://www.govinfo.gov/content/pkg/FR-2000-07-07/pdf/00-14075.pdf#page=2.

188 **"basically a coal-fired victory":** Tom Hamburger, "A Coal-Fired Crusade Helped Bring Crucial Victory to Candidate Bush," *Wall Street Journal,* June 13, 2001, https://www.wsj.com/articles/SB992378085878375783?mg=id-wsj.

188 **a finding published in the *Federal Register* three years into his term:** Mine Safety and Health Administration and the Centers for Disease Control and Prevention, "Proposed Rule: Determination of Concentration of Respirable Coal Mine Dust," *Federal Register* 68, no. 44 (March 6, 2003): 10940–52, https://www.govinfo.gov/content/pkg/FR-2003-03-06/pdf/03-5402.pdf; Mine Safety and Health Administration, "Proposed Rule: Verification of Underground Coal Mine Operators' Dust Control Plans and Compliance Sampling for Respirable Dust," *Federal Register* 68, no. 44 (March 6, 2003): 10784–884, https://www.govinfo.gov/content/pkg/FR-2003-03-06/pdf/03-3941.pdf#page=2.

CHAPTER 16: CROOKED RAFTER

Interviews: Gregory Fino, Edward Terhune Miller, Robert F. Cohen Jr., Patsy Caldwell, Arland Griffith, Daniel Richmond, Tom Johnson

190 **what Jackson Kelly lawyer Mary Rich Maloy did:** Many of the details throughout this chapter about the claims of Charles Caldwell, Calvin Cline, and William Harris are drawn from their respective case files.

191 **a miner had to prove four things:** "Determining Entitlement to Benefits," *Code of Federal Regulations* title 20, chapter VI, subchapter B, part 718, subpart C, https://www.ecfr.gov/cgi-bin/text-idx?SID=fdf45409c8c25b52e1f61f099efa4223&mc=true&node=sp20.4.718.c&rgn=div6.

192 **"evidence of bias":** *Charles Harris v. Shamrock Coal Company, Inc.,* Claim No. 1997-BLA-00611, July 22, 2002, https://www.oalj.dol.gov/Decisions/ALJ/BLA/1997/HARRIS_CHARLES_v_SHAMROCK_COAL_CO_DIR-OWCP_1997BLA00611_(JUL_22_2002)_114112_CADEC_SD.PDF, 5 (footnote).

192 **"bias against the claimant":** *Joe Hurt v. Locust Grove Coal Company,* Claim No. 1998-BLA-00336, September 28, 2001, https://www.oalj.dol.gov/Decisions/ALJ/BLA/1998/HURT_JOE_v_LOCUST_GROVE_COAL_CO_DIR_1998BLA00336_(SEP_28_2001)_091624_CADEC_SD.PDF, 3.

192 **"bias...in failing to attribute changes to coal dust exposure":** *Bobby G. Yates v. Calico Coal Company,* Claim No. 2000-BLA-00097, November 4, 2002, https://www.oalj.dol.gov/Decisions/ALJ/BLA/2000/YATES_BOBBY_G_v_CALICO_COAL_CO_DIR_2000BLA00097_(NOV_04_2002)_132805_CADEC_SD.PDF, 25.

192 **"flip flopping and embracing a discarded theory":** *L. P. v. Consolidation Coal Company,* Claim No. 2004-BLA-05226, June 10, 2009, https://www.oalj.dol.gov/DECISIONS/ALJ/BLA/2004/LP_v_CONSOLIDATION_COAL_C_2004BLA05226_(JUN_10_2009)_140732_CADEC_SD.PDF, 6.

192 **Fino had cowritten a lengthy report for the National Mining Association:** Gregory J. Fino and B. J. Bahl, "Proposed Changes to the Regulations Implementing the Federal Black Lung Benefits Act: Comments on the Medical and Scientific Issues Presented in the Department of Labor's Proposal Dated October 8, 1999," December 1999.

192 **"I don't believe in that anymore":** *William Hercules v. Consolidation Coal Company,* Claim No. 2001-BLA-00469, April 1, 2002, https://www.oalj.dol.gov/Decisions/ALJ/BLA/2001/hercules_william_v_consoldation_coal_co_2001bla00469_(apr_01_2002)_135223_cadec_sd.PDF, 19.

192 **The board had investigated him:** In response to my request under the Kentucky Open Records Act, the Kentucky Board of Medical Licensure provided partially redacted

copies of documents related to the miners' complaints, including correspondence, an investigation memorandum, and photographs.

195 **Dr. Jerome F. Wiot:** Curriculum vitae of Jerome F. Wiot, MD; Otha Linton, "Jerome F. Wiot," *Journal of the American College of Radiology* 9, no. 7 (2012): 523, https://www.jacr.org/article/S1546-1440(12)00005-1/fulltext.

CHAPTER 17: "ALL OF THE EVIDENCE"

Interviews: Mary Fox, Terri Smith, Edward Terhune Miller

207 **"Right-upper-lobe chest mass consistent with lung cancer":** Most details about Gary Fox's health and treatment are drawn from medical records that became part of his Labor Department case file as well as additional medical records that Mary Fox shared with me.

208 **Gary sent the paperwork to the Labor Department office:** Most details about Gary's black lung–benefits claim are drawn from his Labor Department case file, which includes materials such as pleadings, reports, and transcripts.

209 **The connection between Massey and Jackson Kelly ran deep:** Laurence Leamer, *The Price of Justice: A True Story of Greed and Corruption* (New York: St. Martin's, 2013), 17, 119, 361, 368, 378; Michael Shnayerson, *Coal River: How a Few Brave Americans Took On a Powerful Company—and the Federal Government—to Save the Land They Love* (New York: Farrar, Straus and Giroux, 2008), 17; "Personnel File: Harvey Returns to Jackson Kelly," *West Virginia Record,* September 1, 2011, https://wvrecord.com/stories/510601202-personnel-file-harvey-returns-to-jackson-kelly; Ken Ward Jr., "Before Slurry Deal, Records Outlined Massey Pollution," *Charleston Gazette,* June 9, 2012, https://www.wvgazettemail.com/news/special_reports/before-slurry-deal-records-outlined-massey-pollution/article_2bcface7-bb4a-5398-8b8c-5aa07088bc85.html; "Massey Wins Lawsuit Over Coal Dust at School," Associated Press, March 25, 2011.

PART THREE

CHAPTER 18: "CREATIVE APPROACHES"

Interviews: Tammy Cline, Bob Bastress, Forest Bowman

222 **fees to the miner's attorney in a successful claim:** "Information for Attorneys and Representatives: Black Lung Cases," U.S. Department of Labor, https://www.dol.gov/appeals/attorneys_black_lung.htm. The issue of fees for lay representatives is more complicated and has been addressed in specific cases, such as *Charles Kuhn v. Kenley*

Mining Company, decided by the Fourth Circuit Court of Appeals on April 4, 2002 (http://www.ca4.uscourts.gov/Opinions/Unpublished/012255.U.pdf).

223 **He applied to West Virginia University:** John shared with me his academic records and law-school application.

226 ***Great Coastal Express:*** *Great Coastal Express v. International Brotherhood of Teamsters, Chauffeurs, Warehousemen, and Helpers of America,* 675 F.2d 1349 (4th Cir. 1982).

CHAPTER 19: "GO GET 'EM"

Interviews: William DeShazo, Robert F. Cohen

229 **One of John's first new cases:** Most of the details about the claims of William and Laverne DeShazo, Mike and Edna Renick, and Elmer Daugherty are drawn from their respective case files, which include pleadings, medical records, and hearing transcripts, among other documents.

231 **the firm's lawyers took a particularly bold stance:** *Fred B. Wood v. Elkay Mining Company,* Claim No. 2001-BLA-00701; *Billy D. Williams v. Consolidation Coal Company,* Claim No. 2003-BLA-05329.

242 **the district's chief judge, David Faber:** *In Re Jackson Kelly PLLC,* U.S. District Court for the Southern District of West Virginia, Civil Action 2:05-0853, memorandum opinion issued August 30, 2006.

CHAPTER 20: LOST TIME

Interviews: Mary Fox, Terri Smith, Carter Stump, Frank Ratcliff, Marshall Litton, Rick Smith

244 **people living in the nearby town of Sylvester:** Kari Lydersen, "West Virginia Town Fights Blanket of Coal Dust," *New Standard,* May 9, 2006, http://newstandardnews.net/content/index.cfm/items/3140; Kari Lydersen, "Sylvester's Dustbusters," *New Standard,* May 9, 2006, http://newstandardnews.net/content/index.cfm/items/3141; "W. Va. Town Divided Over Coal Plant Suit," Associated Press, January 23, 2003.

245 **"It looked like the sun was even covered":** Lydersen, "West Virginia Town Fights Blanket of Coal Dust."

245 **A trade publication dubbed the structure:** "Don Dome Completed at Elk Run Operations in WV," *U.S. Coal Review,* July 15, 2002; "Dust in the Wind as Massey Elk Run Unit Awaits Verdict," *U.S. Coal Review,* February 10, 2003; "Jury Finds Elk Run Liable for Damages, Not Reckless," *U.S. Coal Review,* February 17, 2003.

245 **"One person's intolerable nuisance":** "Dust in the Wind," *U.S. Coal Review.*

245 **"They're afraid that if Massey loses this":** "W. Va. Town Divided," Associated Press.

245 **jurors found that Elk Run had failed to comply:** Martha Bryson Hodel, "Jury Finds

Massey Subsidiary Liable in Coal Dust Case," Associated Press, February 7, 2003; "Jury Finds Elk Run Liable for Damages," *U.S. Coal Review*.

246 **company agreed to an ongoing dust-monitoring program:** "Elk Run Agrees to Sylvester Dust Protections," Associated Press, February 27, 2003; "Dome over Coal Stockpile at Massey Mine Tears, Deflates," Associated Press, February 7, 2003; "Replacement Dome Rips at Massey Coal Stoker Plant," Associated Press, April 23, 2003; "Judge Orders Massey to Pay Sylvester Residents' Legal Fees," Associated Press, September 19, 2003.

CHAPTER 21: FRAUD ON THE COURT

Interviews: John Cline, Mary Fox, Terri Smith, Gloria Roles, Richard Roles, John Smallwood, Edward Terhune Miller, Tammy Cline, Rick Smith

250 **DID J&K SUBMIT PATHOLOGY?:** John Cline shared with me the detailed notes he kept about his interactions with Gary Fox and his work on Gary's case.

251 **a blueprint of the defense that Jackson Kelly had assembled:** Many of the details in this chapter are drawn from Gary Fox's case file, which includes medical records, pleadings, expert reports, and transcripts, among other things. Some details of Gary's medical care are drawn from additional records that Mary Fox shared with me.

252 **Edsil Keener:** Ruling and Order on Claimant's Motion to Compel and Employer's Motion for Protective Order, *Edsil L. Keener v. Peerless Eagle Coal Co.*, Claim No. 2004-BLA-06265, April 12, 2005, https://www.oalj.dol.gov/DECISIONS/ALJ/BLA/2004/KEENER_SHIRLEY_SWID_v_PEERLESS_EAGLE_COAL__2004BLA06265_(APR_12_2005)_135412_ORDER_PB.PDF; Decision and Order-Denying Benefits, *Edsil L. Keener v. Peerless Eagle Coal Co.*, Claim No. 2004-BLA-06265, August 23, 2005, https://www.oalj.dol.gov/DECISIONS/ALJ/BLA/2004/KEENER_SHIRLEY_SWID_v_PEERLESS_EAGLE_COAL__2004BLA06265_(AUG_23_2005)_172747_CADEC_SD.PDF; Decision and Order on Remand-Denying Benefits, *Edsil L. Keener v. Peerless Eagle Coal Co.*, Claim No. 2004-BLA-06265, December 4, 2007, https://www.oalj.dol.gov/DECISIONS/ALJ/BLA/2004/SK_v_PEERLESS_EAGLE_COAL__2004BLA06265_(DEC_04_2007)_143236_CADEC_SD.PDF.

255 **$876.50 a month:** "Benefit Rates Under Part C, 1973-2019," Office of Workers' Compensation Programs, Division of Coal Mine Workers' Compensation, https://www.dol.gov/owcp/dcmwc/benefits_part_c.htm.

261 **they were simply better qualified:** Curricula vitae of Richard L. Naeye and P. Raphael Caffrey as well as my review of hundreds of decisions by administrative law judges.

262 **Koh had spent much of his career:** Curriculum vitae of Sukjung Gerald Koh; medical-license records from the West Virginia Board of Medicine; Bev Davis, "Pathologist Remembered for Kindness, Compassion," *Beckley Register-Herald*, February

10, 2009, https://www.register-herald.com/news/local_news/pathologist-remembered-for-kindness-compassion/article_fcb5f362-6489-50c5-a21f-9dda4919e23f.html.

263 **Letters from Jackson Kelly to three pulmonologists:** These and other documents later became public as part of the civil lawsuits filed against Jackson Kelly in Raleigh County Circuit Court by the families of miners Gary Fox, Norman Eller, and Clarence Carroll. These cases, filed in 2009, are described later in this book.

265 **the seminal 1944 case:** *Hazel-Atlas Glass Co. v Hartford-Empire Co.*, 322 U.S. 238 (1944).

PART FOUR: INTRODUCTION

Interviews: Francis Green, Donald Rasmussen, John Cline, Craig Robinson

274 **a study he and eight fellow physicians had just finished:** Robert A. Cohen et al., "Lung Pathology in U.S. Coal Workers with Rapidly Progressive Pneumoconiosis Implicates Silica and Silicates," *American Journal of Respiratory and Critical Care Medicine* 193, no. 6 (March 2016): 673–80, https://www.atsjournals.org/doi/pdf/10.1164/rccm.201505-1014OC.

274 **average number of hours a miner worked during a year:** Eva Suarthana et al., "Coal Workers' Pneumoconiosis in the United States: Regional Differences 40 Years After Implementation of the 1969 Federal Coal Mine Health and Safety Act," *Occupational and Environmental Medicine* 68, no. 12 (2011): 912, https://oem.bmj.com/content/68/12/908.

274 **more than twice as much disease as the dust samples suggested there should be:** Ibid., 909–11.

275 **the proportion of films with this particular marker had almost quadrupled:** A. Scott Laney, Edward L. Petsonk, and Michael D. Attfield, "Pneumoconiosis Among Underground Bituminous Coal Miners in the United States: Is Silicosis Becoming More Frequent?," *Occupational and Environmental Medicine* 67, no. 10 (2010), https://oem.bmj.com/content/67/10/652.

277 **article detailed his role as a reluctant hero for miners:** Sam Roberts, "Dr. Donald L. Rasmussen, Crusader for Coal Miners' Health, Dies at 87," *New York Times,* August 2, 2015, https://www.nytimes.com/2015/08/03/health/research/dr-donald-l-rasmussen-crusader-for-coal-miners-health-dies-at-87.html.

CHAPTER 22: "AN AFFRONT TO JUSTICE"

Interviews: John Cline, Mary Fox, Robert F. Cohen, Al Karlin, Dolores Carroll, Norman Eller Jr.

279 **His client was now Mary Fox:** Details about the claims of Gary and Mary Fox are largely drawn from the documents contained in their case files.

281 **accusing the firm of engaging in fraud:** *Mary L. Fox v. Jackson Kelly PLLC,* Circuit Court of Raleigh County, Civil Action No. 09-C-497-K, May 21, 2009.

282 **had also given the lawyers the go-ahead to sue:** *Clarence O. Carroll and Dolores Carroll v. Jackson Kelly PLLC, Rundle & Rundle LC, S.F. Raymond Smith, and Joni Cooper Rundle,* Circuit Court of Raleigh County, Civil Action No. 09-C-614-K, June 24, 2009; *Norman Dale Eller v. Jackson Kelly PLLC, Rundle & Rundle LC, S.F. Raymond Smith, and Joni Cooper Rundle,* Circuit Court of Raleigh County, Civil Action No. 09-C-483-K, May 19, 2009.

282 **Clarence Carroll was eighty-five years old:** Details about Carroll's claim are drawn largely from the documents contained in his case file.

283 **The other case was that of Norman Eller:** Details about Eller's claim are drawn largely from the documents contained in his case file.

286 **"At first blush," Kirkpatrick wrote:** "Order Denying Defendant Jackson Kelly, PLLC's Motions to Dismiss," for consolidated cases *Carroll, Eller,* and *Fox,* issued May 30, 2012.

287 **"So there's going to be some tension in this room":** Most of the details about the two-day hearing are drawn from the official transcript of the proceedings, which took place in Charleston, West Virginia, on June 18 and 19, 2009.

287 **Sanctions up to disbarment were possible:** For more on the West Virginia Office of Disciplinary Counsel's Lawyer Disciplinary Board, see http://www.wvodc.org/comp-packet.pdf.

289 **provisions of the state's ethical standards:** See rules 3.4 and 8.4 of the West Virginia Rules of Professional Conduct at http://www.courtswv.gov/legal-community/court-rules/professional-conduct/contents.html.

290 **he had called the document the "exam report":** These details about Zaldivar's reports and that of another doctor who noted the absence of the "usual summary letter" are drawn from documents in Elmer Daugherty's case file.

291 **They circulated his written opinion by e-mail:** The Department of Labor produced these e-mails in response to my Freedom of Information Act request.

293 **The panel issued its report on March 30, 2010:** Report of the Hearing Panel Subcommittee, *In Re: Douglas A. Smoot,* Before the Lawyer Disciplinary Board of West Virginia, March 30, 2010.

294 **The court issued its decision on November 17, 2010:** *Lawyer Disciplinary Board v. Douglas A. Smoot,* No. 34724, Supreme Court of Appeals of West Virginia, November 17, 2010, http://www.courtswv.gov/supreme-court/docs/fall2010/34724.htm.

296 **The investigative panel declined to file charges:** "Lawyer Disciplinary Board Investigative Panel Findings and Conclusions Regarding Complaints Against William S. Mattingly, Kathy L. Snyder, and Dorothea J. Clark." The panel voted at an October 30, 2010, meeting not to take further action against the three Jackson Kelly attorneys.

297 **"I represent the widow of Gary Fox":** An audio recording of the oral argument is available at the court's website: http://www.ca4.uscourts.gov/OAarchive/mp3/12-2387-20131029.mp3.

297 **Judge J. Harvie Wilkinson III:** "Judge J. Harvie Wilkinson III," U.S. Court of Appeals for the Fourth Circuit, http://www.ca4.uscourts.gov/judges/judges-of-the-court/judge-j-harvie-wilkinson-iii; Neil A. Lewis, "In List of Potential Justices, Many Kinds of Conservative," *New York Times,* July 2, 2005, https://www.nytimes.com/2005/07/02/politics/politicsspecial1/in-list-of-potential-justices-many-kinds-of.html; Elisabeth Bumiller, "White House Letter: The Executive in Chief Likes to Break the Ice," *New York Times,* July 25, 2005, https://www.nytimes.com/2005/07/25/world/americas/white-house-letter-the-executive-in-chief-likes-to-break-the.html.

299 **The unanimous opinion, written by Wilkinson:** *Mary L. Fox v. Elk Run Coal Company,* No. 12-2387 (4th Cir. 2014), http://www.ca4.uscourts.gov/Opinions/Published/122387.P.pdf.

CHAPTER 23: "LACK OF SCIENTIFIC INDEPENDENCE"

Interviews: Steve Day, Nyoka Day, Patience Day, Stepheny Day

302 **he had chosen Dr. George Zaldivar to perform the exam:** Many of the details about Steve Day's medical and work histories and black lung–benefits claims are drawn from the Labor Department case file.

306 **roughly 1,500 cases in which he'd been involved:** To remind the reader: I identified these cases using the full-text search feature on the website of the Office of Administrative Law Judges, which includes final decisions in cases decided since 2000. See https://search.oalj.dol.gov/.

310 **first installment of the series:** Chris Hamby, "Coal Industry's Go-To Law Firm Withheld Evidence of Black Lung, at Expense of Sick Miners," Center for Public Integrity, October 29, 2013, https://publicintegrity.org/environment/coal-industrys-go-to-law-firm-withheld-evidence-of-black-lung-at-expense-of-sick-miners/.

310 **the center released part two:** Chris Hamby, Brian Ross, and Matthew Mosk, "Johns Hopkins Medical Unit Rarely Finds Black Lung, Helping Coal Industry Defeat Miners' Claims," Center for Public Integrity, October 30, 2013, https://publicintegrity.org/environment/johns-hopkins-medical-unit-rarely-finds-black-lung-helping-coal-industry-defeat-miners-claims/.

310 **ABC aired its piece on Wheeler and Johns Hopkins:** Brian Ross et al., "For Top-Ranked Hospital, Tough Questions About Black Lung and Money," ABC News, October 30, 2013, https://abcnews.go.com/Blotter/investigation-johns-hopkins-tough-questions-black-lung-money/story?id=20721430.

311 **final installment of the series:** Chris Hamby, "As Experts Recognize New Form of Black Lung, Coal Industry Follows Familiar Pattern of Denial," Center for Public Integrity, November 1, 2013, https://publicintegrity.org/environment/as-experts-recognize-new-form-of-black-lung-coal-industry-follows-familiar-pattern-of-denial/.

CHAPTER 24: MAKING GOOD

Interviews: Patricia Smith, Joe Main, Dewey Keiper, Al Karlin, John Cline, Dolores Carroll, Robert Bailey Jr., Steve Day, Nyoka Day, Stepheny Day, Patience Day, Francis Green, David Weissman, Jack Parker

312 **the traditionally conservative *Charleston Daily Mail* weighed in:** "Our Views: Black Lung Simply Is Not Acceptable; Nation That Reduced Explosions Can Do Better on Pneumoconiosis," *Charleston Daily Mail,* October 31, 2013.

312 **"shocking and repugnant":** "Black-Lung Victims Cheated; Industry Avoids Paying Miners," *Lexington Herald-Leader,* November 12, 2013.

312 **"We hope this new uproar brings further industry reforms":** "Black Lung; Ugly Allegations," *Charleston Gazette,* November 6, 2013.

313 **"Miners Battle Black Lung, and Bureaucracy":** "Miners Battle Black Lung, and Bureaucracy," *New York Times,* September 7, 2013, https://www.nytimes.com/2014/09/08/opinion/miners-battle-black-lung-and-bureaucracy.html.

313 **Solicitor of Labor Patricia Smith:** Jennifer Frey, "A Labor Force of One," *NYU Law Magazine,* 2010, https://blogs.law.nyu.edu/magazine/2010/m-patricia-smith-77-profile/; "Patricia Smith," National Employment Law Project, https://www.nelp.org/expert/patricia-smith/.

313 **confirmed her as solicitor of labor after a bruising, yearlong fight:** "Nominations: 111th Congress: PN319—M. Patricia Smith—Department of Labor," https://www.congress.gov/nomination/111th-congress/319?q=%7B; "Ranking Member Report: The Nomination of M. Patricia Smith of New York, Nominee to Serve as Solicitor of Labor, U.S. Department of Labor: Accuracy of Senate Testimony," U.S. Senate Committee on Health, Education, Labor and Pensions, February 1, 2010, https://www.help.senate.gov/chair/newsroom/press/ranking-member-report-the-nomination-of-m-patricia-smith-of-new-york-nominee-to-serve-as-solicitor-of-labor-us-department-of-labor-accuracy-of-senate-testimony; James Sherk, "Don't Confirm Patricia Smith," Fox News, October 13, 2009, https://www.foxnews.com/opinion/dont-confirm-patricia-smith; "Editorial: Ms. Smith and the Washington Game," *New York Times,* September 21, 2009, https://www.nytimes.com/2009/09/22/opinion/22tue2.html.

314 **One long-sought reform actually had been accomplished already:** Office of Workers' Compensation Programs, "Regulations Implementing the Byrd Amendments to the Black Lung Benefits Act: Determining Coal Miners' and Survivors' Entitlement to Benefits," *Federal Register* 78, no. 186 (September 25, 2013): 59102–19, https://www.dol.gov/owcp/dcmwc/Final_Byrd_Amendment_Rules_Federal_Register_Print.pdf.

315 **warned that Byrd's proposal would be a "job killer":** Steve Hooks, "Byrd Black Lung Amendment to Health Care Bill Unsettles Industry on Costs," *Platts Coal Trader,*

January 22, 2010; "Our Views: Black Lung Provision Could Do Real Damage; Did Sen. Byrd Fully Consider the Consequences of His Handiwork?," *Charleston Daily Mail,* March 24, 2010; "Byrd Applauds Inclusion of Provisions to Help Victims of Black Lung in Health Care Bill," press release from Senator Robert Byrd's office, March 22, 2010.

316 **stage set up at one end of the atrium:** I attended this April 23, 2014, ceremony.

316 **published a proposed rule:** Mine Safety and Health Administration, "Lowering Miners' Exposure to Respirable Coal Mine Dust, Including Continuous Personal Dust Monitors; Proposed Rule," *Federal Register* 75, no. 201 (October 19, 2010): 64412–506, https://www.govinfo.gov/content/pkg/FR-2010-10-19/pdf/2010-25249.pdf.

317 **the reaction from the coal industry was fierce:** U.S. Government Accountability Office, *Mine Safety: Reports and Key Studies Support the Scientific Conclusions Underlying the Proposed Exposure Limit for Respirable Coal Mine Dust,* GAO-12-832R, August 17, 2012, https://www.gao.gov/assets/600/593780.pdf; Chris Hamby, "GAO Report Supports Science Behind Black Lung Rule," Center for Public Integrity, August 17, 2012, https://publicintegrity.org/2012/08/17/10712/gao-report-supports-science-behind-black-lung-rule; U.S. Government Accountability Office, *Mine Safety: Basis for Proposed Exposure Limit on Respirable Coal Mine Dust and Possible Approaches for Lowering Dust Levels,* GAO-14-345, April 2014, https://www.gao.gov/assets/670/662410.pdf; Chris Hamby, "GAO Report Again Finds Black Lung Proposal Supported by Science," Center for Public Integrity, April 11, 2014, https://publicintegrity.org/environment/gao-report-again-finds-black-lung-proposal-supported-by-science/; Chris Hamby, "GOP Seeks to Kill Black Lung Reform," Center for Public Integrity, July 17, 2012, https://publicintegrity.org/workers-rights/gop-seeks-to-kill-black-lung-reform/.

318 **he had announced that he would retire:** John Bresnahan, "Jay Rockefeller to Retire," Politico, January 11, 2013, https://www.politico.com/story/2013/01/jay-rockefeller-to-retire-086054; Richard S. Grimes, "Jay Rockefeller," *West Virginia Encyclopedia,* https://www.wvencyclopedia.org/articles/110; "John D. Rockefeller IV," Council on Foreign Relations, https://www.cfr.org/expert/john-d-rockefeller-iv.

319 **the reforms his agency was implementing:** Mine Safety and Health Administration, "Lowering Miners' Exposure to Respirable Coal Mine Dust, Including Continuous Personal Dust Monitors; Final Rule," *Federal Register* 79, no. 84 (May 1, 2014): 24814–994, https://www.govinfo.gov/content/pkg/FR-2014-05-01/pdf/2014-09084.pdf.

320 **A coalition led by the National Mining Association sued the Labor Department:** *National Mining Association v. U.S. Department of Labor,* U.S. Court of Appeals for the Eleventh Circuit, No. 14-11942, January 25, 2016, http://media.ca11.uscourts.gov/opinions/pub/files/201411942.pdf.

320 **samples taken since the first two phases went into effect:** Mine Safety and Health Administration, "MSHA Finds Nearly All Respirable Coal Dust Samplings Comply with New

Standards to Lower Levels of Respirable Coal Dust," https://www.msha.gov/msha-finds-nearly-all-respirable-coal-dust-samplings-comply-new-standards-lower-levels-respirable.

321 **their most dire findings yet:** David J. Blackley, Cara N. Halldin, and A. Scott Laney, "Continued Increase in Prevalence of Coal Workers' Pneumoconiosis in the United States, 1970–2017," *American Journal of Public Health* 108, no. 9 (September 2018): 1220–22, https://www.ncbi.nlm.nih.gov/pmc/articles/PMC6085042/pdf/AJPH.2018.304517.pdf.

321 **uncover new clusters of disease:** David J. Blackley et al., "Resurgence of Progressive Massive Fibrosis in Coal Miners—Eastern Kentucky, 2016," *Morbidity and Mortality Weekly Report* 65, no. 49 (December 16, 2016): 1385–89, https://www.cdc.gov/mmwr/volumes/65/wr/mm6549a1.htm; David J. Blackley et al., "Progressive Massive Fibrosis in Coal Miners from 3 Clinics in Virginia," *Journal of the American Medical Association* 319, no. 5 (February 2018): 500–501, https://jamanetwork.com/journals/jama/fullarticle/2671456.

321 **tens of thousands still work underground:** U.S. Energy Information Administration, "Annual Coal Report," https://www.eia.gov/coal/annual/.

321 **a "retrospective study":** Mine Safety and Health Administration, "Retrospective Study of Respirable Coal Mine Dust Rule," *Federal Register* 83, no. 131 (July 9, 2018): 31710–11, https://www.govinfo.gov/content/pkg/FR-2018-07-09/pdf/2018-14536.pdf.

321 **bipartisan backlash from members of Congress:** Letter from Senators Joe Manchin III, Sherrod Brown, Tim Kaine, Mark Warner, and Bob Casey (all Democrats) to Labor Secretary Alexander Acosta, subject line "Re: Regulatory Reform of Existing Standards and Regulations: Retrospective Study of Respirable Coal Mine Dust Rule," December 22, 2017, https://www.manchin.senate.gov/imo/media/doc/DOL%20re%20Respirable%20Dust%20Rule.pdf?cb; letter from Senator Shelley Moore Capito (Republican from West Virginia) to Labor Secretary Alexander Acosta, subject line "Re: Retrospective Study of Respirable Coal Mine Dust Rule," December 22, 2017, https://www.capito.senate.gov/imo/media/doc/12-22-2017%20Letter%20to%20Secretary%20Acosta.pdf.

321 **no immediate plans to roll back the rule:** Mine Safety and Health Administration, "U.S. Department of Labor Seeks Input on Retrospective Study of Respirable Coal Mine Dust Rule," July 6, 2018, https://www.msha.gov/news-media/press-releases/2018/07/06/us-department-labor-seeks-input-retrospective-study-respirable; U.S. House of Representatives, Committee on Education and Labor, Subcommittee on Workforce Protections, *Reviewing the Policies and Priorities of the Mine Safety and Health Administration,* 115th Cong., 2nd sess., February 6, 2018, https://edlabor.house.gov/hearings/reviewing-the-policies-and-priorities-of-the-mine-safety-and-health-administration.

324 **Kirkpatrick rejected Jackson Kelly's argument:** "Order Denying Second Motion to Dismiss of Defendant, Jackson Kelly PLLC" for consolidated cases *Carroll, Eller,* and *Fox,* issued September 4, 2014.

325 **The hearing, titled *Coal Miners' Struggle for Justice*:** U.S. Senate, Committee on Health, Education, Labor, and Pensions, Subcommittee on Employment and Workplace Safety, *Coal Miners' Struggle for Justice: How Unethical Legal and Medical Practices Stack the Deck Against Black Lung Claimants,* 113th Cong., 2nd sess., July 22, 2014, https://www.govinfo.gov/content/pkg/CHRG-113shrg24452/pdf/CHRG-113shrg24452.pdf. Video available at https://www.help.senate.gov/hearings/coal-miners-struggle-for-justice-how-unethical-legal-and-medical-practices-stack-the-deck-against-black-lung-claimants.

325 **a pilot program to raise the quality of the medical reports:** Office of Workers' Compensation Programs, Division of Coal Mine Workers' Compensation, BLBA Bulletin No. 14-05, issued February 24, 2014, https://www.dol.gov/owcp/dcmwc/blba/indexes/BL14.05.pdf.

326 **directed lawyers in regional offices throughout the country to intervene:** Memorandum from the solicitor of labor to regional solicitors, "Subject: Black Lung Benefits Act Program Initiatives," February 24, 2014.

326 **a bulletin to its district directors:** Office of Workers' Compensation Programs, Division of Coal Mine Workers' Compensation, BLBA Bulletin No. 14-09, issued June 2, 2014, https://www.dol.gov/owcp/dcmwc/blba/indexes/BL14.09OCR.pdf.

326 **identified more than a thousand:** "Statement of Christopher P. Lu, Deputy Secretary, Department of Labor, Washington, D.C.," during the July 22, 2014, hearing, https://www.govinfo.gov/content/pkg/CHRG-113shrg24452/pdf/CHRG-113shrg24452.pdf, 8.

326 **"Regulatory changes are not off the table":** Chris Hamby, "Labor Department Unveils Changes to Aid Miners in Black Lung Benefits Cases," Center for Public Integrity, February 24, 2014, https://publicintegrity.org/environment/labor-department-unveils-changes-to-aid-miners-in-black-lung-benefits-cases/.

327 **A brief public notice:** "Black Lung Benefits Act: Medical Evidence and Benefit Payments," Spring 2014 Regulatory Agenda of the U.S. Department of Labor, https://www.reginfo.gov/public/do/eAgendaViewRule?pubId=201404&RIN=1240-AA10.

330 **He died on March 15, 2019:** Robert Bailey Jr.'s obituary from the *Bluefield Daily Telegraph* is available at https://obituaries.bdtonline.com/obituary/robert-bailey-jr-1073374403. Reporter Jessica Lilly of West Virginia Public Radio spoke with Bailey numerous times over the years and produced a wrenching remembrance of his final days; see https://www.wvpublic.org/post/black-lung-patient-fights-miner-benefits-his-final-days#stream/0. In September 2019, local organizers put together the first annual Robert Bailey Jr. Memorial 5K, with the proceeds benefiting miners in need; see https://www.ptonline.net/news/aiding-black-lung-victims-proceeds-from-robert-bailey-jr-k/article_9a1adb50-e13e-11e9-b400-f7d364672502.html.

330 **Steve's breathing impairment was "severe," "irreversible," and "totally disabling":** Details about Steve's claims and medical records are primarily drawn from his case file.

331 **he lay in an open casket with an American flag draped over it:** I attended Steve's funeral, which took place on August 1, 2014.

333 **Fourth Circuit affirmed the dismissal:** *Michael S. Day, Jr., v. Johns Hopkins Health System Corporation,* No. 17-2120, U.S. Court of Appeals for the Fourth Circuit, October 26, 2018, http://www.ca4.uscourts.gov/opinions/172120.P.pdf.

334 **Ramsey later won his claim:** *Larry W. Ramsey v. Eastern Associated Coal, LLC,* Claim No. 2012-BLA-06176, March 30, 2016, https://www.oalj.dol.gov/Decisions/ALJ/BLA/2012/RAMSEY_LARRY_WAYNE_v_EASTERN_ASS_COAL_CO__2012BLA06176_(MAR_30_2016)_123841_CADEC_SD.PDF.

334 **"opinions" and "not facts capable of accurate and ready determination":** *Flemon E. Salmons v. Pontiki Coal, LLC,* Claim No. 2013-BLA-05003, March 30, 2016, https://www.oalj.dol.gov/Decisions/ALJ/BLA/2013/SALMONS_FLEMON_ESS_v_PONTIKI_COAL_CORP_DI_2013BLA05003_(MAR_30_2016)_084144_CADEC_SD.PDF.

334 **"Unlike a criminal conviction":** *John T. May v. Mountaineer Coal Development,* Claim No. 2012-BLA-05618, January 19, 2017, https://www.oalj.dol.gov/Decisions/ALJ/BLA/2012/MAY_JOHN_T_v_MOUNTAINEER_COAL_DEV_2012BLA05618_(JAN_19_2017)_104401_CADEC_SD.PDF.

335 **"Dr. Wheeler's interpretation followed this pattern":** *Thomas Alexander v. Drummond Company Inc.,* Claim No. 2012-BLA-05380, February 10, 2015, https://www.oalj.dol.gov/Decisions/ALJ/BLA/2012/ALEXANDER_THOMAS_v_DRUMMOND_CO_INC_DIR__2012BLA05380_(FEB_10_2015)_102039_CADEC_SD.PDF.

335 **"reason for questioning the credibility of Dr. Wheeler's x-ray readings":** *Catherine D. Bartley v. Union Carbide Corporation,* Claim No. 2011-BLA-05309, May 13, 2015, https://www.oalj.dol.gov/Decisions/ALJ/BLA/2011/BARTLEY_CATHERINE_D_v_UNION_CARBIDE_CORP_a_2011BLA05309_(MAY_13_2015)_102619_CADEC_SD.PDF.

335 **"Dr. Wheeler here continues the narrative":** *Ilene Barr v. Eastern Associated Coal Corp.,* Claim No. 2012-BLA-06060 and 2013-BLA-05428, June 22, 2015, https://www.oalj.dol.gov/Decisions/ALJ/BLA/2012/BARR_ILLENE_0B0_JUNI_v_EASTERN_ASS_COAL_OWC_2012BLA06060_(JUN_22_2015)_093735_CADEC_SD.PDF.

335 **Since the explosion at Massey Energy's Upper Big Branch mine:** Clifford Krauss, "Under Fire Since Explosion, Mining C.E.O. Quits," *New York Times,* December 3, 2010, https://www.nytimes.com/2010/12/04/business/energy-environment/04massey.html; Michael Erman and Ann Saphir, "Alpha Agrees to Buy Massey Energy for About $7.1 Billion," Reuters, January 29, 2011, https://www.reuters.com/article/us-alpha-massey/alpha-agrees-to-buy-massey-energy-for-about-7-1-billion-idUSTRE70S0PC20110130; Ken Ward Jr., "Ex–Upper Big Branch Security Director Found Guilty," *Charleston Gazette,* October 26,

2011, https://www.wvgazettemail.com/news/special_reports/ex-upper-big-branch-security-director-found-guilty/article_cf05d9b2-3a62-5024-a5ee-92077b0a9307.html; Howard Berkes, "Massey Mine Boss Pleads Guilty as Feds Target Execs," NPR, March 29, 2012, https://www.npr.org/sections/thetwo-way/2012/03/29/149639345/massey-mine-boss-pleads-guilty-as-feds-target-execs; Ken Ward Jr., "Former Massey Official Pleads Guilty in Safety Probe, Says He Conspired with CEO," *Charleston Gazette,* February 28, 2013, https://www.wvgazettemail.com/news/special_reports/former-massey-official-pleads-guilty-in-safety-probe-says-he/article_babd2568-0c95-5284-84e3-f508da6dc096.html.

336 **an indictment against Blankenship:** Ken Ward Jr., "Longtime Massey Energy CEO Don Blankenship Indicted," *Charleston Gazette,* November 13, 2014, https://www.wvgazettemail.com/news/special_reports/longtime-massey-energy-ceo-don-blankenship-indicted/article_7c2346a2-b6d6-53f2-9627-53b26af8c7aa.html. Ward posted a copy of the indictment online; see https://www.documentcloud.org/documents/1502683-blankenship-indictment.html.

336 **a trial that spanned most of October and November 2015:** Reporter Ken Ward Jr. and his colleagues at the *Charleston Gazette-Mail* extensively covered Blankenship's trial and the aftermath; see https://www.wvgazettemail.com/news/special_reports/blankenship_trial/.

336 **Blankenship was charged with conspiring to violate:** This was included in a superseding indictment against Blankenship filed on March 10, 2015; see https://www.documentcloud.org/documents/1684603-blankenship-superseding-indictment.html. See also Howard Berkes, "Feds Add Coal-Dust Coverup Allegation to Mine CEO's Indictment," NPR, March 10, 2015, https://www.npr.org/sections/thetwo-way/2015/03/10/392207163/feds-add-coal-dust-allegation-to-mine-ceos-indictment; Ken Ward Jr., "Revised Indictment Filed Against Blankenship," *Charleston Gazette,* March 10, 2015, https://www.wvgazettemail.com/news/special_reports/revised-indictment-filed-against-blankenship/article_cdafe983-3870-58b5-911d-6014d0ba81c3.html.

336 **"if you wore these things the way that you should":** Ken Ward Jr., "Black Lung Sampling at Issue in Blankenship Trial," *Charleston Gazette-Mail,* November 28, 2015, https://www.wvgazettemail.com/news/black-lung-sampling-at-issue-in-blankenship-trial/article_c5b7b1a6-f741-5edb-a3f6-81dd9644e4e4.html.

337 **During his closing argument:** For the transcript of closing arguments on November 17, 2015, see https://www.documentcloud.org/documents/2515678-blankenship-trial-closing-arguments-transcript.html. See also Joel Ebert and Ken Ward Jr., "Jury Begins Deliberations After Blankenship Closing Arguments," *Charleston Gazette-Mail,* November 17, 2015, https://www.wvgazettemail.com/news/jury-begins-deliberations-after-blankenship-closing-arguments/article_6b3ce6ff-cace-5f94-91d5-c8f4557ffdf8.html.

337 **jurors reached their decision:** Ken Ward Jr., "Blankenship Guilty of Conspiring to Violate Mine Safety Rules," *Charleston Gazette-Mail,* December

3, 2015, https://www.wvgazettemail.com/news/blankenship-guilty-of-conspiring-to-violate-mine-safety-rules/article_cc9c8c49-bc00-52c1-a90f-04780ab26bd7.html; Alan Blinder, "Mixed Verdict for Donald Blankenship, Ex-Chief of Massey Energy, After Coal Mine Blast," *New York Times,* December 3, 2015, https://www.nytimes.com/2015/12/04/us/donald-blankenship-massey-energy-upper-big-branch-mine.html?module=inline.

337 **the penalty the judge imposed in April 2016:** Ken Ward Jr., "Blankenship Gets Maximum Sentence: One Year in Prison, $250,000 Fine," *Charleston Gazette-Mail,* April 6, 2016, https://www.wvgazettemail.com/news/blankenship-gets-maximum-sentence-one-year-in-prison-fine/article_c37d6bee-789e-5410-afa1-8629a259d7d8.html; Alan Blinder, "Donald Blankenship Sentenced to a Year in Prison in Mine Safety Case," *New York Times,* April 6, 2016, https://www.nytimes.com/2016/04/07/us/donald-blankenship-sentenced-to-a-year-in-prison-in-mine-safety-case.html.

338 **"In the long march of history, this is a significant change":** Ken Ward Jr., "Ex-Massey CEO Blankenship Now in California Federal Prison," *Charleston Gazette-Mail,* May 12, 2016, https://www.wvgazettemail.com/news/legal_affairs/ex-massey-ceo-blankenship-now-in-california-federal-prison/article_9bbb807a-a9b1-50eb-89cc-6b18d077d079.html.

338 **A federal court later denied his appeal:** *United States of America v. Donald L. Blankenship,* No. 16-4193, U.S. Court of Appeals for the Fourth Circuit, January 19, 2017, http://www.ca4.uscourts.gov/Opinions/Published/164193.P.pdf.

338 **Attorneys from the firm's black lung unit:** "Lawyer Disciplinary Board Investigative Panel Closing," order entered on February 5, 2015. The disciplinary office provided the order and associated correspondence in response to my records request.

339 **Al and John had filed discovery requests:** Because the pleadings regarding discovery in the Raleigh County Circuit Court case file are extensive, I haven't listed each filing here.

340 **At a hearing in Beckley on May 18:** I was not able to attend this hearing. The account of it is based on the official transcript and interviews with participants.

342 **order granting the motion to compel discovery:** "Order Granting Plaintiff's Motion to Compel Production of Documents," *Mary L. Fox v. Jackson Kelly PLLC,* Civil Action No. 09-C-497, Circuit Court of Raleigh County, July 12, 2016.

344 **It told of a man named Gary Fox:** Office of Workers' Compensation Programs, "Notice of Proposed Rulemaking: Black Lung Benefits Act: Disclosure of Medical Information and Payment of Benefits," *Federal Register* 80, no. 82 (April 29, 2015): 23743, https://www.govinfo.gov/content/pkg/FR-2015-04-29/pdf/2015-09573.pdf.

346 **joint submission by the National Mining Association and a group of insurers:** The National Mining Association, American Insurance Association, Old Republic Insurance Company, and the American Mining Insurance Company, "Comments on Proposed Revisions to 20 C.F.R. Part 725—Claims for Benefits Under Part C of Title IV of the Federal Mine Safety and Health Act as Amended," June 29, 2015.

346 **"Lack of information kills people":** Sandra Fogel et al., "Re: Proposed Regulations RIN No. 1240-AA10," June 29, 2015.

347 **The United Mine Workers and the Appalachian Citizens' Law Center:** United Mine Workers of America, "Comments on the Office of Workers' Compensation Programs' Proposed Rule, Black Lung Benefits Act: Disclosure of Medical Information and Payment of Benefits," undated; Stephen A. Sanders on behalf of the Appalachian Citizens' Law Center, "Comments to Proposed Regulations 'Black Lung Benefits Act: Disclosure of Medical Information and Payment of Benefits,'" June 29, 2015.

347 **"the tragic case of Mr. Gary Fox":** H. Morgan Griffith, letter to Secretary of Labor Thomas E. Perez, May 1, 2015.

347 **John's submission rebutted the contention:** John Cline, "Re: Office of Workers' Compensation Programs' Proposed Rule—RIN 1240-AA10, Black Lung Benefits Act: Disclosure of Medical Information & Payment of Benefits," June 29, 2015.

347 **"Black Lung Benefits Improvement Act of 2015":** U.S. Senate, *Black Lung Benefits Improvement Act of 2015,* S 2096, 114th Cong., 1st sess., introduced in Senate September 29, 2015, https://www.congress.gov/bill/114th-congress/senate-bill/2096; U.S. House, *Black Lung Benefits Improvement Act of 2015,* HR 3625, 114th Cong., 1st sess., introduced in house September 28, 2015, https://www.congress.gov/bill/114th-congress/house-bill/3625.

348 **an entry announcing the change John Cline had sought for so long:** Office of Workers' Compensation Programs, "Final Rule: Black Lung Benefits Act: Disclosure of Medical Information and Payment of Benefits," *Federal Register* 81, no. 80 (April 26, 2016): 24464, https://www.govinfo.gov/content/pkg/FR-2016-04-26/pdf/2016-09525.pdf.

349 **announced the finalization of the rule in a press release:** Office of Workers' Compensation Programs, "Black Lung Benefits Act Final Rule Protects Coal Miners' Health," April 25, 2016, https://www.dol.gov/newsroom/releases/owcp/owcp20160425.

EPILOGUE: DECEMBER 2016

Interviews: Arvin Hanshaw, Terri Smith, Gloria Roles, Richard Roles, Mary Fox, Craig Robinson, Debbie Wills, John Cline, Rick Smith

353 **He'd filed his first claim in September 2011:** Many of the details about Arvin's work history, medical records, and benefits claims are drawn from documents in his case file.

360 **the federal government had recently announced a quality-assurance program:** National Institute for Occupational Safety and Health, "Quality Assurance Review of B Readers' Classifications Submitted in the Department of Labor (DOL) Black Lung Benefits Program," NIOSH Docket Number 289, CDC-2016-0020, February 2016, https://www.cdc.gov/niosh/docket/archive/docket289.html.

361 **Overall, the award rate at the initial level:** Office of Workers' Compensation Programs, Division of Coal Mine Workers' Compensation, "Black Lung PDO Claim Decisions at the District Director Level FY 2006-2019," https://www.dol.gov/owcp/dcmwc/statistics/PartCClaimsDecisions.htm.

361 **the proportion of claimants who have a lawyer or lay representative:** Office of Workers' Compensation Programs, Division of Coal Mine Workers' Compensation, "Claims with Attorney or Lay Representation at the District Director Level," https://www.dol.gov/owcp/dcmwc/statistics/AttorneyAndLayRepresentationOfClaimants.htm.

363 **three case files down from his office:** These clients gave John permission to show me their case files.

Index

About the Author

Chris Hamby is an investigative reporter at the *New York Times.* His work has been recognized with the 2014 Pulitzer Prize for Investigative Reporting, Harvard University's Goldsmith Prize for Investigative Reporting, two White House Correspondents' Association awards, and UCLA's Gerald Loeb Award for Distinguished Business and Financial Journalism; he has also received awards from the National Press Club, the National Press Foundation, and the Society of Professional Journalists, among others. He was named a finalist for the Pulitzer Prize in International Reporting in 2017. He has covered a range of subjects, including labor, public health, the environment, criminal justice, politics, and international trade. *Soul Full of Coal Dust* received the J. Anthony Lukas Work-in-Progress Award, given by Columbia University and Harvard's Nieman Foundation. A native of Nashville, Tennessee, Hamby lives and works in Washington, DC.

About the Author